Springer Geophysics

The Springer Geophysics series seeks to publish a broad portfolio of scientific books, aiming at researchers, students, and everyone interested in geophysics. The series includes peer-reviewed monographs, edited volumes, textbooks, and conference proceedings. It covers the entire research area including, but not limited to, applied geophysics, computational geophysics, electrical and electromagnetic geophysics, geodesy, geodynamics, geomagnetism, gravity, lithosphere research, paleomagnetism, planetology, tectonophysics, thermal geophysics, and seismology.

More information about this series at http://www.springer.com/series/10173

Pavel G. Talalay

Thermal Ice Drilling Technology

Geological Publishing House

 Springer

Pavel G. Talalay
College of Construction Engineering
Jilin University
Changchun, China

ISSN 2364-9119   ISSN 2364-9127   (electronic)
Springer Geophysics
ISBN 978-981-13-8850-7   ISBN 978-981-13-8848-4   (eBook)
https://doi.org/10.1007/978-981-13-8848-4

Jointly published with Geological Publishing House, Beijing, China
The print edition is not for sale in China. Customers from China please order the print book from: Geological
Publishing House.

This Springer imprint is published by the registered company Springer Nature Singapore Pte Ltd.
The registered company address is: 152 Beach Road, #21-01/04 Gateway East, Singapore 189721, Singapore

Any sufficiently advanced technology is indistinguishable from magic.
Arthur C. Clarke

Preface

This book is a continuation of the review of ice drilling technology published in 2016, where, in the beginning, a general introduction to ice drilling technology was given (Talalay 2016). The present book includes the design, parameters, and performance of various tools and drills for making holes in snow, firn, and ice via thermal methods. The melting point of ice at normal atmospheric pressure is quite low compared with that of common rocks and is close to 0 °C (Feistel and Wagner (2006) found that the melting point of purified ice is 0.002519 ± 0.000002 °C), and thus melting looks like the obvious solution for making boreholes in ice.

At first glance, thermal drills are simpler and more attractive than mechanical drills because they do not rotate and, for that reason, do not need driving downhole or surface units, anti-torque systems, slip rings, etc. On the other hand, the specific energy required for melting ice (590–680 MJ/m^3) is much higher than the energy required by mechanical systems (1.9–4.8 MJ/m^3) (Koci and Sonderup 1990). Thermal drills also need reliable heating elements that can work for long periods of time at high pressures and a wide temperature range. These drills are difficult to use for penetrating debris-rich ice and recovering subglacial bedrock samples because even though rock melting is possible, it is extremely power-consuming (Armstrong et al. 1962).

On April 1995, T. Folger published a remarkable paper in "Discover" about strange mole-like animals that melt ice holes in the Antarctic ice. Individuals of this newly discovered species were approximately 15 cm long and had a very high metabolic rate—their body temperature was 43 °C. Perhaps their most fascinating feature was a bony plate on their forehead. These animals radiated tremendous amounts of body heat through their "hot plates", which they use to melt holes in ice. That is why they were named as *hotheaded naked ice borers*. News of the ice borers drew more letters from "Discover" readers than any other piece in the magazine's previous 15-year history. Most readers were amused and elaborated on an April's fools hoax.

Joking aside, several years later, hybrid drilling–melting drill heads were designed in Hong Kong Polytechnic University to penetrate through debris-laden ice based on the assumption that the particles would be pushed backward by the hot rotating blades in combination with a solid heating tip. Another "bionic" thermal probe IceMole, was developed in FH Aachen University of Applied Sciences, Germany. Like a mole, the probe is able to "dig" holes in ice horizontally and even vertically upwards (see Sects. 1.4.5 and 1.4.6).

A critical issue in thermal drilling technology is water refreezing in the open hole. The freezing of meltwater in boreholes drilled in temperate glaciers that are at their melting point throughout the year from their surface to their base is relatively slow. However, in polar glaciers, refreezing is very rapid (according to Humphrey and Echelmeyer 1990, 4–23 h with an ice temperature of −25 °C and an initial diameter of 100–240 mm). Thus, refreezing or the removal of meltwater is challenge No. 1.

The diameter of the borehole just above the melting tip is larger than the diameter of the melting tip. Therefore, knowing the refreezing and penetration rates allow for estimating the time for the safe removal of the thermal drill before the hole becomes smaller than the probe. When drilling deep holes in polar ice, the time for the safe removal of thermal drills (except for hot-water drilling systems) is not enough, and the hole needs to be filled with antifreeze drilling fluid. In addition, the melted hole walls tend to be irregular, which could be a problem for some types of logging.

Depending on drilling ability and performance, thermal drilling tools can be divided into the following groups: (1) hot-point drills; (2) electric thermal coring drills; (3) hot-water drilling systems; (4) steam drills; and (5) unconventional thermal drilling systems. Hereinafter, the different drills and tools available for thermal drilling in ice are reviewed and discussed, focusing on particular aspects of drilling operations, drilling problems, and possible solutions. In each section, a description of each type of drill is given in chronological order. Although this review could be useful for learning about ice drilling technology, the experience will ultimately become the best guide, as happens with most equipment.

I am grateful to H. Ueda, V. Chistyakov, R. LeB. Hooke, L. Augustin, V. Zagorodnov, I. Allison, J. Wilson, D. Etheridge, A. Smith, J. Kelty, J. Cherwinka, P. Doran, K. Makinson, A. Pyne, A. Takahahsi, D. Winebrenner, J. Goetz, F. Wilhelms, J. Schwander, X. Fan, G. Da, M. Sysoev, B. Liu, Y. Liu, D. Pomraning, and other experts for providing pictures, reports, and other materials. I would like to thank Y. Sun for his continuing support when preparing the book and Y. Yang, who helped me a lot with the publication procedure. The completion of this manuscript would not have been possible without the financial support of the Program for Jilin University Science and Technology Innovative Research Team (Project No. 2017TD-24), the Fundamental Research Funds for the Central Universities, and the Ministry of Science and Technology of China (Project No. 2016YFC1400300).

Drilling in ice is often a nontypical and challenging process. The deepest hole drilled in ice via thermal methods was bored at Vostok station, East Antarctica, to a depth of 2755.3 m (Vasiliev et al. 2007). The deepest hole drilled in ice via mechanical methods was bored to a depth of 3769.3 m also at Vostok station (Lukin and Vasiliev 2014). However, the thickest Antarctic ice, which is at Terre Adélie (69° 54' S, 135° 12' E), is 4776-m deep (Riffenburgh 2007). Therefore, the deepest polar ice has not been drilled yet. Ice drills that worked perfectly in some ice intervals or sites failed under different conditions. Whenever this occurred, designers and engineers would have more questions than answers. I selected Clarke's third law from his book "Profiles of the Future", 1961, as an epigraph to this book because the separating line between state-of-the-art engineering and magic can frequently be blurred.

Changchun, China   Pavel G. Talalay
December 2018

References

Armstrong DE, Coleman JS, McInteer BB et al (1962) Rock melting as a drilling technology. Los Alamos National Laboratory Report, LA-3243

Feistel R, Wagner W (2006) A new equation of state for H_2O ice Ih. J Phys Chem Ref Data 35(2):021–1047

Folger T (1995) Strange molelike animal melts ice tunnels with its head. Discover 16(4):14–15

Humphrey N, Echelmeyer K (1990) Hot-water drilling and bore-hole closure in cold ice. J Glaciol 36(124):287–298

Koci BR, Sonderup JM (1990) Evaluation of deep ice core drilling system. Polar Ice Coring Office, University of Alaska Fairbanks, PICO Tech. Rep. 90–1

Lukin VV, Vasiliev NI (2014) Technological aspects of the final phase of drilling borehole 5G and unsealing Vostok Subglacial Lake, East Antarctica. Ann Glaciol 55(65):83–89

Riffenburgh B (ed) (2007) Encyclopedia of the Antarctic: Vol I (A–K), New York

Talalay PG (2016) Mechanical ice drilling technology. Geological Publishing House, Beijing and Springer Science + Business Media Singapore

Vasiliev NI, Talalay PG, Bobin NE et al (2007) Deep drilling at Vostok station, Antarctica: History and recent events. Ann Glaciol 47:10–23

Contents

Hot-Point Drills

Abstract

Hot-points are designed for producing boreholes without a core by melting ice and used to install ablation sticks, to determine ice thickness, to locate englacial and subglacial streams, to deploy sensors and tools under ice shelves, to measure temperatures and closure rates of glaciers, to study subglacial environment in Earth and other planets. The energy for the melting tip heater is provided by a hot fluid that is pumped down using hoses in a closed circuit or by electricity transported via electrical cable.

Keywords

Thermal tip • Ice melting • Fluid heating medium • Electric heaters • Meltwater refreezing • Drill's vertical stabilization

The first ice thermal drills, which were used for glacier research, were designed for producing boreholes without a core by melting ice. It has been affirmed that these types of ice drills are known as hot-point drills, or just hot points, after a company that mass-produced electric heaters and electric stoves (Hotpoint n.d.). There are also other names for non-coring thermal drills, such as thermal/melting needle/probe/sonde.

Hot points are used to install ablation sticks, to determine ice thickness (though hot points do not provide absolute assurance that bedrock has been reached because they cannot distinguish between bedrock and englacial debris), to locate englacial and subglacial streams, to deploy sensors and tools under ice shelves, to measure temperatures, closure rates and other internal properties of glaciers, such as the variation of ice flow velocity and density with depth, to identify dust layers, to perform video observations, and so on. Recently, interest in the use of hot points was rekindled in connection with certain proposed projects for subglacial environment exploration in the Earth and other planets.

The main advantages of hot points compared with other types of drills used in glacier research are their rather high penetration rate of up to 22 m/h, their lightweight drilling setup, and their relatively low power consumption when used for making small-diameter (20–60 mm) boreholes. Many hot points are designed to be transported on ice by only one person. In view of these circumstances, their weight is reduced to a minimum; not only their total weight, but also the weight of the heaviest piece that cannot be dismantled. The energy for the melting tip heater is provided by a hot fluid that is pumped down using hoses in a closed circuit or by electricity transported via electrical cable.

The first electrically heated hot point was designed by Russian physicist and glaciologist Weinberg (1912). The main purpose of his device was to measure the thickness of mountain glaciers. The drill consisted of a wooden rod (1 m in length) with a 5-kg piece of lead at the lower end. The thermal tip was made of a flat spiral band composed of manganin and was isolated from the lead by a sheet of mica. The individual turns of the spiral were also insulated using thin mica tape. The hot point was tested in a 10-m-high ice tower constructed at the yard of Tomsk Technological Institute (now called Tomsk Polytechnic University) (Fig. 1.1), and an average rate of 1.5 m/h was routinely achieved at a power of 1.63–1.72 kW. Although Weinberg's drill was never used in the field, his idea was successfully realized thirty years later in 1940–1941 at Ross Ice Shelf, Antarctica during a US Antarctic Service Expedition where a hot point was used for the first time to create a borehole with a depth of 41 m (Wade 1945).

Hot points have been used mainly for drilling in temperate glaciers where boreholes filled with fresh water stay open for days, which allows for interruptions in the drilling process and reaming. However, a few applications of special hot-point drills in cold ice have been tested and some of them succeeded (Fig. 1.2): (1) using a heating power cable that partly prevents meltwater refreezing (Classen 1970; Suto et al. 2008); (2) antifreeze-assisted drilling, in which hydrophilic liquid is added into the hole and mixed with meltwater (Hooke 1976; Grześ 1980; Morev et al. 1988); (3) bailing of meltwater from the lower portion of the borehole (Vasilenko et al. 1988; Zeibig and Delisle 1994);

© Geological Publishing House and Springer Nature Singapore Pte Ltd. 2020
P. G. Talalay, *Thermal Ice Drilling Technology*, Springer Geophysics,
https://doi.org/10.1007/978-981-13-8848-4_1

(4) replacing meltwater with a non-freezing hydrophobic drilling fluid (Gillet et al. 1984); and (5) freezing-in instrumented probes carrying a tethering power cable (Philberth 1976; Hansen and Kersten 1984). The latter ones (the so-called Philberth probes) reached a depth of 1005 m in cold ice in Greenland.

The axial asymmetry of the heat flux from the melting tip of the hot point, nonhomogeneous dust and air bubbles concentration in the ice, and drill and borehole axis misalignment tend to divert the drill from its vertical path. In order to drill a hole as vertically as possible, Aamot (1967a) suggested using "pendulum pivot" heaters on the top of the drill (Fig. 1.2e). In such cases, the sonde is inherently stable and its attitude is plumb at all times. If deflected by external disturbances, the sonde returns to a vertical attitude like a pendulum. Other methods to perform the vertical stabilization of freezing-in hot-point drills are discussed in the summary presented in Sect. 1.3.

There were numerous patent inventions devoted to hot points that were never used. One of the first examples of such interesting inventions was designed by Koechlin (1945) and is shown in Fig. 1.3. The figure shows a drilling device comprising an elongated electrically heated shell with a continuous circulation of fluid and suspended by a supply cable. Remenieras and Terrier (1951) reported that Koechlin's drill had been probably used several times in Switzerland, but gave no details. In the following review, we mainly include designs that were practically used or tested at glaciers as well as new hot point concepts that are currently in progress.

1.1 Hot-Point Drills with a Fluid-Heating Medium

These hot-point drills use an intermediate heat medium to transfer thermal energy from the surface heater to the downhole melting tip. Water is the simplest fluid heating

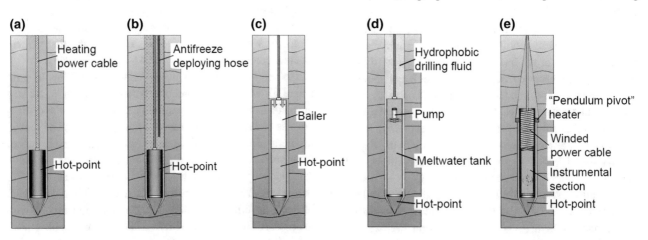

Fig. 1.2 Prevention of the refreezing of boreholes drilled with hot-point drills in cold ice: **a** with heating cable; **b** by adding antifreeze; **c** by bailing; **d** by replacing meltwater with a non-freezing hydrophobic drilling fluid; **e** using a freezing-in hot point with on-board tethering cable (modified from Zacny et al. 2016)

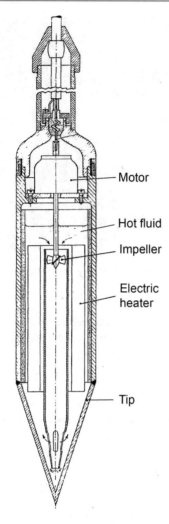

Fig. 1.3 Hot point of Koechlin (1945)

Fig. 1.4 Thermal drilling at Miage glacier, Alps, 1942 (Calciati 1945)

medium readily available in glaciers. However, ethylene glycol, propylene glycol, or other antifreeze liquids for preventing freezing during idle times and transport operations are safer options.

1.1.1 Calciati's Hot-Point Drill

The first hot point with a hot-water heating medium was developed by Italian engineer Calciati (1945), who drilled 20 holes in the Hosand and Miage glaciers (Formazza valley, Alps) to a maximal depth of 125 m to determine the bedrock topography during the summer and autumn of 1942 (Fig. 1.4). A downhole conical tip was fixed to the lower end of a 5-m-long steel pipe that contained two concentric chambers for water circulation. The steel pipe was connected to a wood-heated boiler by means of two hoses. The circulation of hot water was produced by a hand pump mounted on the boiler. The hot water flowing inside the tip heated it and was then pumped back to the boiler.

The diameter of the melted borehole was ∼180 mm and the penetration rate varied from 4 to 10 m/h, including the time required for the attachment of the extensions. The penetration rate significantly decreased with the presence of sand or small pebbles in the ice and, when the tip met pebbles of considerable size or bedrock, the advance stopped. A total of 1470 m of holes were drilled by a team consisting from 5–6 men over 36 days, including the work interruptions for moving from one hole to another, installation, bad weather, and breakdowns. The average consumption of firewood was estimated at 1.5 kg per meter of hole. In all cases, the melted water remained in the hole even when bedrock was reached.

1.1.2 ETH Lightweight Hot Point

Kasser (1960) from the Hydrology Department of the Swiss Federal Institute of Technology (ETH), Zurich, designed a light thermal ice drill similar to Calciati's hot point for the installation of ablation sticks. Heat was supplied from a gasoline blowtorch to heat up water to approximately 80–90 °C (the volume of the blowtorch fuel tank was 3 L). A double-acting piston pump was mounted inside the heating pot and was operated by hand with an external lever. The pump drove the water with a rate of 1.7 L/min at pressures as high as 0.4 MPa in a closed circuit through an insulated hose line with a diameter of 11 mm (10, 20, and 30-m-long pieces were available) down a 1.5-m-long pipe with an outer diameter (OD) of 34 mm with two tubes inside and then to a 0.38-m-long paraboloid hot point. Hot water entered the central tube of the hot point and jetted through a nozzle with an OD of 2 mm to the tip, where it deflected at 180° (Fig. 1.5) and returned via the annulus between the middle pipe and the jacket back to the blowtorch.

The equipment was designed to be transported via back-packing. The heaviest single item is the heater assemble weighed 17.3 kg. The hot point weighed 1.2 kg;

Fig. 1.6 Drilling with an ETH hot point at Grosser Aletschgletscher, July 1959 (Kasser 1960)

its rigid guide tube, which was made heavier with lead, weighed 2.7 kg; and the two-conduit rubber hose weighed 1.0 kg/m to exactly compensate for buoyancy. Tools, fuel, and occasionally water and ethylene glycol antifreeze also had to be carried.

This drill has been tested out since 1950 on numerous sites at the Swiss Alps to depths as great as 63 m at Grosser Aletschgletscher in temperate ice (Fig. 1.6) and to a maximum depth of 30 m at the Jungfraujoch in ice at temperatures near −2 °C. During 1959–1960, this system was used in rather cold ice with a temperature of approximately −10 °C in the International Greenland Expedition (EGIG) in Western Greenland (Shreve 1962b). Drilling to the depth of 30 m took ∼2 h. The borehole diameter was ∼40 mm. The average temperature of the water upon leaving the heater was 75 °C, whereas it was 42 °C when returning. Under normal conditions, the penetration rate varied from 12 to 22 m/h. The typical fuel consumption was ∼65 mL/m. To prevent water from freezing inside the closed circuit during drilling interruptions, it was replaced on several occasions by a 50% ethylene glycol aqueous solution, which has a freezing point of −39 °C.

1.1.3 Glacier Girl Recovery

This story falls slightly beyond the scope of the present book but shows a remarkable application of thermal drills. On July 15, 1942, a squadron of six P-38 Lightnings and two B-17 Flying Fortress bombers was flying from Greenland to Iceland when they ran head-on into an Arctic blizzard (Hayes 1995). As conditions deteriorated, they decided to turn back and were forced to make emergency landings on the Greenland ice sheet. All the crew members were subsequently rescued. However, the aircrafts were eventually

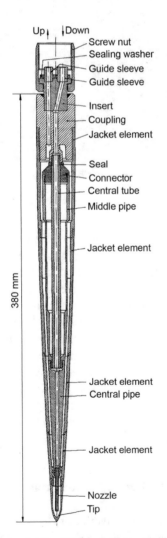

Fig. 1.5 ETH hot point (modified from Kasser 1960)

buried under the snow and ice that built up over the subsequent decades.

A combination of historic photos, the understanding of ice movements, and subsurface sensing systems, such as ground-penetrating radar and magnetometry, led to discovery of the Lost Squadron's location in 1988. The aircrafts were found at a depth of 82 m approximately 4 km away from the crash-landing site. The Greenland Expedition Society decided to recover one of the buried planes. In 1990, a temporary base was constructed on the ice sheet surface with a hoist to suspend a gigantic hot point, called the Thermal Meltdown Generator, which had a diameter of approximately 1.2 m and a height of 1.5 m (Fig. 1.7). The hot point weighed ∼250 kg and was suspended by a chain. It had a cone-shaped tip that was wrapped in copper tubing through which hot water circulated. Water from the solar collector was fed into the generator-powered boiler, and hot water was pumped into the hot point.

The Thermal Meltdown Generator could melt a shaft with a rate of 0.6–1.2 m/h and a submersible pump pushed the

meltwater up to the surface. When the drill touched the wing of a B-17 bomber, a worker was lowered into the hole, where he used a hot-water hose to melt a cavern around the plane's remains. Water was pumped to the surface as the ice melted, and slowly the bomber became exposed. It soon became apparent that the B-17 was very badly crushed, far beyond worthwhile salvage.

It occurred that the smaller, more rugged P-38 s would probably be in much better conditions than the B-17 had been. Two years later, in 1992, the team once again melted a shaft in Greenland's ice and cut away ice around another buried fighter with pressurized steam. As anticipated, the P-38 was in much better conditions (Fig. 1.8).

A team of technicians was lowered into the ice cavern to begin the process of disassembling the aircraft so that it could be shuttled to the surface piece by piece. To lift the last, biggest piece—the 2.7-t center section—additional shafts were drilled and were connected in-line; a total of five shafts were excavated (Fig. 1.9). All the parts were shipped to the United States and, after nine years of restoration, the salvaged Lockheed P-38 Lightning, dubbed as the Glacier Girl, was exhibited at the Lost Squadron Museum in Middlesboro.

1.1.4 IceCube Firn Drill

When drilling ice sheets with hot-water drilling systems, the firn does not hold water until it reaches a sufficient density (e.g., according to Benson et al. 2014, at the South Pole 35–40 m below the surface). In firn, the hot water just seeps away through the snow without melting it and the methods for drilling large holes and water circulation do not work. During the first stage of the construction of the IceCube Neutrino Observatory (or simply IceCube; the hot-water drill system used in this project at the Amundsen–Scott Station in Antarctica is described in Sect. 3.4.2), a large cone-shaped aluminum vessel filled with hot water was used to melt the firn. Although this worked, it was a very slow process, even with substantial water flow out of the nozzle at the bottom. To improve heat transfer, the aluminum vessel was replaced with coils of a copper tube with an OD of 13 mm wound into the shape of a cone. Because of the shape, drillers had nicknamed the system as Carrot Drill. The high flow rate through the tube ensured good heat transfer and improved the drilling rate enough that it could be used in a closed-loop configuration. At the tip, a fast stream of water impinged against the inside of a hollow copper cap. This allowed for good heat transfer to the cap and enabled it to keep melting the snow. This method, however, still took a significant amount of time because the system used hot-water resources from the main drill system and plumbing had to be reconfigured between firn and deep drilling.

Fig. 1.7 Thermal meltdown generator, Eastern Greenland, summer of 1992 (*Photo* L. Sapienza from Hayes 1995)

Fig. 1.8 Lockheed P-38 Lightning at a depth of 82 m, Greenland, summer of 1992 (*Photo* L. Sapienza from Epps Aviation 2009)

Fig. 1.9 Drilling of on-line shafts, Eastern Greenland, summer of 1992 (*Photo* L. Sapienza from Hayes 1995)

An independent firn drill was built to allow for predrilling the firn holes in parallel with hot-water deep drilling operations and was first used at the South Pole in 2006–2007 (Fig. 1.10). This reduced the time constraints on firn drilling while simultaneously increasing the time available for deep drilling. The independent firn drill required a reel to hold the supply and return hose, a shelter to house electronics and people during drilling operations, and a catch pan to catch any spills of the drilling fluid. The drill used electric heat and a propylene glycol water mixture as the heat-transfer fluid instead of plain water so that the liquid would not freeze when the drill was not being used. Heating was performed with five 30-kW electric heating elements. The system was left running unmanned with periodic checks, and a penetration rate of 2 m/h was reliably achieved while consuming approximately 1100 L of fuel per firn hole. Moving the independent firn drill from hole to hole required two or three persons, but could be done in as little as 30 min.

1.1.5 Coiled Hot Points for Drilling in Sea or Lake Ice

In the same way as the IceCube firn drill, these hot points use coiled copper tubes with different shapes heated via the loop circulation of hot antifreeze usually provided by commercial washers. The penetration rate of these systems is quite low ($\sim$0.1 m/h), but they can create large-diameter holes to deploy divers, samplers, and autonomous

Fig. 1.10 Firn-drill copper coiled cone (*Credit* J. Cherwinka)

underwater vehicles into the sea or lakes covered by multi-year ice with a thickness of many meters.

The "plain" hot point was designed to create holes of several meters in depth and up to 0.8 m in diameter in order to study Antarctic lakes (Fig. 1.11). For this application, a Hotsy 500 Hot Water Pressure Washer was modified to be a closed circuit containing an ethylene glycol aqueous solution (Doran, personal communication, 2016). The copper coil melted down and drillers had to be constantly adjusting the ropes to level the hot point and keep it going straight down. To increase the load, a few stones were put on top of the coil. The melted water had to be pumped out or otherwise the hole would acquire a bell shape at the bottom and melting would become less efficient. Because drilled ice contains many mineral inclusions that gravitate to the bottom, it was essential to shut down and pull the coil out of the hole to clear the sediment from under the coil, which slowed the melting process. From then on, diving holes were drilled using other more effective methods.

Another coiled hot point was used in November 2005 and October 2006 to drill holes with a diameter of 24″ through Ross Sea ice for the deployment of mooring systems and to perform ocean current observations needed for the installation of the ANDRILL (ANtarctic DRILLing Project) drilling platform (Limeburner et al. 2006). Two pressure hoses connected to a pump and a boiler (Fig. 1.12) were attached to a coil. The hole melting process was slow and equipment failures were frequent owing to the cold and harsh operating conditions. It took more than three days to reach the underlying seawater at a depth of 7 m.

A small-diameter coiled hot point Stinger was designed in 2007 in the SCINI (Submersible Capable of under Ice Navigation and Imaging) project to study the marine environment of the Southern Ocean in the McMurdo area (Project S.C.I.N.I. n.d.). Stinger was used to deploy instruments under the ice and included a pressure washer driving heated glycol through a coil (Fig. 1.13). It required refueling of the power washer every 5 h. It took ∼24 h to melt 2-m-thick sea ice and three days to melt 7-m-thick ice.

1.1.6 Summary

Hot point drills with a fluid heating medium use electric heaters, gasoline blowtorches, or gas-fired burners on the surface to warm up the heat-transfer fluid and have to pump it down downhole hot point, which is heated and melts ice at the bottom. Circulation usually goes on in a closed circuit. The efficiency of these drills is lower than the efficiency of drills using downhole heaters because some of the energy is eventually lost in the hoses and pipes during heat transfer. Nevertheless, hot-point drills with a fluid heating medium could still be a good pick for shallow drilling, for example, for ablation stick installation or creating large-diameter holes in snow and firn. Especially noteworthy is the fact that these hot points, unlike electrically heated ones, cannot burn out down the borehole.

1.2 Electric-Heated Hot Points

Electric-heated hot points were actively used in glacier research in the 1950–1970s, and were then replaced in many applications by faster and less power-consuming electromechanical drills suspended on cable (Talalay 2016) or by faster and more power-consuming shallow hot-water ice drilling systems (see Sect. 3.1). Nevertheless, electric-heated hot points were recently used to identify snow–firn density distributions, to determine the thickness of the grounded accumulation of sea ice (such as in ridges and stamukhas), for deployment of small sensors under ice shelves, and for other purposes (Kharitonov and Morev 2011; Zagorodnov

(a)   **(b)**

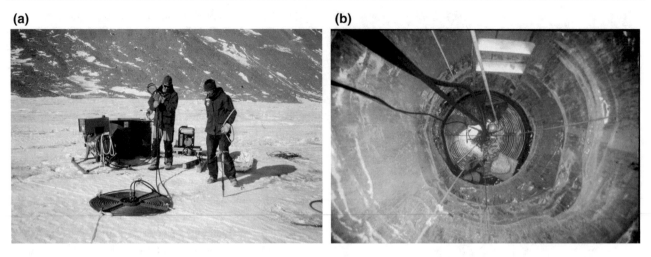

Fig. 1.11 "Plain" hot point: **a** ready to melt; **b** at the base of the lake's ice cover (*Credit* P. Doran)

(a)   **(b)**

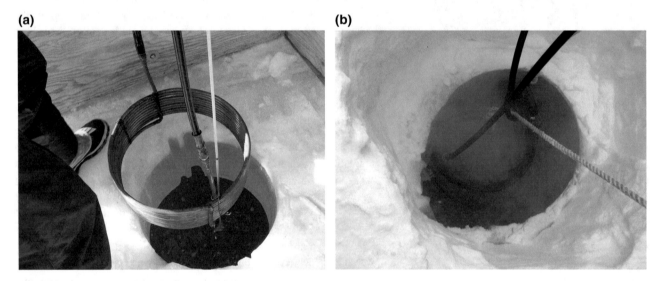

Fig. 1.12 Coiled hot point at Ross Sea ice: **a** removing it from the hole, November 2005 (Antarctica: ANDRILL 2005); **b** drilling, October 2006 (Limeburner et al. 2006)

et al. 2013, 2014). In recent years, interest in electric-heated hot points raised because of subglacial lake explorations and extraterrestrial investigations. Many of them are reviewed in the following two sections. Some parameters of electric-heated hot points are shown in Table 1.1.

1.2.1 NRC Ice Thickness Hot-Point Kit

Several hot points were designed for the rapid manual drilling of small-diameter holes to measure thickness of lake, river, and sea ice. To the best of our knowledge, the first thickness hot-point kit was designed by the National Research Council (NRC) of Canada in the middle of the 1940s (Northwood 1947). This battery-operated (6 V) lightweight hot point was capable of drilling a hole with a diameter of 0.5″ at a penetration rate of 36 m/h. It consisted essentially of a rod long enough to penetrate the desired thickness range with an electrically heated tip (Fig. 1.14). Its electrical circuit had a low resistance, with that of the tip heater being only 0.05 Ω. A cylindrical brass cover, which served mutually as an outer case and the principal conductor, became warm enough to prevent the drill from freezing in. The heating element was a piece of projector carbon fitted between an outer and an inner tubing at the lower end of the drill. A standard automobile starting relay was used for switching.

When operated in free air, the carbon heating element tended to become so hot that much caution had to be used to prevent it from burning out. The drill was also only capable of penetrating a few meters, but its rate of penetration was high and it could easily be hand-carried.

Fig. 1.13 Preparation of a heated-coil Stinger for drilling near McMurdo Station, Antarctica, 2007 (Project S.C.I.N.I. n.d.)

1.2.2 Hot Point of the US Antarctic Service

The first hot point drill in Antarctica was used for temperature measurements at Ross Ice Shelf during the US Antarctic Service Expedition in 1940–1941 (Fig. 1.15) (Wade 1945; Miller 1953). Using a hand auger, initial drilling reached 10 m. From there up to the maximum depth attained, which was 41 m, the hot point was employed. The heating element was a 1-kW heating coil. Unfortunately, if heated in air for more than a few minutes, the element would burn out. The hot point was designed with an outer diameter of 57 mm on a tube that was 0.61 m long. It was attached to a guide pipe that was 2 m long.

1.2.3 Hot Point of the University of Cambridge

The original electric-heated hot point was designed in the University of Cambridge in order to measure vertical velocity distributions in the interior of mountain glaciers

(Gerrard et al. 1952). This hot point consisted of a steel tube 40 mm in diameter and 120 mm in length, which contained three parallel coils of tantalum wire insulated by heat-resistant clay that had been baked in situ (Fig. 1.16). The electric heating element could output 2.5 kW at 330 V. It was welded onto a 3″ steel tube that was in turn screwed onto the drill pipe. The tip was designed to stand the full weight of a 180-m steel tube behind it.

A drilling site was selected at the Jungfraufirn glacier, Swiss Alps at an altitude of 3350 m and a distance of ∼500 m from the Jungfraujoch Research Station (Fig. 1.17). Drilling started in July 1948 and bedrock was reached after two weeks at a depth of 137 m. The hot point yielded a maximum penetration rate of 1.5 m/h in ice and worked as long as it was surrounded by ice and water. In porous firn, melt-water tended to seep away from it, causing it to run dry and burn out.

The hole was lined with mild steel tubes with an inner diameter (ID) and OD of 2.5″ and 3″, respectively, which were used in sections with a length of 7 m and screwed together with couplings. The final tube was lifted 50 cm and clamped to a large steel plate resting on the glacier surface so that it would not scrape against the glacier bed. After approximately two weeks, the ice in the borehole closed in around the tube and gripped it firmly. Great care had been taken to make the couplings between the successive sections of the tube watertight so that water would not seep into the tube and subsequently freeze. These precautions proved successful. In the following few years, a few sets of inclinometer readings of the tube were carried out.

1.2.4 Nizery Hot Point

A very effective hot point was designed by French engineer A. Nizery (Remenieras and Terrier 1951; Nizery 1951; Nizery and Terrier 1952) for test drilling operations in mountain glaciers carried out by the Service des Etudes et Recherches Hydrauliques, Electricité de France. This thermal drilling system consisted of a hot point with an open resistive heater, a flexible rubber cable supporting the hot point, a derrick with a cable winding system, and an electricity-generating set (Fig. 1.18a).

The hot point body was made of a hollow brass cylinder with an OD of 50 mm and a length of 1 m (Fig. 1.18b); its weight, including the lead ballast, was ∼15 kg. The heating resistance was made of a Tophet A (nickel-chromium resistance alloy synonymous with nichrome) wire with a diameter of 1.63 mm and a length of 1 m, wound on a conical helical path, the maximum diameter of which was 50 mm. The ends of the resistance were attached to the electrodes inside the bit body using connecting screws in order to allow for quick replacement of the resistance in case

Table 1.1 Electric-heated hot points

Drilling site	Year	Glacier type	Power/kW	Hot point diameter/mm	Penetration rate/(m h^{-1})	Max depth reached/m	References
Ross Ice Shelf, Antarctica	1940–1941	Polar glacier	1	57	NA	41	Wade (1945)
Jungfraufirn, Alps	1948	Temperate glacier	2.5	76	1.5	137	Gerrard et al. (1952)
Mer de Glace, Alps	1948–1949	Temperate glacier	7.8	50	20–25	195	Nizery (1951)
Upper Seward Glacier, Canada	1948–1949	Temperate glacier	1.7[a]	32	11.4 in firn; 2.2 in ice	62.2	Sharp (1951)
Traverse Camp IV – Station Centrale, Greenland	1949–1950	Polar glacier	8	80	9	50	Negre (1950)
Barnes Ice Cap, Canada	1950	Polar glacier	1.8	76	0.8–0.9	21.4	Ward (1952)
Malaspina Glacier, Alaska	1951	Temperate glacier	NA	NA	NA	305	Sharp (1953)
Bråsvellbreen, Svalbard	1951	Sub-polar glacier	NA	NA	2.4	30	Miller (1953)
Saskatchewan Glacier, Canada	1952–1953	Temperate glacier	NA	45	4.3-4.9	120	Meier et al. (1954)
Taku Glacier, Alaska	1952–1953	Sub-polar glacier	1.5	63.5	1.3	52	Miller (1953)
Salmon Glacier, Canada	1956	Temperate glacier	1.8	76	1.8	756	Mathews (1957, 1959)
Austerdalsbre, Norway	1956–1959	Temperate glacier	2.6	81	3.3–3.8	157	Ward (1961)
McCall Glacier, Alaska, USA	1957	Temperate glacier	NA	NA	3.4	91.4	Orvig and Mason (1963)
Blue Glacier, USA	1957–1964	Temperate glacier	2.5	51	Up to 9.0	260	Shreve and Sharp 1970
Vostok Station, Antarctica	1958–1959	Polar glacier	2.0	150	0.25–1.6	52	Ignatov (1960b)
Athabaska Glacier, Canada	1959–1960	Temperate glacier	1.8	50.8	6.5	322	Stacey (1960), Savage and Paterson (1963)
Isfallsglaciären, Sweden	1961	Sub-polar glacier	2.5	50	6	NA	Ekman (1961)
Blue Glacier, USA	1961–1962	Temperate glacier	0.22	18	5.5–6.0	142	LaChapelle (1963)
Black Rapids Glacier, Alaska	1967	Sub-polar glacier	0.25	20.3	5	62	Aamot (1968c)
Rusty Glacier, Canada	1969	Sub-polar glacier	2.5	76	1.0–1.5	48.8	Classen (1970)
Jankuat Glacier, Central Caucasus, USSR	1970–1973	Temperate glacier	1–2	40	5–6	111	Sukhanov et al. (1974)
Abramov Glacier, Pamir, USSR	1971–1972	Temperate glacier	1.5	40	7–11	172	Ryumin et al. (1974)
Blue Glacier; South Cascade Glacier, USA	1971–1975	Temperate glaciers	1.3	50.8	5–6	210	Taylor (1976)
Severnyi Polyus-19, Arctic Ocean	1972	Sea ice	1.7	40	7–9	34	Morev (1976)
Mer de Glace, Glaciers of the Sommet de Bellecôte	1973	Temperate glaciers	0.65	18	12	180	Gillet (1975)
Obruchev Glacier, Polar Urals, USSR	1973–1974	Temperate glacier	2.0	40	12–14	137	Zagorodnov et al. (1976)

(continued)

Table 1.1 (continued)

Drilling site	Year	Glacier type	Power/kW	Hot point diameter/mm	Penetration rate/(m h^{-1})	Max depth reached/m	References
Vavilov Glacier, Severnaya Zemlya, USSR	1974–1976	Polar glacier	NA	40	NA	310	Morev and Pukhov (1981)
Barnes Ice Cap, Canada	1974–1978	Polar glacier	2.0; 4.4	40; 80	7.6–8.2; 4.9	276	Hooke (1976) Hooke and Alexander (1980)
Barnes Ice Cap, Canada	1975	Polar glacier	2.24	76	4.0	200	Classen (1977)
Svalbard	1975–1982	Temperate glaciers	1–3	40	10–18	Interval 368–586[b]	Kotlyakov (1985)
Hans Glacier, Svalbard	1979	Sub-polar glacier	0.7	27	2–4	NA	Grześ (1980)
Dome C, Antarctica	1981–1982	Polar glacier	2.25	43	5	Interval 180–235	Gillet et al. (1984)
Davydov Glacier, Internal Tien Shan	1985	Temperate glacier	NA	70	3	109	Vasilenko et al. (1988)
Victoria Land, Antarctica	1992–1994	Polar glacier	2.5	76[a]	2.2	102	Zeibig and Delisle (1994)
Garabashi Glacier, Central Caucasus, USSR	1987	Temperate glacier	NA	40[a]	NA	77.7	Bazhev et al. (1988)
McMurdo Ice Shelf, Antarctica	2011	Sub-polar glacier	1.8	40	7.2	190.4–192.9	Zagorodnov et al. (2014)

Note: *NA* not available; [a]author's estimated values; [b]hot point drilling started from 368 m depth, drilled with electro-thermal drill ETB-3

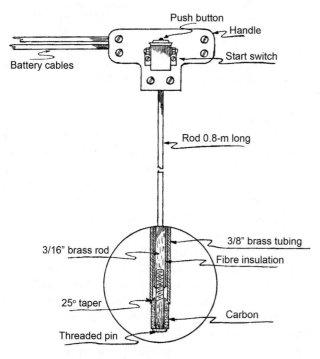

Fig. 1.14 Battery-operated lightweight hot point for determining the thickness of sea or lake ice (Northwood 1947)

Fig. 1.15 Makeshift drill tower erected near the rear of the Snow Cruiser, a large vehicle intended to facilitate transport in Antarctica, 1940–1941 (Wade 1945)

of it burning out. The resistance, which was approximately 0.54 Ω, took 120 A in normal service under a voltage of 65 V. Under these conditions, the power of the hot point was 7.8 kW.

To suspend and to supply power to the hot point, two flexible 30-mm^2 copper cables insulated with rubber were used. The weight of each cable was 0.50 kg/m. A three-legged metallic derrick supported the winding system of the cable. The cable was wound in one or two coils on the capstan pulley controlled by a worm gear through a direct-current variable-speed 0.37-kW electric motor. The driving of the cable was automatically stopped through lack of contact of the coils on the capstan as soon as the drilling bit rested at the bottom of the hole and pull was no longer

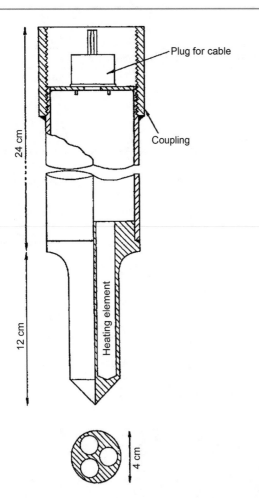

Fig. 1.16 Hot point made for an expedition at Jungfraufirn by Edur A. G., Zurich (Gerrard et al. 1952)

Fig. 1.17 Tripod used for thermal drilling at Jungfraufirn, Swiss Apls, July 1948 (Clarke 1987 after Gerrard et al. 1952)

exerted on the cable. The power needed for the drill and the auxiliary equipment (capstan, lighting) was ~ 10 kW.

The first trial tests on the Mer de Glace in 1948 demonstrated the good working order of the rig. A drilling program was undertaken on that glacier between July and October in 1949 as a part of reconnaissance work for the subglacial water supply of a hydroelectric plant. The objective of the drillings was to investigate two cross-sections of the glacier. The first cross-section, named Montenvers, was approximately 570 m long at the surface (altitude 1815 m asl). The second one, Alltée Verte, was located 1 km up from the valley and was approximately 900 m long (altitude of approximately 1950 m asl). On the Montenvers cross-section, 14 holes totaling 950 m were drilled. The deepest hole was stopped by the bedrock 195 m below the surface of the glacier. On the Alltée Verte cross-section, seven holes of approximately equal depth and totaling 450 m were drilled.

Two electric generating sets arranged in parallel each consisting of a four-cylinder air-cooled gasoline engine coupled by a belt to an 80-V 60-A direct-current generator were used. The weight of each one of those electricity generating sets was 150 kg; its gasoline consumption was approximately 5 L/h. The total weight of the rig was 1200 kg and it required two to three persons for field work. The average rate of penetration was from 20 to 25 m/h; the diameter of the holes varied between 60 and 80 mm.

Normally, the hole stayed full of water, which did not refreeze even several days after drilling. Glacier cracks had running streams of water in them, which sometimes caused overflowing of the drill hole at the surface of the glacier. When the hole reached the bedrock and emptied, the subglacial stream could be heard. Drilling operations were sometimes stopped by isolated stones which appeared irregularly in the glacier but became extremely rare 20 m below the surface. When small stones were met, the hot point could sometimes either push them aside or go around them. Getting through thin sand beds with the hot point was

Fig. 1.18 Thermal drilling system used at Mer de Glace near Chamonix, France, 1948–1949: **a** general configuration of the drill rig; **b** hot point (Nizery 1951)

(a)

(b)

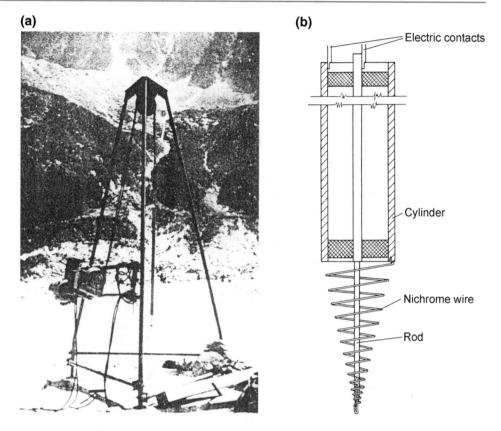

sometimes possible with a little deterioration of the heating resistance.

1.2.5 AINA Hot Points

In the Snow Cornice expedition of the Arctic Institute of North America (AINA) to the St. Elias Mountains in 1948, a hot-point drill was employed to drill a set of boreholes into the upper Seward Glacier lying within the Yukon Territory, Canada (Sharp 1951). The drilling site was set up at Airstrip Station at an elevation of 1791 m asl, ~4.8 km west of the eastern border of the glacier.

The equipment consisted of a portable 2-kW generator, a hot point, a 60-m-long small-diameter aluminum drill string, and a conductor cable (Fig. 1.19). The generator weighed 47.6 kg, including a two-cycle gasoline-driven engine, and could deliver a maximal DC of 70 A at 28.5 V. The drill pipe was used as one of the conductors. A separate and rather heavy cable was run down inside the pipe and served as the other conductor. The reason for using the pipe string was to ensure that the drilled borehole would be vertical.

The hot point consisted of an enclosed 24-V heating coil in a sealed brass cylinder with a rounded lower end. The tube was 32 mm in diameter and 0.28 m long. At its upper end, it was threaded into the conductor-guide pipe with a water-tight fitting.

Fig. 1.19 Hot-point drilling at upper Seward Glacier, Yukon Territory, Canada (Sharp 1951)

In the summer seasons of 1948 and 1949, dozens of boreholes were drilled, with a maximum depth of 62.2 m. The rate of penetration varied considerably owing to the stratification of the firn and to the ice layers encountered. The average rate in firn at a depth of 19.5 m was 11.4 m/h; below that depth, penetration rate dropped to a nearly constant value of 2.2 m/h. The lowest temperature recorded in the Seward firn was −1.1 °C.

Fig. 1.20 Schematic of the
AINA hot points used at Taku
Glacier, Alaska in 1952 and 1953
(all dimensions are given in mm):
a hot point with a tubular heating
element casted into a copper body
(dotted line shows the altered
length of the tip); **b** hot point with
a cartridge heater; **c** hot point with
a monel housing conformed to the
shape of an internal Calrod
heating tube (inset shows a helical
resistance wire centered in a
sheath of dense magnesium
oxide) (Miller 1953)

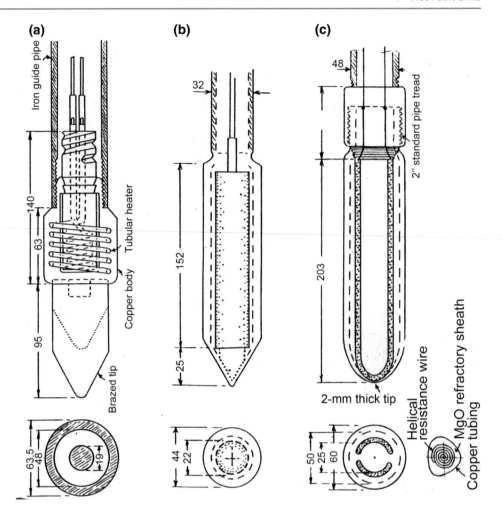

The second set of AINA hot points were designed for carrying out temperature measurements in each season of the year and in different sectors of the Taku Glacier, which is the largest ice stream leading from the huge highland ice known as the Juneau Icefield, Alaska (Miller 1953). The first hot point used a 1.25–2 kW Calrod tubular heating element that was casted into a copper body, and heavy-duty soldering iron was placed at the tip to improve the drill's gravity feed characteristic and to reduce the tendency of the upper end to lean against the sidewall (Fig. 1.20a). The wedged tip plus heater was 250 mm long and had a diameter of 48 mm at its lower end and of 63.5 mm at the cast-in section. A 1.7-m-long guide pipe with a 51-mm inner diameter was screwed to the upper end of the copper body. The hot point (without the guide tube) weighed 23.6 kg. The total length of the drill was ~2 m. The metal haulage loop was welded to the water-tight coupling at the upper end of the guide tube. A manila rope with a diameter of 6.4 mm was attached to the loop for hoisting operations. Power was supplied via a double-conductor cable with a diameter of 15 mm via a 1.5-kW 120-V portable generator that weighed 64 kg. In 1952, a 52-m-deep hole was drilled using this hot point over

40.5 h of almost continuous work; however, a shortage of gasoline forced the termination of the drilling operation.

In 1953, work was resumed. Two new drills were constructed. One had a 500-W cartridge heater installed on a sharp conical tip (Fig. 1.20b). This drill had an overall length of 2.7 m and was used only for shallow drilling in seasonal snowpack with an average rate of 4.5 m/h. The second one was a 2.1-m-long drill with a 1.5-kW internal heater of the Calrod type that was brought in close contact with the ice by the fine machining of a thin monel casing that, in turn, was tapered and rounded to conform with the heater's shape (Fig. 1.20c). The tip had a spherical shape; the diameter of this hot point was 60 mm. Unfortunately, the drill was not as effective as expected; its penetration rate in snow and upper firn was 1.5–4.5 m/h, and it dropped to 0.8 m/h in old firn.

1.2.6   E.P.F. Hot Point Drills

In addition to the two holes drilled using a conventional drill rig, several boreholes that went down to ~50 m were drilled using thermal hot points at site Camp IV, which was 40 km from the edge of the ice sheet, in summer 1949, and at a point ~170 km inland from the edge of the ice sheet and Station Centrale in summer 1950 within the frame of the Expéditions Polaires Françaises (E.P.F.) in Greenland (Miller 1953). The hot points were designed after Nizery (1951) (see Sect. 1.2.4) and consisted of a brass cylinder at the base of which was a bare wire cone-wound heating head with an OD of 80 mm. These hot points were supplied via a cable with a diameter of 45 mm by a single-phase 116-V generator.

At Camp IV, the first hole reached 45 m in 5 h at an average rate of 9 m/h. However, the cable had to be abandoned due to the refreezing of water on the walls 15 m down. The diameter of this hole was only 1–2 mm larger than that of the hot point. In the second hole, another technique was tried. By repeated raising and lowering of the hot point, this borehole was reamed enough so that the cable would not freeze during operations. As a result, a hole of 100 mm in diameter was obtained. The equipment was then withdrawn each night to avoid losing it.

In a third hole, after 8 h of drilling, a depth of 25 m was reached. After interruption of the drilling at night, it was found that the hole had closed at a depth of 10 m because of meltwater refreezing. The drilling continued, and after 30 more hours of continuous and uninterrupted work, a depth of 50 m was finally attained. Below this level, the hot point stopped, even though it was still working well and the power dissipated was the same as it had been at the shallower depths near the surface (~8 kW). What caused the arrest of the hot point at this depth was not known, but because no further progress was possible, the drilling was terminated.

1.2.7   Hot Points of the Baffin Island Expedition

Two types of hot points were used for drilling at the Barnes Ice Cap located in central Baffin Island, Nunavut, Canada with the aim to install an aluminum pipe string and to measure the vertical velocity and temperature distribution of the ice (Ward 1952). The first one consisted of a commercial sealed-cartridge heater (115 V, 1.2 kW) that was ~150 mm long and had a diameter of 25 mm (Fig. 1.21a). The heater was fitted into the center of a thick copper bit that was slid inside the aluminum tip.

The other hot point consisted of a copper core wound externally with a single spiral of resistance wire that was ~7.6 m long (2 kW at 115 V) and insulated with ceramic beads (Fig. 1.21b). The extended tip of the copper core slid into a hole in the aluminum tip to provide good thermal conductivity. Both heaters could be removed if necessary.

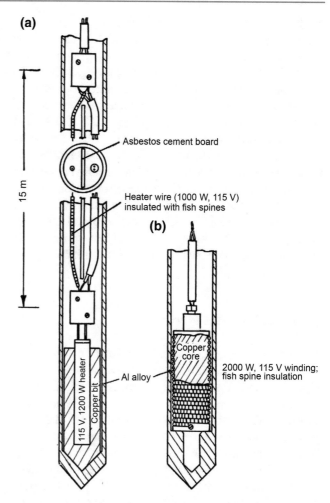

Fig. 1.21 Hot points used at Barnes Ice Cap: **a** with a cartridge heater; **b** with resistance wire (Ward 1952)

The hot points were suspended from 3-m-long aluminum pipes with an ID and an OD of 63 and 73 mm, respectively, which were screwed together with a bituminous jointing compound to form watertight joints. Power was provided by a 2-kW gasoline generator via a heavy conductor cable.

Both hot points were tested early in June 1950 for the purpose of permanently setting a few thermistors in the ice. The hot point equipped with the 1.2-kW heater froze at a depth of 3 m and was freed by slowly moving the heater up and down the pipe. By continually rotating and moving the string up and down, a much greater depth was reached with the 2-kW heater. Further progress became impossible at a depth of 9.1 m and the hot point was rapidly lifted out of the ice. Meltwater froze in the hole within a few minutes of the hot point being extracted. This unfortunately prevented the thermistor from being lowered further than 4 m.

The heater arrangement was then modified to prevent the refreezing of meltwater using an additional 15.3-m-long resistance wire (1 kW at 115 V) threaded with ceramic beads and installed inside the suspending pipes. This long heater was stretched out in the tube alongside the rubber-insulated cable, but was separated from it by an asbestos cement board of equal length (see Fig. 1.21a). The hot point and the pipe heater were connected in parallel.

This new arrangement was quite difficult to erect. It sunk into the ice to a depth of 21.4 m at Camp A.2X after 25 h of continuous work, conducted early in July 1950. The unexpectedly low ice temperature unfortunately prevented the hot point from reaching a great depth, presumably because the string had frozen in the ice above the top of the long heater. The actual distribution of power used in the final stages of penetration amounted to ∼1.8 kW at the tip and 100 W/m in the heating pipes over a length of 15.3 m. In the end, it was possible to lower thermistors into the drill string to a depth of only 9 m.

Fig. 1.22 Drilling operations at Saskatchewan glacier, Alberta, Canada, August 1952 (Meier et al. 1954)

1.2.8 Caltech Hot Points

In the 1950–1960s, the California Institute of Technology (Caltech), Pasadena, USA, designed several types of electrically heated hot points to use in different projects. In the summer of 1951, a vertical hole ∼50 mm in diameter was drilled to a depth of 305 m near the center of the Malaspina glacier, Alaska, using the first Caltech hot point (Sharp 1953). Unfortunately, the technical details of the drilling device and the drilling technology used were not published. Drilling stopped at that point because the hot point ceased to function. The glacier was 595 m thick. The borehole was cased with an aluminum pipe with an ID and an OD of 35 and 40 mm, respectively. The orientation of the pipe was determined using a small-diameter inclinometer.

The second Caltech 45-mm-diameter hot point was designed to study the magnitude and distribution of velocity within the Saskatchewan glacier, Alberta, Canada (Meier et al. 1954). This hot point operated at a current ranging from 8.0 to 9.5 A, and maximum penetration rates of 4.3–4.9 m/h were attained under normal conditions. Power was supplied by a portable 2.5–kW generator driven by a small gasoline engine. In 1952, drilling was attempted at two sites near the center of the glacier (Fig. 1.22). Both attempts were unsuccessful owing to burnt out or shorted hot points; the greatest depth attained was 45 m. An aluminum pipe with an ID and an OD of 35.1 and 41.9 mm, respectively, was selected for the drill stem because this allowed for the use of a small-diameter inclinometer. However, while setting down, the drill stem got stuck. In 1953, drilling was continued at a site higher up the glacier. The operation got off to a reasonably good start but ended in failure when the deepest hole, which was 120 m deep, was lost during the process of replacing a shorted hot point. The failure of the five available hot points resulted in the early end of the field season.

Another electrical hot point series was designed in Caltech for drilling in McCall Glacier, a small valley glacier in the Romanzof mountains of the Brooks Range (Alaska), and in Blue Glacier, a large temperate glacier located to the north of Mount Olympus, Washington, USA (Shreve 1961; Orvig and Mason 1963; Shreve and Sharp 1970). This 51-mm-diameter hot point was attached to the lower end of a string consisting of 3-m-long aluminum pipes with an ID and an OD of 35 and 42 mm, respectively, threaded with commercial aluminum taper-thread couplings and sealed with various commercial thread sealants. The power line passed through the inside of the pipe down to the hot point, and the pipe served as the ground return.

In July 1957, a 91.4-m-deep hole was drilled in the McCall Glacier (Orvig and Mason 1963). Power was supplied by a 4.9-kW Witte diesel generator. The drilling was carried out over a 27-hour period, and the average rate of penetration was 3.4 m/h. The average temperature measured during the 11-month period after drilling was −0.9 °C at a depth of 13.7 m and −1.1 °C at a depth of 91.4 m.

Another borehole experiment was done at Blue Glacier (Shreve 1961). From 1957 to 1964, 15 holes with depths in the range of 119–260 m were drilled using Caltech hot points. Drilling continued around the clock until the completion of each hole. Power was supplied by a commercial portable 2.5-kW, 230-V AC generator driven by a two-cycle gasoline engine that required ∼4 L/h of fuel when operated under full load conditions. The entire power plant weighed 55 kg. Aluminum pipes with a diameter of 40 mm were then placed in the holes, and annual inclinometer surveys were carried out (Shreve and Sharp 1970).

The first two boreholes at Blue Glacier were drilled with a hot-point prototype that burned out at power inputs greater than 1.6 kW. With that input power, these hot points could provide a maximum rate of penetration of 3 m/h, but had an

Fig. 1.23 Hot point design used in the late stage of the project at Blue Glacier (Shreve and Sharp 1970)

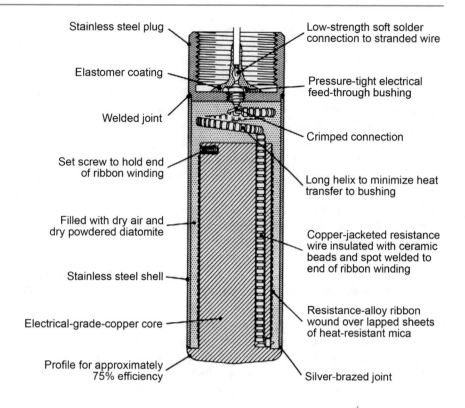

Stainless steel plug

Elastomer coating

Welded joint

Set screw to hold end of ribbon winding

Filled with dry air and dry powdered diatomite

Stainless steel shell

Electrical-grade-copper core

Profile for approximately 75% efficiency

Low-strength soft solder connection to stranded wire

Pressure-tight electrical feed-through bushing

Crimped connection

Long helix to minimize heat transfer to bushing

Copper-jacketed resistance wire insulated with ceramic beads and spot welded to end of ribbon winding

Resistance-alloy ribbon wound over lapped sheets of heat-resistant mica

Silver-brazed joint

average operating lifetime of only ~24 h. In subsequent drillings, improved hot points were used (Fig. 1.23). A copper-jacketed resistance wire isolated with ceramic beads was inserted into the hole of the short and thick copper core, which was insulated by overlapping sheets of heat-resistant mica. The helical winding was made of flat-strip wire and was grounded to the core at the high-temperature end to minimize the mechanical, thermal, and electrical stress on the mica. The front of the hot point was blunt, which yields the highest efficiency according to Shreve (1962a). The space between the copper core and the stainless-steel shell was filled with insulating powdered diatomite that had been baked for 24 h at high temperatures to remove water vapor.

Under normal operation conditions at a power input of 2.1 kW, the improved version of the hot point could drill at a rate of 9 m/h, making a hole that was ~60 mm in diameter. Two holes were terminated by hot-point failure, and the rest were terminated by very slowly penetrable (less than 0.3 m/h) material at the bottom of the hole. On rare occasions during drilling, the borehole pipe fell free for distances from 0.1 to 2 m, sometimes at depths exceeding 150 m. Simultaneously, the water level in the hole usually dropped by as much as several meters or sometimes overflowed; however, within an hour, it returned to its normal level of a few centimeters below the surface. When a hole was completed, the power wire, which was made to break at the lower end, was removed, and the inclination of the emplaced pipe was measured, usually at 8-m intervals.

A major source of difficulty in subsequent surveys of the Blue Glacier boreholes was the ice that formed inside the pipes despite all preventive measures. The casing pipes were initially filled with an antifreeze solution, but by the following summer and every summer thereafter, they were invariably blocked with ice, sometimes at depths as great as 200 m. Mechanical removal of the ice was possible near the surface but impractical at greater depths. Special 33-mm pipe-thawing hot points were therefore constructed following the principles developed for the main hot points. In these pipe-thawing hot points, however, the frontal surface had larger diameter than the body of the hot point. This allows the hot point to melt its way back up the pipe if the ice forms above it. In addition, the temperature of the core relative to the shell was thermostatically regulated, thereby temporarily protecting the heating element when the hot point entered air-filled sections of the pipe.

A new method to measure glacier deformation using a hot-point drill was suggested by Kamb and Shreve (1966). First, a steel cable was suspended in an uncased hole drilled through the ice. The deformed hole is recovered with a hot point that follows the steel cable. This method allows for a greater resolution of detailed flow features than that possible in holes cased with pipes, and it avoids the loss of holes due to pipe buckling or breakage. Using this method, three holes in the ice base were served 50 m apart in Blue Glacier.

During the summers of 1969 and 1970, a conventional 51-mm-diameter Caltech hot point was used again at Blue Glacier near the equilibrium line, where the ice thickness is approximately 120 m (Harrison and Kamb 1976). The holes, which tended to refreeze slowly, were maintained at a diameter of 60 mm using a special conically-shaped thermal drill. A thin layer of debris accumulated beneath the thermal drill up to a thickness of only 2–3 mm, which was detected via borehole photography. This layer was enough to slow the thermal drilling rate, which was normally ~ 7 m/h, by three orders of magnitude.

1.2.9 Hot Point of Oxford University

In the summer of 1951, an Oxford University expedition to Northeast Land in Svalbard was carried out to drill 16 holes in Bråsvellbreen, an advancing polar glacier, in order to measure englacial temperatures and to attain a better understanding of the behavior of this glacier which, in 1938, suffered a spectacular sudden slip-type of advance, causing it to push forward into the Barents Sea over a distance of more than 15 km.

The drill had a tail attachment consisting of a 19-mm steel tube extending upward from 0.5 to 1.3 m above the drill head (Miller 1953). The tube was weighted with lead rings at its lower end. These had outer diameters of 25 or 31 mm, which allowed the drill weight to be altered from 0.7 to 2.5 kg. A faster and smoother drilling rate (0.46 m/h) was achieved with the weight of the ring with the smaller diameter. Changes in the length of the guide pipe, however, had little effect on the speed of advance.

Serious freezing occurred in the borehole. It was found that a heater could be extracted from a partially frozen hole by slowly jiggling it up and down and by allowing heat from the hot point to melt away projecting ice obstructions. An effort was also made to employ an extraction heater at the upper end of the hot point's tail assembly. This was tested both as a separate resistance and as one connected in series with the lower heater; however, using a separate cable for the extractor was more satisfactory.

Heating the drill with one third of the power in the body heater and two thirds at the tip heater yielded a much slower rate but a smoother hole. The deepest hole drilled in Bråsvellbreen reached a depth of 30 m and required approximately 13 h of work. One third of this time was spent on reaming the upper part of the hole. The overall rate on each hole averaged 0.3 m/h. By subtracting the time necessary for raising, lowering, and so forth, a practical rate of ~ 2.4 m/h was attained. Six different heating elements

burned out during the course of these efforts—five in dry holes and one in free air when the current was accidentally left on. The coils were robust enough themselves, but problems seemed to arise in the asbestos and silica insulation cylinders on which they were mounted.

1.2.10 Granduc Mines Hot Point

A series of deep holes was drilled using hot points in the spring of 1956 in Salmon Glacier in the Coast Mountains of British Columbia near the south-eastern tip of Alaska (Mathews 1957, 1959). Drilling was done by Granduc Mines Ltd., a company engaged in the exploration of a large copper deposit in a nunatak, 40 km northwest of Stewart. Tunneling between Stewart and the mine had been considered as one possible means of providing access. To ensure that the tunnel would pass beneath Salmon Glacier, after an unsuccessful attempt at a seismic survey the company decided to drill holes in the ice to establish the bedrock profile along the tunnel line.

Drilling was conducted by means of a Calrod thermal element sheathed in copper and 76 mm in diameter, powered by a 220-V 10-kW portable generator with a variable voltage transformer. The hot point was suspended from the surface by means of a 4.7-mm wire rope to which the electric cable was clamped. When drilling through firn, the element operated at a reduced voltage. After the hot point entered the ice, the power input was increased to 220 V and 8 A, which allowed for an average penetration rate close to 1.8 m/h. Operations were conducted through the floor of an insulated plywood hut, which also served as the living quarters for the drill crew, consisting of two men. Apart from brief halts for maintenance or for the replacement of burnt-out elements, drilling was conducted around the clock until each hole was completed.

The first hole was abandoned at a depth of 323 m without reaching the bottom of the glacier, but five other holes (the deepest holes ever drilled using hot points) were terminated at depths ranging from 495 to 756 m before obstacles were reached that could not be penetrated after 24 h of drilling. These obstacles were assumed to be the bedrock floor of the glacier, although it was recognized that large boulders embedded within the ice near the glacier base might similarly prevent the penetration of the hot point.

Each hour, the hot point would be lowered to the bottom of the hole and ~ 3 m of slack cable were paid out; at the end of the subsequent hour, the remaining slack was withdrawn, the depth of penetration was noted, and 3 m of slack cable were again paid out. The first hole, which was drilled using this procedure, became so crooked that it had to be abandoned. In later works, a pipe 6 m in length was rigidly attached behind the heating element to guide it in its descent.

This device greatly reduced the crookedness of the holes, but the survey of hole No. 4 (496 m) indicated that once the drill unit was deflected from the vertical path, it tended to continue off plumb. Difficulties were encountered in hole No. 3 at approximately 460 m because of rock fragments, one of which was retrieved via fishing. However, in all holes other than this one and the first one, drilling continued with no significant problems until the bottom was reached.

To measure the vertical distribution of velocity in Salmon Glacier, a 494-m-long string threaded using 38.1-mm aluminum pipes was inserted in hole No. 4. Surveys in this hole carried out over a 94-day period show that movement diminished downward from a maximum value of 0.225 m per day at the surface through a slightly lower value at mid depth and down to 45% of this value at the bottom of the pipe.

Following the completion of the drilling operations in Salmon Glacier, the crew and the gear were moved to the neighboring Leduc Glacier, where by the end of summer 1956 four additional holes had been completed using the hot point, with a maximal depth of 366 m (Mathews 1957). Glaciological research was continued in 1957, completing five holes in the west fork of Leduc Glacier and reaching

what was apparently bedrock at depths of 201, 368, 384, 398, and 177 m (Mathews 1958).

1.2.11   Hot Points of the Cambridge Austerdalsbre Expedition

Two cased and two uncased boreholes were drilled using electric hot points in the temperate glacier of Austerdalsbre, Norway in 1956–1959 (Ward 1961). The first 127-mm-diameter hot point had a heating element with a power of 2.5 kW at 230 V (Fig. 1.24a). A commercial heating tubing element was wound into a tight spiral and casted into a separate aluminum cone that was fitted inside the outer nose cone. This arrangement allowed the heater to be replaceable in case of failure or when the casing pipe following the hot point had been set to its final position. The bottom face of the hot point was formed into a 90° cone; it had a series of circular grooves with a triangular section. A thermostat fitted on the top side of the heater casting allowed for opening the circuit to protect the heating element from overheating and burning out. The hot point was connected to a string composed of the 4.6-m-long pipes with an OD of 118 mm and

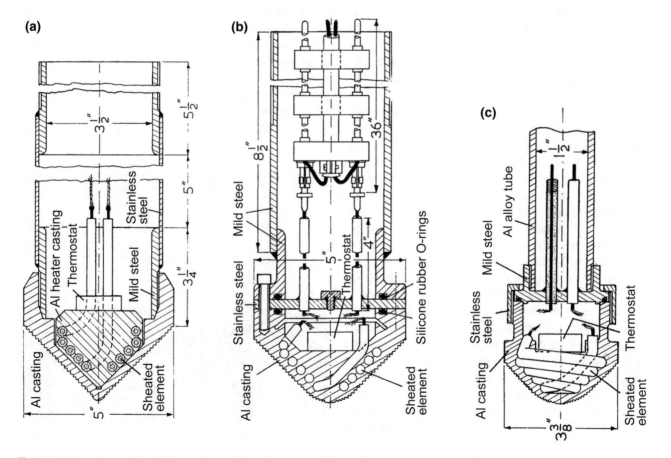

Fig. 1.24 Hot points used for drilling at Austerdalsbre, Norway: **a** 127-mm–diameter hot point used in 1956; **b** 127-mm-diameter hot point used in 1958; **c** 86-mm-diameter hot point used for drilling holes H1 and H2 in 1959; all sizes are given in inches (Ward 1961)

an ID of 105.2 mm, which were fitted with screwed-socket joints with an OD of 121 mm. Power was supplied by an electrical generator coupled to a 230-V 3-kVA self-exciting alternator.

This hot point was used for drilling at Austerdalsbre in the summer of 1956. In the first 5 m, the drill entered the ice at the rate of 1.8 m/h. However, the penetration rate at higher depths was halved, probably because of a thin layer of rock debris that accumulated beneath the melting tip owing to dirty ice. The operation was finally stopped when no effective penetration was obtained after dissipating 2.4 kW in the drill for 14 h, and it was presumed that bedrock had been reached at a depth of 39.3 m. The hole around the pipes remained full of water throughout the drilling operations. The engine regularly consumed 2.3 L/h of gasoline.

A second hot-point design had the same diameter of 127 mm (Fig. 1.24b). The 16.3-Ω tubing element was cast directly into the aluminum-alloy nose cone and the thermostat was mounted directly behind it. The connections to the external cable, which had to be separated when the drilling was complete, were made by means of a special plug and socket with a travel of 0.9 m. This hot point was used at the Austerdalsbre glacier in 1958. All available pipes were sunk into the glacier to a depth of 121 m without any serious problems and without reaching bedrock. Surprisingly, the average penetration rate (1.8 m/h) in the first 60 m was less than that for the rest of the drilling operation (2.0 m/h). The average rate of gasoline consumption was ~ 0.9 L/h, which was slightly higher than that in 1956.

The third hot-point design made was similar to the previous model; the same tip form was used but the head diameter was reduced as much as possible without substantially changing the power of the heating element (Fig. 1.24c). Two drill heads were made: one with a diameter of 81 mm and another one with a diameter of 86 mm. They could be loaded with as high as 2.6 kW. A total length of 9.9 m of aluminum tube was screwed behind the hot point. Metal discs with a diameter of 81 mm were fitted to the tube at its mid-point and its upper end to serve as guides.

In the summer of 1959, two vertical uncased holes were drilled in Austerdalsbre with these hot points. The rate of penetration in the first hole down to a depth of 117 m varied in the range of 3.3–3.8 m/h. Between depths of 117 and 140 m, the rate of penetration gradually decreased. At a depth of 140 m, the drill head with a diameter of 81 mm was changed to the drill head with a diameter of 86 mm on account of a partial short between the conductors within the compartment. Unfortunately, after the drill head was changed, the rate of penetration dramatically dropped owing to some debris retarding the melting of the ice. Finally, the hole was abandoned at a depth of 157 m on account of very slow progress in the lower 17 m.

The drill was removed and used to drill the second hole without any modification. Apart from one thin resistant zone at a depth of 18 m, the rate of penetration remained almost steady throughout the drilling at ~ 2.9 m/h. At a depth of near 100 m, the drill got suddenly stuck at the bottom of the hole. The heater was left operating for ten minutes without any significant penetration, and the drill was removed from the hole with some difficulty—only the upper part of the drill head was recovered.

1.2.12 Hot Point Experiments in 4th–8th SAEs

During 1958–1959, in the 4th Soviet Antarctic Expedition (SAE) of the Arctic and Antarctic Research Institute (AARI, Leningrad, now renamed to St. Petersburg), four boreholes were drilled with a hot-point thermal drill to a maximum depth of 52 m at Vostok Station (Fig. 1.25). Nine types of hot points were tested (Ignatov 1960b, 1962). One of the devices is shown in Fig. 1.26. The body was made from a 1-mm-thick zinc-coated steel sheet rolled into a tube with an OD of 150 mm. Sixteen heating elements divided into four sets with a power of 1 kW per set were fixed inside the body and molded with an aluminum alloy. During drilling, two sets (2 kW) of heating elements were used, whereas the others served as backup. The power lines went through a 1.4-m-long tube with an OD of 40 mm and with a stabilizer at the upper end. The hot point included a lead dead-weight, resulting in a total drill weight of ~ 40 kg. A hand winch was used for tripping operations. Drilling of the upper snow was going on quite smoothly until impermeable firn water refroze on the borehole walls, causing the hoisting of the hot

Fig. 1.25 First experience of thermal drilling at Vostok Station during the 4th SAE (Ignatov 1962)

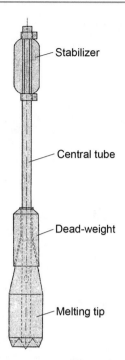

Fig. 1.26 Hot point used at Vostok Station (Ignatov 1960b)

point from depth of 40–50 m took up to 5–8 h. This hot point allowed for the study of the rate of penetration of drills versus depth for various firn densities (Ignatov 1960a). The penetration rate decreased from 1.3–1.6 m/h in snow to 0.25 m/h in dense firn at a power of 1.5 kW dissipated by the melting tip.

During the 5th SAE, thermal drilling with an electric hot point was conducted down to a depth of 44 m in the central part of the ice-cap-covered Drigalski Island, 78 km north of Mirny Station (Korotkevich 1965). Two other types of hot point were used at Lazarev Station (Dubrovin 1960; Korotkevich 1965). The first one was made from a 110-mm-diameter tube with two electric heaters (the power of each heater was 0.5 kW). During March–June 1960, four holes with depths ranging from 42 to 44 m were drilled using this simple device, with a mean penetration rate of 0.43–0.58 m/h. The melt water was removed using a special container. The second thermal drill was made from a steel fuel drum. The diameter of this thermal drill was 700 mm, and the total power of the heaters was 1.8 kW. It was used for the deepening of a 9-m well down to 36 m. The uninterrupted drilling of this hole took almost one month, with a mean penetration rate of only 0.05 m/h.

In 1961, two types of hot points with conical and flat face melting tips were tested at Mirny Station during the 6th SAE (Morev 1966). The diameter of both hot points was 130 mm. The conical tip was a copper vessel with a spherical point containing six tubular electric heaters embedded in lead with a total power of 6 kW. A penetration rate of 1.03 m/h was

achieved at a power consumption of 2.7 kW. When an asymmetric connection of the heating elements was tried, it was noted that the hole deflected in the direction of the more intensely heated side of the hot point. The hot point with a flat face consisted of two 3-kW tubular heaters coiled into a spiral and embedded in lead in a steel cylinder. The tubular heaters protruded by a third of their diameter. A penetration rate of 3.6 m/h was achieved with power consumption of 3.6 kW. The diameter (~ 170 mm) of the borehole produced by the conical hot point was larger than that produced by the drill with the flat face (~ 150 mm).

In 1963, the original hot point was tested at Mirny Station in ice with a temperature of $-10\ ^{\circ}$C, and it produced a 25-m-deep borehole. This 6.5-m long drill consisted of a heater with a flat face melting tip with an OD of 130 mm, a centrifugal pump, and a 22-L water collection tank with a length of 4.5 m (Morev 1966). Water was removed from the bottom and placed into the water collection tank through a heated tube. When the tank was full with melt water, the drill was lifted and water was removed from the tank. The power consumed by the drill was 4.5 kW, including power of the melting tip (3.6 kW), the power of the heater in the water-lifting tube (0.5 kW), and the losses in the power cable (0.4 kW). The drill weighed 55 kg. The diameter of the hole was ~ 150 mm, the average rate of penetration was 2.2 m/h, and the length of a run was 1.3 m. Drilling was discontinued because of the breakdown of the hydraulic winch.

1.2.13 Stacey Hot Point

This project was organized by the Universities of British Columbia and Alberta (Stacey 1960). Calrod tubing heating was used as the element of a hot point rated at 230 V and 1.5 kW and designed for running at a maximum temperature of 816 °C (Fig. 1.27). It employed a stainless-steel sheath that was 7.6 mm in diameter and had an overall length of 1.22 m. Three hot points were produced for the first drilling campaign, each with a slightly different design. In all three models, the heater was cut to a suitable length, wound into a helix, and cast into copper.

Stainless-steel tubing formed the hot point's body and was silver-soldered at its ends to the copper casting and the brass upper section. Heat flow through the sidewalls was prevented via Sil-o-cel (a diatomaceous earth material) and rock wool insulation. The body of the hot point served as one electrical line to the heater. Insulation of the input terminal was achieved by filling the surrounding volume with Araldite, which is a synthetic resin. A weak joint was provided to the electric cable to allow it to be withdrawn from the hole after drilling.

Fig. 1.27 Hot point used at
Athabaska Glacier in 1959
(Stacey 1960)

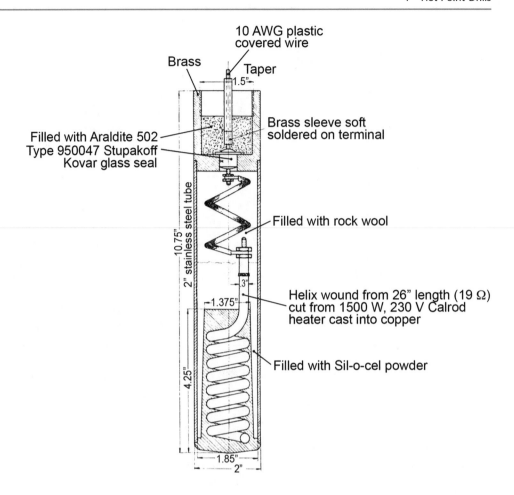

During the summer of 1959, five holes were drilled at Athabaska Glacier, Canadian Rockies to depths of 18.3, 76, 198, 228, and 312 m. The hot point was fixed to the end of a pipe string comprising 3.05-m-long aluminum pipes with an ID and an OD of 34.8 and 50 mm, respectively. The pipes were joined by threaded pipe coupling (OD of 55 mm) and the joints were sealed with an Alcoa lubricant compound. The power for drilling was provided by a two-stroke gasoline-driven motor-generator rated at 230 V and 3 kW; its total weight was 64 kg. The electrical output was made variable from 0–230 V via the inclusion of a rheostat in the field circuit of the generator.

The first hot point, which had a 16-Ω element, burned out on an obstruction in a hole 18.3 m below the surface because the heater overheated under these conditions. Another hot point with a 19-Ω heater failed at the lower termination of the filament. The last hot point, which had the same heater resistance as the previous one, was used for drilling the two remaining holes, of which the second one reached the base of the glacier 312 m below the surface. The hot point was assumed to have reached the bottom of the glacier when the drilling speed did not exceed a few centimeters per hour for a period of 2 h (Savage and Paterson 1963). The casing was then raised 1 m and clamped in that position until refreezing caused the ice to seize the casing securely. It took 48 h to drill this hole, with an average rate of penetration of 6.5 m/h.

During the summer of 1960 drilling at Athabaska Glacier was continued and several holes were drilled, including three holes deeper than 100 m (322, 209, and 116 m). The maximum rate of penetration was ~6 m/h. The greater portion of the ice was penetrated at the maximum rate, but layers of ice in which the drilling rate dropped to a few centimeters per hour for periods in the order of an hour were occasionally encountered. It was assumed that these layers of ice were contaminated with dirt, which formed an insulating film at the hot point. Not infrequently, the casing would fall freely for distances as great as 1 m. In such cases it was assumed that the hot point had penetrated a water-filled cavity within the ice and fallen through it. Whenever the hot point failed (four such failures occurred in 1959 and five in 1960), it was necessary to pull all of the casing to the surface to replace the hot point.

The temperature of at least the upper 100 m of ice appeared to be somewhat below the melting point. During drilling operations, it was necessary to introduce antifreeze (ethylene glycol) into the borehole outside the casing to prevent the

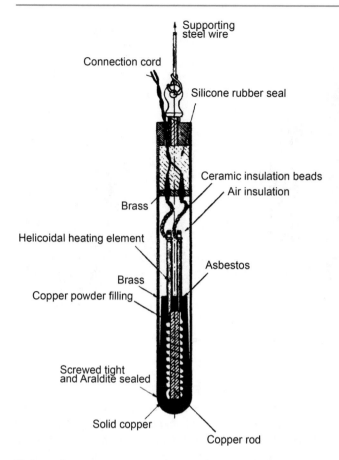

Fig. 1.28 Hot point used at Isfallsglaciären in 1961 (Ekman 1961)

casing from being seized by the ice forming at the borehole walls. The antifreeze was introduced through a 10-m auxiliary borehole beside the cased hole. Moreover, it was necessary to keep the casing filled with a solution of 5% antifreeze in water to prevent ice from forming within the casing.

The configuration of each borehole was determined via an inclinometer survey. The survey was repeated in subsequent years; three surveys (1960, 1961, and 1962) were carried out in most of the boreholes. During these years and between inclinometer surveys, the antifreeze was flushed from within the casing and the water that replaced it froze. Before another inclinometer survey could be carried out it was necessary to clear this ice by means of a small-diameter hot point, which operated inside the casing. During such clearing operations, there is a continuous hazard of ice reforming above the hot point.

1.2.14 Hot Point of Stockholm University

As a preparation for digging a tunnel along the bottom of Isfallsglaciären, Kebnekajse, Swedish Lapland, hot-point thermal drilling was performed for ice thickness measurements in the frontal zone of the glacier in March 1961

(Ekman 1961). One hot point was provided by Columbia University (Stacey 1960) and another one was manufactured and modified in Stockholm (Fig. 1.28). The heating elements (2.5 kW) were wound into a double spiral such that both ends could be taken out through the upper sealing. In the center of the heating elements, a copper rod was inserted. All outer surfaces were nickel-plated.

A Homelite gasoline-driven generator (2.5 kW, 115/230 V, 63 kg) was used as power source, and the current was transmitted to the hot point through an auto transformer. A number of drillings were carried out with both hot points, and it was found that they had both stopped 0.5–0.6 m short of the glacier bed, where the till content was too high for further penetration. The rate of penetration was approximately the same for both systems: ~6 m/h at 1.7 kW. The diameter of the hot points was 50 mm, and the hole diameter was 55–60 mm. The mean temperature of the ice was −4 °C.

1.2.15 LaChapelle Silicon-Carbide Hot Point

After numerous lab and field tests carried out in 1961 and 1962, a small-sized electrically powered hot point of only 18 mm in diameter was designed in the Department of Atmospheric Sciences, University of Washington (LaChapelle 1963). This hot point differed fundamentally from the more common type of hot points in the material of its resistance element as well as in its method for melting ice. In order to reduce the power consumption and weight of the hot point, silicon carbide (a ceramic material with a relatively high electrical conductivity) was used as a heating element (Fig. 1.29a). This element was a short section of a standard silicon carbide Globar Type A-S heating rod that was 8 mm in diameter rated at 115 V. Heaters with higher resistances for use at 220 V had very short operating lives.

The silicon carbide element had metalized ends for the electrical connections. Aluminum foil was used in the final design; a single layer of aluminum foil was pressed firmly against each end of the carbide element until it adhered, and then additional layers of foil were added until the element could be seated in the tip. To reduce current density by creating an increased contact area, the silicon carbide rod had cone-shaped surface ends instead of plane ends. This also provided maximum length to the element and hence maximum circuit resistance.

The first versions of silicon carbide elements with plane silver ends seldom drilled more than 5–8 m of ice before failing. The simplest way to extend the life of silicon carbide heating elements was to maintain the operating current below the critical level at which deterioration became rapid. This critical current density at the metal-carbide junction appears to be in the range of 7.0–7.5 A/cm for the type of

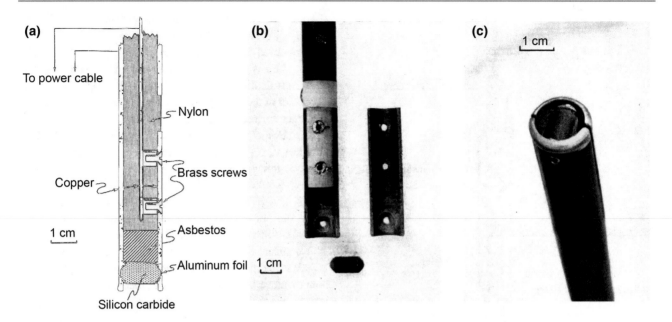

Fig. 1.29 Silicon-carbide hot point: **a** schematic; **b** hot point with opened split-tube (asbestos shield is not shown); **c** view from the bottom end of the hot point (LaChapelle 1963)

silicon carbide used, which normally operated at a contact current density of 6.0–6.5 A/cm. In these cases, the maximum operating power of the drill was approximately 220 W. This yielded a rate of penetration ranging from 5.5 to 6.0 m/h. The drill with the final improved version of the carbide elements configuration drilled from 40 to 70 m before failing. Failure usually occurred when gradual contact erosion reduced the effective contact area and thus raised the current density above the critical level.

The silicon carbide was placed in between a thick-walled copper split tube that could be easily split into two pieces for quick replacement of the heating element (Fig. 1.29b, c). The cover of the split tube was electrically insulated from the tube body by a nylon plug, which was thermally shielded from the heating element by a layer of asbestos. Electrical connections were made through the copper tube and its insulated section. A 2.5-m-long galvanized steel pipe (which is not shown in the figures) behind the copper tube provided alignment in the borehole and weight to ensure uniform drilling.

The silicon carbide element heated the surrounding water, which in turn melted the ice. This heat transfer method limited the hot point to be used only in impermeable ice where water remained in the hole; the heating element was very quickly destroyed via over-heating in permeable firn. To drill through firn, an auxiliary and simple thermal firn hot point was constructed from copper tubes and a high-density 165-W electric cartridge heater. This hot point had the same diameter as the carbide hot point and penetrated firn at a rate ranging from 6 to 8 m/h down to impermeable ice at depths of 15–20 m.

During the summer field season of 1962, twenty boreholes were drilled in the Blue Glacier accumulation zone, with a combined length totaling 983 m. A total of 718 m of this length were drilled using the silicon carbide hot point; the rest was drilled in firn layers using a hot point with a resistance-wire heater. Of the twenty holes, six were abandoned at depths ranging from 5 to 27 m when englacial crevasses or cavities were struck, which were too wide to allow for spanning with an aligned borehole. Five holes were abandoned at depths ranging from 2 to 58 m owing to technical difficulties (on three occasions, the drill got stuck in the hole and was lost). The remaining nine holes reached presumed bedrock at depths ranging from 38.6 to 142.0 m.

1.2.16 Aamot Buoyancy-Stabilized Hot-Point Drill

To ensure the vertical attitude of the drill and the plumb hole, Aamot (1968c, d) from the US Army Cold Regions Research and Engineering Laboratory (CRREL), Hanover, USA suggested using a heavy hot point and a light upper section that floats in the surrounding meltwater. The buoyant force is less than the weight of the drill in air, but its rectifying moment around the fulcrum is greater than the tilting moment of the drill's weight. This solid copper hot point was long in order to achieve a large contact pressure between the tip and the ice (Fig. 1.30). A 50-Ω electric resistance cartridge heating element was heated only in the lower half. It was completely soldered to the copper with tin-lead solder.

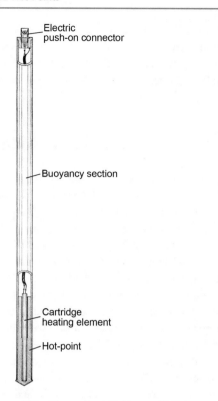

Fig. 1.30 Buoyancy-stabilized hot-point drill (Aamot 1968c)

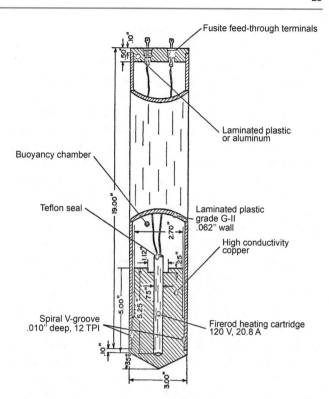

Fig. 1.31 Hot point design used at Rusty Glacier, 1969 (Classen 1970)

The buoyancy section consisted of a tube of laminated plastic (glass cloth with epoxy resin). It was bonded and sealed against the hot point and cap with an epoxy resin adhesive. The cable had a push-on connector to allow for the recovery of at least the cable if the drill could not be retrieved. The diameter of the hot point was 20.3 mm and the drill length was approximately 0.5 m. Power was transmitted from a 117-V AC generator through a 100-m-long coaxial cable. The total power consumption was ~0.25 kW.

In August 1967, five of these drills were used for ice thickness measurements in the Black Rapids glacier, Alaska. The first and second holes were 51 and 53 m deep, respectively. The third hole reached an empty cavity at a depth of 22 m; the drill made contact again with solid material at 60 m. The fourth hole was started near the third one and reached a depth of 62 m. In the fifth hole, the drill froze in place at 10 m after having been stopped overnight. The penetration rate was approximately 5 m/h. None of the drills were recovered, but the cable was pulled back out of each hole.

1.2.17 Classen Hot Points

Classen electric-heated hot point had a diameter of 76.2 mm and a copper melting tip with a 110° cone (Fig. 1.31)

(Classen 1970; Classen and Clarke 1972). An electric heating Firerod cartridge (maximum power of 2.5 kW at 240 V, 20.6 A) with a diameter of 19 mm and a length of 152 mm (effective heated length of 102 mm) was soldered onto the copper body of the tip. Originally, the hot point included a buoyancy chamber to ensure its verticality in the same way as suggested by H. W. C. Aamot (see the previous section). However, tests carried out prior to field use indicated that the wall thickness of the buoyancy chamber was not enough to withstand the anticipated pressures. To correct this fault, six probes were completely filled with an epoxy resin. In two others, the buoyancy chamber was perforated and allowed to fill with water. Fusite feed-through terminals in the cap were connected to a power cable. Later on, the end cap was redesigned to allow for the attachment of a 2.1-m-long steering rod.

To reduce the hole closure and prevent the anchoring of the power cable during drilling, the power cable was heated. It was expected that this arrangement would last 40 h before the hole would close to a diameter of 22 mm, which is slightly larger than the cable package (the probes were not recovered after drilling completion). Power was supplied by two 16AWG cables dissipating ~14.1 W/m. The 160-m-long line had a resistance of 2.5 Ω, dissipating 1.1 kW power each. If the water in the hole froze faster than estimated, heating could be increased by changing one of the 16AWG cables to a 22AWG cable with a resistance of 8 Ω and dissipating 3.3 kW. The power cable package was laid out on ice to prevent it from overheating. A 5-kVA Kohler generator of the 5RMS65 type was equipped with a voltage regulator that adjusted the engine speed, eliminating the need for a variable transformer.

During the summer of 1969, a thermal-drilling and ice-temperature measurements program was carried out on Rusty Glacier, a small valley glacier in the Yukon Territory, Canada. Eight additional hot points were purchased from the University of Toronto based on the prototype model developed by Stacey (1960) with tubular heating elements molded into the copper tip (see Sect. 1.2.13). Thermal drilling resulted in a total of seven instrumented holes at six locations on the glacier. The depth of the holes varied from 14.6 to 48.8 m; three holes reached bedrock. The rate of penetration was usually in the range of 1.0–1.5 m/h, with substantial sidewall melting. Only one element burn-out occurred during drilling owing to accidental overloading. Persistent generator difficulties, however, prevented the completion of three holes. A thermistor cable was attached to the power line package at 5-m intervals as the hot point descended. Temperature measurements indicated that the glacier was below the pressure-melting point throughout.

Basically, the same Classen 76-mm-diameter hot point was used in spring of 1975 for temperature measurements along a flow line in the surge area of Barnes Ice Cap (Classen 1977). Minor modifications to the drilling equipment included soldering of the heating element (2.24 kW at 20 A) into the copper tip as opposed to the use of a thermal compound, the insertion of the heating element to within 6 mm of the tip, increasing the probe length to 0.6 m, and employing a four-conductor power cable for variable line heating. The thermistor cable was taped to the hot-point power-supply cable as the drill descended. Hole closure due to the refreezing of meltwater in the hole prevented the retrieval of the hot point, and the entire instrument package was allowed to freeze in place once drilling had ceased.

A total of seven boreholes were drilled, with a cumulative depth of 547 m. This required 156 h of drilling time and ~34 m³ of fuel. These seven attempts yielded three instrumented sites with depths of 200, 122, and 88 m. Drilling of the first hole was stopped because of insufficient manpower (two-man crew) at the drilling site to maintain a workable shift schedule over a long period of time. Two other boreholes were terminated owing to the sudden reduction of the drilling rate, which supposedly occurred in englacial debris-laden ice. One hole (10 m) was ceased because of freeze-in due to the obstruction of the thermistor pod on the edge of the drill hole. The drilling of two other holes (17 and 40 m) was stopped owing to shorts in the power cable in the presence of high currents. In addition, the PVC-jacket insulation of the cable was not suitable for the air temperatures encountered, which ranged between −34.0

and −1.3 °C. Another hole (70 m) was lost when the hot point burned out because of a circuit overload owing to wrongly increased generator speed. Penetration rates generally approached 4.0 m/h and seldom decreased below 3.4 m/h.

1.2.18 ETI-1 Hot Point and Modifications

At the end of the 1960s and the beginning of the 1970s, V. A. Morev (AARI) developed a lightweight hot point Elektroigla (electric needle) of the ETI-1 type (Korotkevich and Kudryashov 1976; Morev et al. 1984). These drills, which were 1.5–2.0 m long, consisted of a melting tip of 40 or 80 mm in diameter attached to a pipe via a coupling nut and a connector (Fig. 1.32a). A cable termination and a spring centralizer were attached to the upper end of the pipe. Power was transmitted through a central electric wire and the armor of the electromechanical cable. The 40-mm-diameter drill weighed only 5–7 kg.

The melting tip had a conical shape with a spherical pointed end (Fig. 1.32b). An electric heater coil (nichrome wire) was winded on a copper core and covered by a steel jacket. The heater connections were sealed with rubber. The rated power of the smaller hot point was 1–2 kW, and that of the larger one was 3–4 kW. The ETI-1 hot point and its later modifications were used for drilling several hundreds of boreholes in high-mountain glaciers and sub-polar ice caps.

The first experiments were carried out on the edge of the northern part of Marukh Glacier, Central Caucasus, where a 20-m-deep hole was drilled (Sukhanov et al. 1974). In 1970–1973, a series of boreholes, including seven holes that reached bedrock, were drilled in Jankuat Glacier (Sukhanov et al. 1974; Golubev et al. 1976). The deepest hole reached 111 m. The speed of penetration gradually decreased from 10–12 m/h in the near-surface permeable firn zone to 5–6 m/h in the pure ice below 17–20 m. In the interval near bedrock, the penetration rate dropped to 2–3 m/h owing to debris in the ice.

In 1971–1972, the glacier thickness and ice structure of Abramov Glacier, Alai Ridge, Pamir (South Kyrgyzstan) were investigated using modified ETI-1 hot points (Ryumin et al. 1974; Morev 1976). To make penetration through dirty ice easier, the conical part of the hot point was enlarged two times. The total cumulative depth of these holes was nearly 1500 m, the deepest one is 172 m deep. The mean penetration rate was 7–11 m/h at a drill head power of 1.5 kW.

In 1972, dozens of holes were drilled using ETI-1 hot points for a total cumulative depth of nearly 300 m and a maximum depth of 34 m at drifting station Severnyi Polyus-19 in the ice island in the Arctic Ocean (Morev 1976). The average penetration rate was 7–9 m/h at a power of 1.7 kW at the melting tip. In 1973–1974, 28 holes were

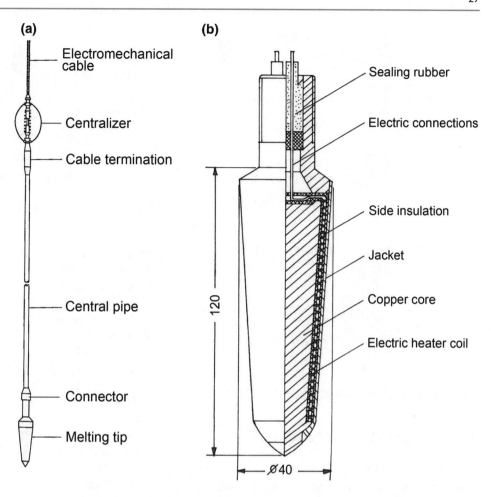

Fig. 1.32 ETI-1 hot point: **a** general schematic (Morev et al. 1984); **b** melting tip (Korotkevich and Kudryashov 1976)

drilled, having a total cumulative depth of 1426 m (maximum depth of 137 m) in Obruchev Glacier, Polar Urals (Morev 1976; Zagorodnov et al. 1976). Eleven holes were drilled down to bedrock. The other holes penetrated only the upper 20–30 m.

In 1974–1976 and in 1980, eight boreholes were drilled by an AARI drilling team using an ETI-1 hot point at Vavilov Glacier, Severnaya Zemlya (Morev and Pukhov 1981; Morev et al. 1988). To study drilling fluid stability, two holes drilled in 1974 were filled with low-temperature drilling fluids. In 1976, a 310-m-deep borehole was drilled with a modified ETI-1 hot point employing an antifreeze-assisted drilling technique. The borehole was kept open via the injection of antifreeze (ethanol). Because of the accumulation of dirt at the bottom of the hole, which required excessive thrust loads, the axis of the hole deviated considerably from the vertical direction up to 45° [sic]. The true vertical depth of the hole was only 180 m.

In 1975–1982 more than 20 boreholes were drilled at the Svalbard glaciers using modified ETI-1 hot points (Kotlyakov 1985; Zagorodnov 1981, 1988; Zagorodnov and Zinger

1982; Zagorodnov et al. 1984). The rate of penetration varied in the range of 10–18 m/h. The deepest borehole was drilled in the central part of the Amundsen Plateau in 1980. It was started from a depth of 368 m, drilled using a coring electro-thermal drill, and was terminated at 586 m (Zagorodnov and Zotikov 1981). Approximately 20 m above the bottom, the hot point was accidentally disconnected from the cable. A new drill and a whipstock were built from available materials and a spare heating head. The whipstock was installed just above the lost hot point, which was successfully bypassed. After the end of the drilling operations, a deviation survey was carried out. The inclination in the main part of the hole did not exceed 1°–2°, but it abruptly increased to 30° near the bottom.

In 1987, two holes were drilled at Garabashi Glacier in the southern Elbrus slope using a hot point, reaching 73.2 and 77.7 m (Bazhev et al. 1988; Zagorodnov et al. 1991). Both holes reached bedrock. Based on the penetration rate, the thickness of the firn zone was estimated to be 24 m. The glacier was temperate all the way down and was close to the melting point.

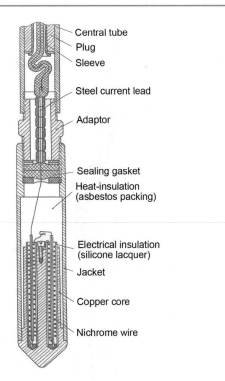

Fig. 1.33 Mikroteb-1 hot point (Sukhanov et al. 1974)

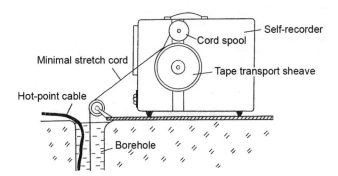

Fig. 1.34 Device for recording hot-point penetration rates (Zagorodnov et al. 1976)

1.2.19 Mikroteb-1 Hot Point

The Mikroteb-1 hot point was designed by L.A. Sukhanov, Moscow State University, USSR in the beginning of the 1970s and had a melting tip with the shape of a blunt cone (Fig. 1.33) (Sukhanov et al. 1974; Golubev et al. 1976). Four nichrome spirals were fixed inside quartz tubes with a diameter of 3 mm and installed into the blind holes of an 18-mm-diameter copper core. The total power consumption of the heaters was only 0.2–0.37 kW. This hot point was used together with an ETI-1 hot point to estimate the thickness of Jankuat Glacier, Central Caucasus, in 1970–1973. The Mikroteb-1 hot point had a slower drilling speed than ETI-1: 4–5.5 m/h in the upper firn and 1–1.5 m/h in pure ice.

1.2.20 Speedograph and ETI Hot Point with Penetration Rate Recording

The penetration rate of a hot point depends on the density of the snow–firn–ice layers that are melted. Numerous experiments showed a linear relationship between snow/firn density and the rate of penetration. The concept of using a hot point's penetration rate as a measure of the snow-firn density distribution was first tested at Taku Glacier, Juneau Icefield, Alaska in 1952–1953 (Miller 1953). Experiments in which

the penetration rate was recorded and the density changes of the snow-firn layers were interpreted were carried out at Vostok station, Antarctica in 1958–1959 (Ignatov 1960a) and Jankuat Glacier, Central Caucasus in 1970 (Sukhanov et al. 1974; Sukhanov 1975). Although these tests validated the technical concepts, the calibration of the tools used proved very challenging.

In 1973, V. Zagorodnov developed a hot-point penetration rate recorder called Speedograph based on a self-recording voltmeter N-390, which was connected to a hot point and registered the penetration rate (Fig. 1.34) (Zagorodnov et al. 1976). A few dozen penetration rate profiles with a resolution of 5.6 mm were obtained from ice fields in the Polar Urals and Svalbard. The deepest profile obtained was 27 m deep in Lomonosov Fonna Glacier in Svalbard (Zagorodnov and Zotikov 1981). However, these measurements were never properly calibrated.

In order to determine the porosity and thickness of grounded accumulations of sea ice, such as ridges and stamukhas, the ETI-1 hot point (see Sect. 1.2.18) was modified (ETI-3M2 and ETI-4 types) in the AARI to drill shallow holes and was equipped with a computer system to record penetration rates (Morev and Kharitonov 2001). This hot point consisted of a heating head in the form of two mirror-like cones and a drill body tube (Fig. 1.35). The lower cone was actually the thermal head, with a resistive heating element inside that consumed 85–90% of the supplied power. The remaining 10–15% of the supplied power was consumed by the lateral surface of the upper cone to heat meltwater during drilling and to melt refrozen ice jams during hot-point lifting. The ETI-4 hot point was different in that an additional heater was installed in the drill body tube to avoid icing and difficulties when raising the hot point from the borehole (Kharitonov and Morev 2005).

The outer diameter of the thermal head was 35 mm and the height of the lower conical tip was only 5 mm, which allowed for determining voids with a vertical size of not more than 10 mm. The total hot-point length was 1.8 m, and

Fig. 1.35 ETI hot point with penetration rate recording (Morev and Kharitonov 2001)

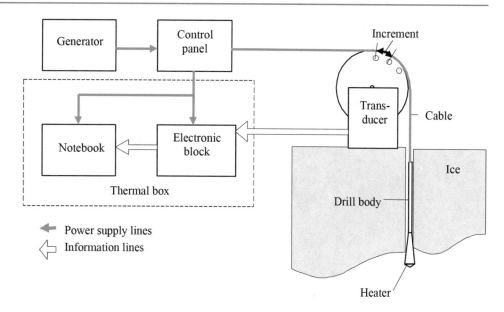

it weighted 2.6 kg. Power was provided by a 2-kW (220 V) generator. The total weight of the drill set (without the generator and fuel) was 19 kg.

This hot point and the registration system were first tested on the northeast shelf of Sakhalin Island in the spring of 1998 (Kharitonov 2008). A total of 89 records of the penetration rates were made on 14 ice ridges, and 18 records were made on two stamukhas. The thickness of these stamukhas was in the range of 14–15 m. The penetration rate in hard clear ice with a temperature of approximately −5 °C was 16–18 m/h with a power consumption of 2 kW. In the ensuing years, hot-point drilling experiments with rate recording on ice ridges and stamukhas were performed in Pechora Sea in 1999; Caspian Sea in 2003 and 2005; drifting stations in the Arctic Ocean Severnyi Polyus–32, Severnyi Polyus–33, and Severnyi Polyus–35 in 2004, 2005, and 2008, respectively; and Sea of Azov in 2005 (Kharitonov and Morev 2005, 2009, 2011; Kharitonov 2005, 2008).

In the 2000s, V. Zagorodnov totally modified his first Speedograph (Zagorodnov et al. 2013). The current version (Fig. 1.36) contains a small 32-mm-diameter hot point (1.0 kW, 0.95 m in length, weight of 5 kg), a mast with a wheel, a winch with a 50-m cable, a PC data acquisition system, and a power generator. The data acquisition system allows it to record depth, penetration rate, hot-point load, tip power, and temperature. The weight of Speedograph (including the shelter and power supply system) is ∼150 kg. Hot-point penetration rates vary typically in the range of 9–12 m/h. Speedograph needs careful calibration: under stable power supply conditions, the hot point penetrates sections 5–10 cm long of ice samples or cores with known densities.

An improved Speedograph with high depth resolution (0.9–13 mm) has been used in Crawford Point, southwestern Greenland in 2007 and the Elbrus western plateau in 2009 to a maximum depth of 33 m. The Speedograph continuous profiles obtained in Greenland were compared with the density profiles obtained in snow pits. However, the profiles were poorly correlated, which was perhaps caused by the significant horizontal variability of the snow–firn–ice densities in this area.

Fig. 1.36 Speedograph in Greenland, 2007; the hot point is in the hole (*Credit* V. Zagorodnov)

Fig. 1.37 Hot point designed by
Taylor (1976)

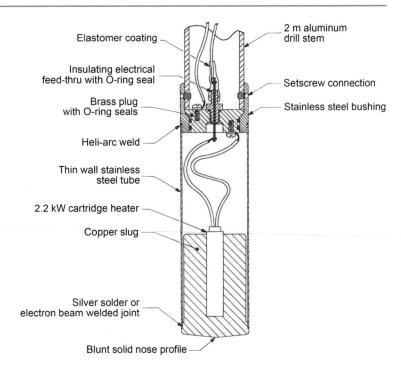

Elastomer coating

Insulating electrical
feed-thru with O-ring seal

Brass plug
with O-ring seals

Heli-arc weld

Thin wall stainless
steel tube

2.2 kW cartridge heater

Copper slug

Silver solder or
electron beam welded joint

Blunt solid nose profile

2 m aluminum
drill stem

Setscrew connection

Stainless steel bushing

1.2.21 Taylor Hot Points

Taylor (1976), from the University of Washington, designed a hot point with a basic diameter of 50.8 mm that used a cartridge heater rated at 2.2 kW and 220 V (Fig. 1.37). Hot points with a similar design with diameters of 25.4 and 31.8 mm were also constructed using a cartridge heater rated at 1 kW and 110 V or 220 V and measuring 12.7 mm in diameter and 38 mm in length. To prevent refreezing in the drilled boreholes and to ensure the safe passage of the instrumentation, a tapered borehole reamer 63 mm in diameter and 610 mm long was designed. This reamer was machined from an aluminum bar and had a heating cartridge that was 12.7 mm in diameter and 510 mm in length and rated at 2.2 kW and 220 V inserted in a long hole on the cylinder axis. The termination of the drill cable was standardized so as to allow for swapping between hot points with different sizes and the reamer.

The hot points could be opened easily for maintenance and for replacement of the heater in case of burn-out (Fig. 1.38). The heater could be inserted into the copper slug with a light push. The contact surface between the heater and copper was treated with a commercial silver plating compound to improve and maintain heat transfer. The heater was wired through a watertight plug. The nose was blunt and a thin-wall stainless-steel tube was chosen for the body. The hot point was guided by a 2-m-long drill stem made of aluminum tubing weighted with ~7 kg of lead (Fig. 1.39). The surface set contained a drum with a 150-m-long cable, a meter wheel sheave, and a power control box.

Fig. 1.38 Hot point assembled and with the plug and heater assembly removed with a small pulling tool (Taylor 1976)

Performance checks of these Taylor hot points over several field seasons and hundreds of meters of boreholes with a maximum depth of 210 m yielded typical drilling rates of 5–6 m/h at a power of approximately 1.3 kW for basic diameter probes. Interesting experiments were carried out by Rogers and LaChapelle (1974) in Blue Glacier at Mount Olympus, Washington in 1971. To measure vertical strain of the glacier, three holes were drilled to a depth of ~90 m and were thermally reamed to a diameter of 62 mm. Then, a total of 40 markers were placed in each hole, starting from the bottom and proceeding to the top. The markers were electrically conducting rings made from thin (0.75 mm) phosphor-bronze

Fig. 1.39 Hot-point platform and operator holding the drill stem (Taylor 1976)

spring strip and were placed concentric to the axis of the hole. The location of these rings was detected by passing a resonant electrical circuit housed in a polyvinylchloride pressure capsule.

In the same year, two other holes, which were not cased, were made using 25-mm-diameter hot points with an average rate of ~8 m/h in Blue Glacier to depths of 192 and 76 m (Harrison 1975). Upon the completion of these holes, the drill and drilling cable were recovered and replaced by a single-thermistor cable assembly. Again in 1971, two boreholes were drilled to a depth of ~70 m to measure the ice thickness near the centerline of Nisqually Glacier, which is the largest glacier on the south side of Mount Rainier, Washington (Hodge 1974).

In 1973–1975, Taylor 50.8-mm-diameter hot points were used to connect with active water systems at or near the bed of South Cascade Glacier, Washington (Hodge 1976, 1979). Typical drilling rates were 5–6 m/h using full generator output (1.7–1.8 kW). Drilling was maintained as long as possible in the debris-laden ice near the glacier bed. Of the 13 holes that were ~100 m apart drilled in 1973 and 1974, seven connected with englacial or subglacial water systems while drilling. In the early spring and early summer of 1975,

14 boreholes were drilled in the lower half of the accumulation area with the same hot-point design. However, connection with a water system was not successful until the fourteenth hole was drilled.

1.2.22   Hot Points of the Laboratoire de Glaciologie, CNRS

A small-sized hot point with a diameter of only 18 mm for measuring the thickness of glaciers was designed in the Laboratoire de Glaciologie, Centre National de la Recherche Scientifique (CNRS), Grenoble, France (Gillet 1975). A tubing heating element was twisted up and casted into a silver tip (Fig. 1.40). The casting of silver was done in a vacuum furnace to guarantee a perfect thermal contact between the heating element and the silver without air bubbles or inclusions. The hot point operated with a single-phase voltage signal varying from 0 to 220 V supplied by a 1.5-kW generator.

A larger hot point (45 mm in diameter) was supplied with a three-phase voltage of 380 V to drill holes for measuring the deformation of ice. A load sensor was attached to the cable termination of the hot point. An additional heating resistance was placed at the top of the hot point to be used in case of jamming on the way back.

During lab tests, the 18-mm-diameter hot point drilled in pure ice with a rate of 18.6 m/h for a head electric power of 825 W and a rate of 15 m/h at 650 W. In 1973, numerous drilling operations were carried out in Mer de Glace and the glaciers of the Sommet de Bellecôte (French Alps). The speed obtained there (12 m/h at 650 W) was lower than in lab tests. When drilling in a glacier, the speed decreased very sharply at times for a few centimeters or decimeters, probably

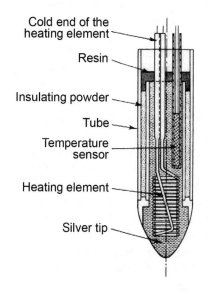

Fig. 1.40 Hot point used at the French Alps (Gillet 1975)

Fig. 1.41 The drill set up at Barnes Ice Cap; from left to right: box with the core barrel, tool box, drum with glycol, control panel, hand pump, and cable reel, June 1976 (*Photo* R. LeB. Hooke)

because of debris layers. In the Mer de Glace, in a place where bedrock was reached at 180 m, this phenomenon was observed at 60 m and was so strong that drillers believed for a moment that they had reached bedrock.

1.2.23 Hot Points of the University of Minnesota

Two hot points with diameters of 40 mm (2 kW at 9.1 A) and 80 mm (4.4 kW at 9.8 A) were designed and built at the University of Minnesota (Hooke 1976). The drill set consisted of a 5-kW generator, an autotransformer, a transformer, cable, the hot points, a core thermal drill, and an ethylene-glycol injection system (Fig. 1.41). The complete set weighed ∼500 kg, excluding the fuel and glycol. The downhole part of the drill consisted of the hot point itself and a buoyancy section (Fig. 1.42). The length of the buoyancy section was adjusted so that when submerged with the hot point attached, it assumed a vertical attitude and thus kept the borehole plumb, as proposed by Aamot (1968c, d).

A cylindrical cartridge heater was inserted into a hole in a solid brass or copper hot-point tip with a parabolic surface profile (Fig. 1.43). A small cylindrical cap was screwed at the end of the tip for easy cartridge installation. The shape of the parabola was designed to satisfy the condition that each tip section had the melting capacity just necessary to melt the ice encountered as the hot point penetrated the ice at a uniform velocity. The weight of the 40-mm-diameter hot point was only 1 kg, whereas the 80-mm-diameter one weighed 8.7 kg.

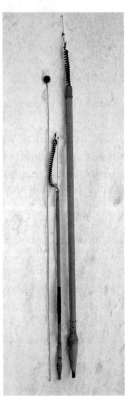

Fig. 1.42 Hot point and buoyancy sections (*Photo* R. LeB. Hooke)

Power for these hot points was supplied through four conductors. Two conductors were connected together and used as a single line. The third conductor was grounded to the buoyancy section and, hence, to the hot point. The fourth

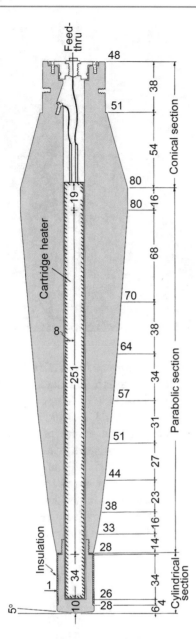

Fig. 1.43 Cross-section of an 80-mm hot point showing the diameter of points at various positions and the distances between these positions (all dimensions are given in mm) (modified from Hooke 1976)

conductor was connected to the hot-point body through a sliding switch. As long as current flowed through this switch, the signal lamp on the control box remained on. Whenever the circuit was broken and the lamp went out, an alert operator would let out more cable.

The switch included a 1-m-long stiff, straight, and springy wire sliding down into a stainless-steel tube with an ID of 4 mm passing entirely through the buoyancy section from one end to the other. A cylindrical brass sleeve with a rounded tip was slipped over the end of the wire. This sleeve made electrical contact with the tube and, hence, with the

buoyancy section. When retracted from the stainless-steel tube, the sleeve entered a short thick-walled leucite tube and the electrical contact was broken.

Power was provided by a 5-kW gasoline-powered Onan (model 5CCK) generator with 120 and 240-V outlets. Power from one of the 240-V outlets was delivered to a variable autotransformer and thence through a step-up transformer, which doubled the voltage. Using this arrangement, the voltage at the control-box outlet could be varied continuously from 0 to ~560 V.

To prevent the melted water produced by the hot point in cold ice from freezing, it was diluted with ethylene glycol and left in the hole. The glycol system consisted of a reservoir, made from a 45-L drum, from where glycol was drawn by a hand-operated piston pump developing pressures of 0.5–1 MPa and forced down the borehole through polyethylene tubing with an ID of 9.6 mm. Between the reservoir and the pump, there was a filter to remove foreign particles from the glycol. A transparent tube and scale were installed on the side of the drum to monitor glycol use. The tubing was taped to the electrical cable, and both were handled simultaneously on the cable reel. This cable system proved to be poorly adapted for use owing to differences in the coefficients of thermal expansion of the cable and tubing, and perhaps also to the plastic stretching of the tubing.

At the top of the buoyancy section, a piece of ~1-m-long latex tubing was used to connect the plastic tubing to a stainless-steel tube with an ID of 4 mm that passed entirely through the buoyancy section. This tube was parallel to the tube mentioned earlier for the switch. Thus, the glycol was injected immediately above the hot point at the junction between the hot point and the buoyancy section.

In 1970–1978, 18 boreholes were drilled using hot points of both diameters in the south dome of the Barnes Ice Cap, Baffin Island, Canada, the deepest one being 300 m deep (Hooke and Alexander 1980). Drilling continued until debris in the basal ice effectively prevented further penetration. However, the depth of the deepest borehole was limited by the length of the drill cable.

In field tests, the penetration rates obtained were 7.6–8.2 m/h with the 40-mm-diameter hot point and ~4.9 m/h with the 80-mm-diameter hot point. The actual borehole diameter was ~53 mm when using the small hot point and ~90 mm when using the larger one. In some instances, a thermal core drill was used interchangeably with the hot point to clean up the accumulation of sediment at the bottom of the hole. For example, in one of the holes drilled in 1975, debris-laden ice was encountered at a depth of 41 m, but the use of the thermal core drill allowed the drillers to penetrate an additional 11.5 m of this dirty ice before drilling rates became unreasonably low (<0.1 m/h).

The holes were cased with round aluminum tubing with an ID of 28 mm or square tubing with an inside dimension

Fig. 1.44 Simple hot point of the Japanese Glaciological Expedition to Nepal, Khumbu Glacier, East Nepal Himalayas, 1974 (Mae et al. 1975)

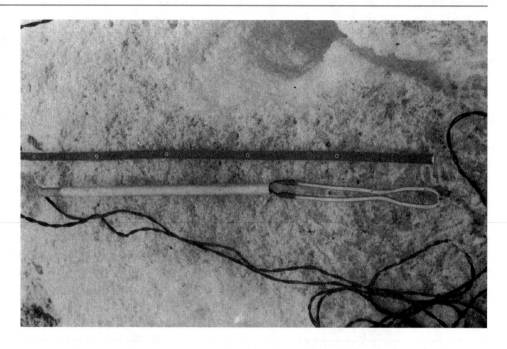

of 32 mm (Hooke and Hanson 1986). In the latter holes, deformation was measured in addition to temperature. To sink the casing into the holes filled with the dilute water-glycol mixture and to avoid buoyant forces, the casing was filled with diesel fuel. As the casing sank into the hole, it displaced the water-glycol mixture. Thus, when the hole was fully cased, there was diesel fuel inside the casing and water glycol mixture outside the casing.

1.2.24 The Simple Hot-Point of the Japanese Glaciological Expedition to Nepal

In the Japanese Glaciological Expedition to Nepal, researchers decided to use a 0.5-kW electric heating cartridge as a hot point drill (Fig. 1.44) (Mae et al. 1975). The heater was connected directly to a 1.2-kW gasoline generator. In August 1974, this simple hot point was used for drilling in Khumbu Glacier, flowing from Mt. Everest in the East Nepal Himalayas. One hole was drilled to a depth of 20.3 m with an average penetration rate of 0.56 m/h. The diameter of the drilled hole was ∼150 mm near the surface. The heater was still functional after stopping, but the feeding cable had frozen in the hole (the temperature at a depth of 2.7 m was measured ten weeks after drilling to be −5.3 °C).

1.2.25 Hot Point of the Geological Survey of India

A series of hot points were designed for the Geological Survey of India in the mid-1970s in connection with the

glaciological research project at Zemu Glacier, which is the largest glacier in the Eastern Himalaya with ∼26 km in length and located at the base of Kangchenjunga, the third highest mountain in the world, in India (Datta 1980; Dutta 1980). The working version of this hot point (Fig. 1.45) consisted of two parts: a heating chamber and a dead weight on the top of a probe. The 340-mm-long and 75-mm-diameter body of the heating chamber was made of gunmetal. The tip was machined in the shape of a cone with a hole in the central part. This hole served for fixing a 90-mm-long copper heating rod that was 16 mm in diameter.

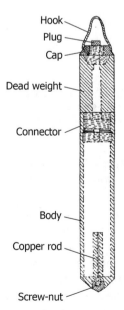

Fig. 1.45 Schematic of the hot point designed in the Geological Survey of India (Datta 1980)

A thin brass sleeve (not shown) with an outer thin mica insulation was slipped over the rod. A heating coil was wound over the sleeve and separated turn-to-turn by thick asbestos rope. The space between the heating rod and the body was filled with asbestos powder. All threaded connections were sealed with araldite.

The probe was suspended on a polythene rope connected to the hook. The electric cable and the polythene rope were coupled with one another using clips. To hoist the probe, a wooden coil was used along with a small tripod with three 0.56-m-long legs and a double pulley system to operate both the electric cable and the polythene rope at the same time. Power was provided by a 3-kVA single-phase generator through a voltmeter, an ammeter, a safety fuse, and a heavy-duty switch.

The hot-point probe was tested several times in a lab in Calcutta, and drilling at Zemu Glacier was performed during 1976. A rate of penetration of 1.56 m/h was achieved with a 1-kW heating element, resulting in a hot-point efficiency of 57%. Unfortunately, the drilled depth in the glacier was only ~2.5 m. Then, the operations were terminated because of problems with the generator and breakdown of the probe body after it was dropped on the ground by a porter during carrying.

1.2.26   Grześ Hot Point

A portable hot point was designed in the Institute of Geography and Spatial Organization, Polish Academy of Sciences, Toruń (Grześ 1980). A copper tip was wound with resistance wire (0.7 kW) and insulated with ceramic beads and mica (Fig. 1.46). The OD of the tip was 27 mm. The hot point was covered by a stainless-steel jacket threaded with the copper tip. Upward heat flow was prevented by ceramic inserts and a layer of asbestos rubber. In addition, silicon gum filled the top of the hot point to make it water tight. The total hot point length was 155 mm and it weighed 1.4 kg.

To control the verticality of the drilled hole, the hot point was connected to a 10-m-long guiding brass pipe with an OD of 16 mm, which was threaded to the upper end of the hot point. The guide included an oval sleeve with the same diameter as the tip. Power was supplied by a small generator (220 V, 1.5 kW) through a self-transformer, which was used for adjusting the voltage, and two-wire cable.

Drilling was done at Hans Glacier, Svalbard in 1979 to install ablation stakes at depths of 5–10 m and to carry out temperature measurements at depths of a few tens of meters. To prevent the hot point from getting stuck owing to refreezing of the melted water, methyl alcohol was poured into the borehole in amounts of 0.25–0.33 L/m. The diameter of the melted boreholes was ~40 mm, with the maximal rate of penetration achieved being 8.5 m/h. However,

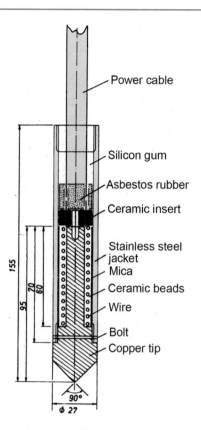

Fig. 1.46 Hot point used at Hans Glacier, Svalbard, 1979 (Grześ 1980)

the rate of penetration was usually kept in the range of 2–4 m/h.

1.2.27   Climatopic Thermal Probe

The aim of the Climatopic thermal probe designed in the Laboratoire de Glaciologie et Géophysique de l'Environnement (LGGE), CNRS, Grenoble, France was to obtain continuous stable-isotope recordings and also to measure, at a later stage, the temperatures in the hole (Gillet et al. 1984). It was expected with runs of up to 6 m to drill deeper than 3000 m in a single summer season. To prevent borehole closure, it was filled with a mixture of diesel fuel of arctic blend DF-A and densifier trichlorofluoromethane CFC-11. The drill had an OD of 43 mm, weighed 45 kg, and was 15.6 m long. It consisted of five sections having the following lengths (Fig. 1.47): hot point (0.4 m), pump and flow measurement section (0.7 m), water tank (10.1 m), electronic and suspension section (0.4 m), and cable termination section (0.4 m).

Two different types of hot point were used. The first one was developed for temperate glaciers. An insulated nickel-chrome wire was cast in pure silver. This hot point was very efficient and reliable, but relatively expensive. The

Fig. 1.47 Schematic of the Climatopic thermal probe (Gillet et al. 1984)

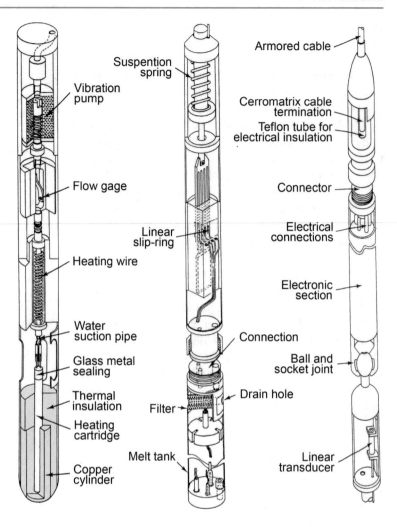

second one used a high-power-density cartridge with a 46-Ω resistance element soldered to a copper cylinder (Fig. 1.48). Both types had a power rating of 2.25 kW at 320 V. To ensure water tightness, a glass-metal seal was soldered on the cold part of the heating element.

Meltwater was sucked at 0.35 m above the bottom of the hole using a vibration pump. The flow of this pump (20–30 L/h) was higher than the production of water (9–11 L/h) to make sure that all the water produced was pumped up. The power delivered to this pump section was relatively high (0.4 kW). The flow was measured using a special flow gage, whereas a differential pressure gage indicated the water level in the tank. The water tank was made from two stainless-steel tubes screwed together. To prevent the full refreezing of water, a heating wire placed along the axis of the tube irradiated a power density of 35 W/m. This heat was not sufficient to prevent the formation of ice. After each run, the tank was placed in a heated enclosure before all the water was recovered. A floating piston gave an alarm signal whenever the tank became full.

In order to reduce the size of the cable, the telemetering system implemented used a bifilar line for four measurements: flow rate, water level, alarm, and drill load. The transmitter was located in the upper part of the drill in a 3.5-m-long steel pressure chamber. The drill load was monitored via the elongation (50 mm) of a spring moving a magnet near a cell by exploiting the Hall effect.

The 4000-m-long cable used was 8.9 mm in diameter and weighed 1050 kg. It consisted of two outer layers of steel armor and four conductors; two 0.93-mm^2 conductors used in parallel and the armor provided a 7-A current path for the hot point. The surface voltage was approximately 900 V; two 0.34-mm^2 conductors were used for telemetering. Voltage was supplied by a three-phase generator using a specially designed inverter, the output voltage from which could be varied from 0 to 1000 V.

The cable was spooled using the LeBus system on a duralumin drum connected by a chain drive to a gear reducer. A 12-kW three-phase 380-V variable speed motor (120–2400 rpm) provided a maximum hoisting speed of 1.43 m/s.

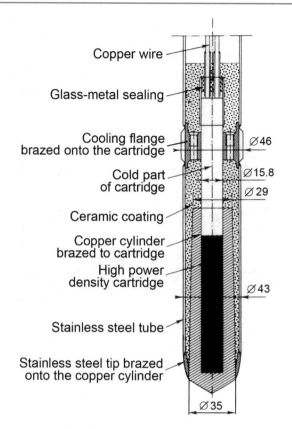

Copper wire

Glass-metal sealing

Cooling flange brazed onto the cartridge ⌀46

Cold part of cartridge ⌀15.8

⌀29

Ceramic coating

Copper cylinder brazed to cartridge

High power density cartridge ⌀43

Stainless steel tube

Stainless steel tip brazed onto the copper cylinder

⌀35

Fig. 1.48 Hot point of the Climatopic probe with a cartridge heater (Gillet et al. 1984)

A disc brake was placed between the motor and the gear reducer. In each run, a small hydraulic variable-speed motor reducer was connected to the main motor via an electromagnetic coupling and yielded a speed between 2 and 70 m/h. The mast was 11.4 m high and consisted of two 250-mm-diameter duralumin tubes. It was surrounded by four air-heated polyethylene tubes designed to receive the water tank for recovering the water after each run. At the top, the pulley was equipped with an encoder (200 pulses per revolution) to measure the depth of the drill in the borehole.

The drill was tested during the 1981–1982 season at Dome C, Antarctica starting from the bottom of the 180-m-deep hole cored using a shallow electromechanical drill in 1978–1979. For the first 28 m, the penetration rate was up to 5 m/h. It was lower than the rate obtained in laboratory tests (8 m/h), perhaps because of impurities encountered at the bottom of the hole. To prevent any loss of fluid in the permeable firn, a 130-m-long polyethylene tube casing was installed in the hole. As the drill passed, small chips of polyethylene were torn off. These particles dropped to the bottom of the hole and had to be cleaned via coring with a small electromechanical drill designed for that purpose as well as for the recovery of small ice cores. The length of the melting runs was no more than 4.6 m. This was

due to problems with the flow sensor and the refreezing of water in the water tank.

From depths of 208–217 m, it became increasingly difficult to penetrate the ice. Increasing friction of the drill against the wall of the hole was observed and, at a depth of 217 m, it was not possible to penetrate further. One of the possible explanations for this problem was the deformation of the water tank tubes. The recovery of water in the water tank set in a vertical position in the heated polyethylene tubes was not satisfactory. With an initial temperature of 80 °C, the final temperature at the top was not higher than 15 °C and, in adverse weather conditions, it could be as low as −15 °C. The tanks therefore had to be placed in a shelter after each run. The strength of the stainless-steel tubes was not high enough and, after a few days of such handling, they became permanently deformed, making it impossible for the drill to pass through the borehole.

Because the bulk of the weight of the drill was located in the upper section (mainly because of the thick pressure chamber) it was not easy to obtain a vertical hole even if the drill was carefully kept in suspension. Because the pressure chamber was rigid, it probably faced difficulties when trying to follow the bending radius of the hole without considerable frictional drag.

At a depth of 217 m, the hole was reamed. Restarting from 207 m, the probe advanced very slowly (0.5–1.0 m/h) in order to obtain a convenient borehole diameter. Friction was immediately reduced and, at 220 m, the penetration rate increased to 2 m/h. Usually, drilling took ~45 min. At a rate of 5 m/h, 10 L/h of meltwater was obtained. The water level in the tank was not measured because of the breakdown of the differential pressure sensor. Finally, drilling was stopped at a depth of 235 m.

1.2.28 Leningrad Mining Institute Hot Point

In 1985, a 109-m-deep borehole was drilled in the middle part of Davydov Glacier (Fig. 1.49), northern slope of Akshirak, using a 70-mm-diameter thermal point designed by the Leningrad Mining Institute (LMI) (Vasilenko et al. 1988). At depths ranging from 0 to 15 m, several narrow sections caused by refrozen water on the hole wall were encountered by the hot point during hoisting. As a result, drilling runs (to a depth of 40 m) were alternated with water bailing. The rate of penetration down to 102 m was rather stable at 3 m/h. From there on, the rate rapidly decreased, apparently because of solid inclusions in the ice and, at 109 m, penetration stopped. An inclinometer survey showed that the hole was almost vertical and that the inclination did not exceed 2°. Temperature was measured only down to 30 m. The hole temperature first decreased from −0.2 °C

Fig. 1.49 Hot-point drilling setup at Davydov Glacier, Tien Shan, summer of 1985 (*Credit* D. N. Dmitriev)

near the surface to −5.8 °C at 6 m and then uniformly increased to −0.2 °C at 30 m.

1.2.29 BGR Hot Point

This hot point was built in the Federal Institute for Geosciences and Natural Resources, Hannover, Germany (Bundesanstalt für Geowissenschaften und Rohstoffe, BGR) as a part of research project to measure terrestrial heat flow in boreholes drilled into almost stagnant Antarctic ice fields and specifically along a profile across the Transantarctic Mountains (Zeibig and Delisle 1994).

This hot point contained a cartridge heating element with a power output of 2.5 kW (Fig. 1.50). The tip was connected to a copper pipe (water tank) to collect the melted water, which was recovered periodically as the hot point was lifted out. Four one-way valves were fixed at the upper end of the pipe, through which the meltwater entered. The inner part of the pipe was equipped with a heater (160 W) to prevent the premature freezing of the meltwater within the pipe and of the outer surface of the pipe to the ice wall. A second heating element (400 W) was placed on the top section of the hot point to melt the ice on the way out of the borehole to prevent the hot point from becoming jammed by refrozen ice. The total weight of the hot point was 12 kg.

A tripod, an electrically driven winch, and a control box were mounted on a 2 m × 3 m platform installed on two sledges with a clearance of 0.3 m above the snow surface (Fig. 1.51). The working area at the platform was protected

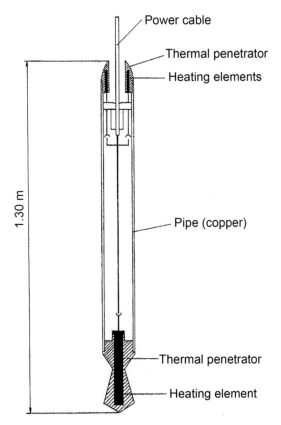

Fig. 1.50 BGR hot point (Zeibig and Delisle 1994)

by a 2-m-high shelter with an aluminum frame covered by canvas. The maximum working load of the 9.8-mm-diameter

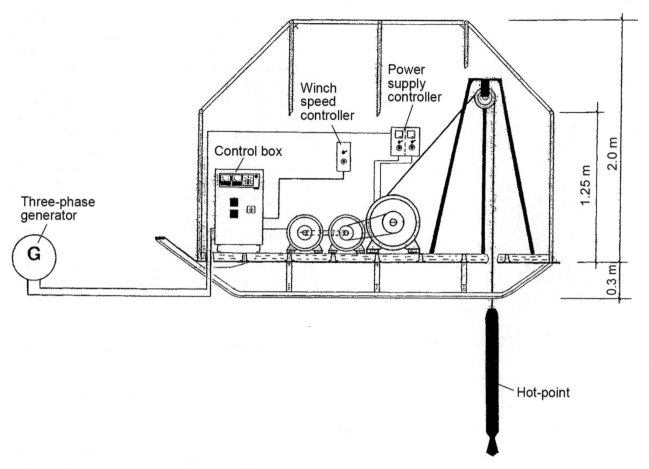

Fig. 1.51   BGR hot point's shelter (Zeibig and Delisle 1994)

two-line cable with Kevlar armoring was 200 kg. The winch could provide maximum tripping speed of 0.6 m/s.

A total of 48 boreholes in Antarctic ice were drilled over two seasons using this system. However, during the field season of 1992–1993, the first test of this hot point with a 3-m-long water-tank pipe in blue ice near Ambalada Peak failed. Down to 4 m, melted water was recovered with a normal water discharge rate of 5.7 L/m, which corresponds to an average borehole diameter of 80 mm. However, for some unknown reason, only 1 L/m of meltwater was recovered below this depth. At a depth of 10 m, the hot point froze in the hole and was lost. To prevent further accidents, the water-tank pipe was shortened to 1.3 m. Then, the equipment, which was housed on sledges, was transported from Gondwana Station (Terra Nova Bay) by helicopter into the field at Victoria Land, and three boreholes with depths of 51, 102, and 83 m were drilled. Shelter and other equipment were transported between drill sites by snowmobile at an average speed of 11 km/h.

During drilling, the greatest problems occurred at the firn–ice interfaces. The bottom of firn layers tended to move inward and reduce the borehole diameter. These sections had to be re-drilled from time to time. This problem was most severe at the last site, where the firn and ice were warmest (−24 °C). The rate of penetration gradually decreased from an initial 5 to ∼2.8 m/h. The typical rate in dense ice averaged 2.2 m/h. The actual production rate was, however, closer to 0.9 m/h owing to the time needed for the bailing trips.

This hot point was used the second time on Antarctic ice in the season of 1993–1994 within the Italian Antarctic Program. A total of 45 short holes of depths varying between 18 and 40 m were drilled at four sites to determine the seismic refraction profile across the Transantarctic Mountains south of the Drygalski Ice Tongue in Victoria Land.

1.2.30   Tohoku University Thermal Sonde

Suto et al. (2008) from Tohoku University, Japan proposed an electro-thermal sonde (Fig. 1.52) that can penetrate through thick ice using a hot point and continuously analyze meltwater in a similar way as the Climatopic thermal probe (see Sect. 1.2.27). The proposed diameter of the sonde was

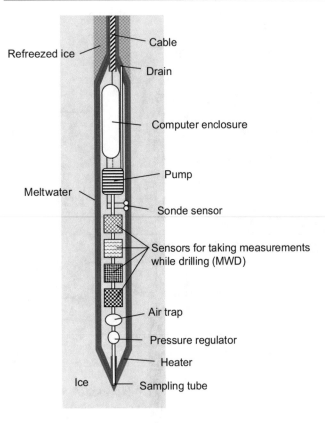

Fig. 1.52 Conceptual model of the Tohoku University thermal sonde (Suto et al. 2008)

150 mm, with a length of 1 m. A thermal tip was used to heat and melt the ice at the bottom and a side heater was used to retain meltwater around the sonde. Electric power

was supplied through a cable that would also heat the surrounding area to protect it from refreezing. Experiments with a one-third scaled conical thermal tip showed the it could melt ice with a temperature of −20 °C at a maximal rate of 0.15 m/h with a load of ∼20 N.

According to theoretical assumptions, the required power supply for the heaters in the sonde and for the heating cable greatly increased with depth: (1) to penetrate through 300 m of temperate glacier (temperature ∼0 °C at all depths), the sonde would require 4.8 kW with a non-heated cable; (2) to penetrate through 1000 m of ice with a temperature of −25 °C at the surface and 0 °C at the bottom, the sonde would require 10 kW plus 19–32 kW to heat the cable; and (3) to penetrate through 3000 m of cold ice with a temperature of −55 °C at the surface and 0 °C at the bottom, the sonde would require 19 kW plus 140–235 kW to heat the cable. The implementation of the last case is not realistic, but drilling down to 1000 m could be considered an expedient.

1.2.31   Hot Point of the Byrd Polar Research Center

Zagorodnov et al. (2014) from the Byrd Polar Research Center (BPRC), Ohio State University suggested installing small-diameter sensors under ice shelves via dry-hole electromechanical ice coring as close as possible to the shelf glacier base and then drilling through the shelf glacier down to sub-ice shelf cavities with a hot-point drill. The proposed hot point had a 40-mm-diameter melting tip and was 1.5 m long (Fig. 1.53). The top anchor mechanism allowed the hot

Fig. 1.53 BPRC hot point:
a general structure;
b cross-sections of the centralizer and the melting tip; **c** anchor in fixation state (modified from Zagorodnov et al. 2014; Zacny et al. 2016)

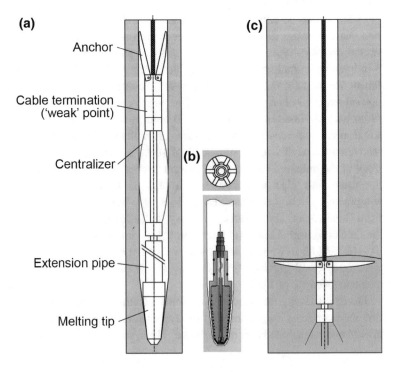

point to be jettisoned below the shelf glacier base after penetration and made retrieval of the electromechanical cable to the surface possible. This jettison feature was designed to reduce the probability of the drill and cable of becoming stuck in the borehole after penetrating into the ocean. The shape of the anchor blades allowed the drill to move up and down in the borehole.

The melting tip had a long conical shape. A long (1.575 m) Watlow coiled cable heater with a small diameter (1.58 mm) was mounted close to the melting surface of the tip in spiral grooves in the copper core. To ensure heat dissipation from the coiled cable heater, it was molded in pure silver. The housing and the protective shell of the melting tip were made of stainless steel. The final tests for each of the 11 melting tips (one was lost during silver casting) included: non-electrified high-pressure test at 5.5 MPa for 2 h; operation at maximum power in water at 3 kW for 5 min; and 3 h of operation at a power of 2 kW. Only five tips passed all three tests.

In November–December 2011, these hot points were used at the Windless Bight site, McMurdo Ice Shelf, Antarctica, within 18 km of the ice shelf barrier, where two boreholes were predrilled using an electromechanical drill to depths of 185.7 and 190.4 m. Then, using hot points, they were completed down the sub-ice shelf cavity to depths of 192.7 and 192.9 m, respectively (Tyler et al. 2013; Zagorodnov et al. 2014). Drilling and sensor installation took 110 working hours. The total weight of the drilling equipment, including the power system and fuel, was less than 400 kg.

Vertical stabilization of the hot point was achieved via pendulum steering. The drill was partly (80% of its weight) hung and its penetration rate was limited by the rate at which the cable was fed. This rate was controlled by the winch motor and was less than the hot-point penetration rate with the full weight of the drill applied to the melting tip. Vertical stabilization was also assisted using a centralizer, which kept the top of the drill in the center of the borehole. The tip power of the hot point was set to 1.8 kW, and the cable-feeding rate was set to 7.2 m/h lower than the maximal penetration rate (7.6 m/h). At this rate, the hot point produced a borehole 56 mm in diameter. The hot point passed freely down to the sub-ice shelf cavity, and the drilling cable was disengaged from the hot point and hoisted to the surface.

1.2.32 Large-Diameter Electric-Heated Hot Points

To drill large-diameter holes through snow and firn, which may be necessary for pre-drilling with hot-water ice-drilling systems or for the construction of water wells in ice sheets, electric-heated hot points can be used. The Raytheon Polar

Services Company (RPSC), which is a division that provided logistics, operations, and staffing for the National Science Foundation's operations in Antarctica and Antarctic waters, designed large-diameter hot point that was used to melt a hole to a depth of 61.5 m through snow and firn to establish the initial well shaft for the Rodwell-3 water well at Amundsen-Scott station in 2006–2007 (Fig. 1.54) (Haehnel and Knuth 2011). However, that hole would need to be re-drilled years later before the water well could finally be started in 2011–12 and put online a year later.

Another large-diameter hot point was developed in the Polar Research Center, Jilin University (JLU), China. This system included (Fig. 1.55) a stainless-steel base sized 0.97 m × 3.5 m and weighting 256 kg, a winch with a 100-m-long steel rope with a diameter of 12 mm, a 1.97-m-high T-shaped mast, driving units, and a thermal drill head.

The winch had two driving units that operated alternatingly, depending on the specified mode. During tripping, the winch was geared by a 3-kW electric motor (YE-100L2-4, 3 × 380 V, 50 Hz) through a double-strand chain (1:2) and a worm reducer (1:100). This system allowed for a rope-winding speed of 0.4 m/s. For drilling, the driving unit was changed with the help of two electromagnetic clutches into a 0.25-kW YC-71C-4 motor with an additional reducer (1:184). This system allowed for precise control of the rope-winding speed within the range of 0.2–8.5 m/h. The self-locking feature of the worm gear reducer served as a brake.

The drill head consisted of a lower hot point, in which a tubular heating element was casted into an ogival copper body, and three 20-mm-diameter tubular heating elements wound into a conical spiral. The total weight of the drill head was 62.5 kg; its OD was 476 mm and its height was 760 mm. Power was supplied through three conductor cable with silicone rubber isolation (the diameter of each conductor was 10 mm). To prevent overheating, the power of the drill head was adjusted using a variable transformer. During drilling in snow and incompact firn, the supplied voltage was kept in the range of 75–80 V, at which the temperature of the tubular heating elements reached maximal values as high as 140 °C. When the firn became impermeable to water, the voltage could be adjusted to the rated 220 V. Thus, the total power of the drill head changed from 20 to 48 kW depending on the drilled formations. Preliminary tests showed penetration rates up to 2–3 m/h in incompact firn.

1.2.33 Summary

Initially, electric-heated hot points were fixed at the lower end of a pipe string that, in many cases, was left in the

Fig. 1.54 Hot-point drill used to establish the initial well shaft for the Rodwell at the South Pole (Haehnel and Knuth 2011)

Fig. 1.55 JLU large-diameter firn-drilling system during laboratory tests, February 2018 (*Credit* B. Liu)

borehole as casing. Later, they were modified to be suspended on a rope or armored cable and did not require extension pipes. To maintain their vertical stability, some hot-point drills were equipped with centralizers. However, most of them were attached to long guiding tubes (in certain instances, up to 10–15 m long) to reduce hole deflection.

In general, electric-heated hot points were used in temperate glaciers, which are at their melting point throughout the year from their surface to their base and in which refreezing of meltwater takes a long time. In colder ice, the closure of holes due to the refreezing of meltwater prevents the retrieval of the hot point. For such cases, some hot points had quick-disconnect couplings to the external cable, which had to be separated when drilling was complete. In some occasions, the entire instrument package was allowed to freeze in place once drilling had ceased.

To drill deeper and safer, four possible hot-point technologies were proposed to penetrate cold ice: (1) using a heating power cable that partly prevents the refreezing of meltwater (Classen 1970; Suto et al. 2008); (2) antifreeze-assisted drilling, in which a hydrophilic liquid is added into the hole and mixed with meltwater (Grześ 1980; Hooke 1976; Morev et al. 1988); (3) bailing of meltwater (Vasilenko et al. 1988; Zeibig and Delisle 1994); and (4) replacing meltwater with a non-freezing hydrophobic drilling fluid (Gillet et al. 1984).

The main element of hot-point drills that makes penetrating ice possible is their electrically heated tip. Two types of electric hot-point melting tips were developed: (1) those with an open high-resistance heater (nichrome wire or silicon carbide) and (2) those with a closed resistive heater. The latter are also called hot points with a heated solid penetration tip.

It is generally accepted that hot points with an open heater have less internal heat loss and can therefore penetrate ice faster. As reported by Nizery (1951), a melting rate of 20–25 m/h was achieved using a thermal head with an open nichrome wire. On the other hand, nichrome wires are difficult to support unless they are quite thick. The small air or steam bubbles generated in the melting process causes local hot spots, which easily lead to the fusing of the nichrome element. In addition, the shape of the drilled hole is difficult to control. Because of operational limitations and although melting tips with open nichrome wires have a simple design, they were used only once since their first deployment in 1948–1949.

The original hot-point drill was designed by LaChapelle (1963), in which an open heater was made from a silicon carbide thermal element. This element got in contact with the melted water near the bottom in the space surrounding the heater, and then the heat was transferred to the ice. The main

disadvantage of this hot point is the quite short lifetime of heaters: it was necessary to change the carbide thermal elements every 40–70 m of drilling.

Three main types of heating elements were used in hot points with a closed resistive heater: (1) cartridge(s), (2) nichrome wire isolated by ceramic beads, and (3) sheathed tubular elements. The last two types of heaters are wound in cylindrical or spiral patterns and provide more efficient heat transfer to the working faces, which results in reduced heater temperature and hence more reliable units. However, the use of a cartridge heater allowed for ease of design when scaling to different hot-point diameters and for ease of field substitution for different wattages and voltages depending on the available electrical generators. Careful attention must be paid to produce a symmetrical heating head in order to prevent leaning of the drill. Using metal with high thermal conductivity helps in this regard.

The melting rate of hot points depends on the following variables: (1) input power, (2) cross-sectional area of the thermal tip, (3) axial load (in a limited range), (4) temperature of the drilled ice, and (5) certain design features, such as the material and shape of the tip, properties of the thermal element, etc. The first two are included as some of the main hot-point parameters, namely the *specific power*, which is the input power per unit area of the cross-section of the thermal tip. All the available information of the hot points designed over the past few decades is summarized in Fig. 1.56 from the point of view of the achieved rate of penetration vs specific power. In this case, the influence of other parameters on the melting rate was neglected.

In general, the specific power was in the range of 40–160 W/cm^2, with two exceptions. The specific power of the hot point designed in the Laboratoire de Glaciologie,

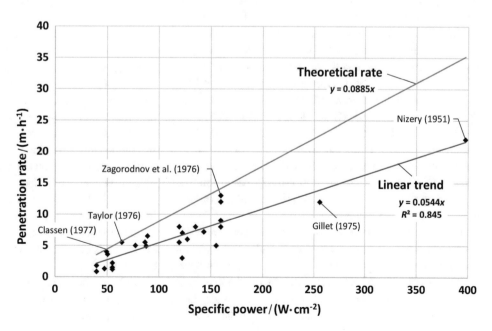

Fig. 1.56 Penetration rate of hot points versus specific power

Grenoble (Gillet 1975) was ∼255 W/cm², which was attained because of the very small size of this hot point (only 18 mm). The second exception is Nizery's hot point with an open nichrome wire (Nizery 1951). The specific power of this device according to reported information was almost 400 W/cm², which explains the extremely high achieved rate. On the other hand, Mellor (1986) noted that, according to practical experience, the maximum specific power for a long-life heater is ∼300 W/cm². Higher specific powers are not worth pursuing, as they only lead to film boiling at the ice interface, thus inhibiting a more rapid penetration.

The melting rate scales with the inverse of the cross-sectional area of the hot point and is directly proportional to the melting tip power. That is one of the reasons for many of these devices having a quite small diameter in the range of 18–40 mm with a power of 1–2 kW. Assuming that all the input power melts ice at a temperature of −10 °C (the method used for estimating the power required for borehole melting is described by Talalay et al. 2014), the theoretical rate can be calculated according to the following simple equation:

$$v_t = 0.0885\, W_F,$$

where W_F is the specific power of the hot point in W/cm².

This theoretical rate line is drawn in Fig. 1.56 as a potential hot point's limit. The direct proportionality between melting rate and specific power has been clearly confirmed by the data presented in more than 25 references. Based on statistics processing, the experimental melting rate can be estimated by a linear trend with a correlation factor of 0.845 as follows:

$$v_e = 0.0544\, W_F,$$

The thermal efficiency of a hot point is the ratio of its true rate of penetration to its theoretical one. Thus, 100% efficiency is achieved when all the available electrical energy is consumed for melting a hole in the ice with a diameter equal to the thermal tip diameter. The estimated efficiency of high-performance hot points was approximately 0.75–0.85 (Shreve 1961; Ward 1961; Hooke 1976; and some others). Many hot points had thermal efficiencies lower than that, in the range of 0.50–0.65 (Gillet 1975; Taylor 1976). The efficiency of the fastest mystery Nizery's hot point (0.63) lies within the average trend. The average efficiency of all the hot points shown in Fig. 1.56 is 0.65. The estimated efficiency of a few hot points (0.91–0.97) based on data reported by Classen (1977), Taylor (1976), and Zagorodnov et al. (1976) is likely overstated.

Low efficiency results in the wasting of energy and time because it creates bumpy holes with irregular cross-sections owing to the larger percentage of energy that heats water, which moves up and sideways and melts the ice in an uncontrolled manner. On the other hand, it does more harm than good to increase the thermal efficiency because the drills become more likely to get stuck in the hole and be lost. The probability of a drill getting stuck in the borehole for a given stem length seems to be largely a function of the clearance between the drill and the hole. Taylor (1976) suggested that a difference of ∼5 mm between the hot-point diameter and the hole diameter should be a good design guideline to achieve a useful, efficiently drilled holes with a smooth and constant cross-section in temperate glaciers with minimum risk.

Two common shapes of solid penetration tips are usually used. The first one has a cylindrical core with a conical or blunt end (Stacey 1960; Shreve and Sharp 1970; Classen 1970; Taylor 1976). The outer surface of the core is covered by a material with low thermal diffusivity (stainless steel) to decrease lateral heat losses. The second tip design has a parabolic shape to satisfy the condition that each section of the tip along its height should have the melting capacity just necessary to melt the ice encountered as the hot point penetrates at a uniform velocity (Gillet 1975; Hooke 1976). The parabolic design has several advantages over the cylindrical design. First, in the parabolic design, the temperatures in the hot point are substantially lower, thus reducing the possibility of failure of the heater. Secondly, construction is simpler because of the lower temperatures and because insulation is not required to prevent radial heat loss. Thirdly, small debris particles can be removed with the melted water from the bottom sidewards, improving drilling efficiency in debris-laden ice. Finally, the elongated shape of the melting tip provides better vertical stability during penetration. Talalay et al. (2019) tested seven tip shapes (cylinder, cosine, ogive, sphere, catenary, parabola, and cone) at the same power level, axial load, ice temperature, and ice contact area. The 60° cone-shaped tip demonstrated the highest rate of penetration.

Gillet (1975) suggested the use of a hot-point tip made from silver because the thermal diffusivity of silver, 417 W/(m °C) at 100 °C, is higher than that of copper, 382 W/(m °C) at 100 °C. Although the rate achieved during drilling at a temperate glacier reached 12 m/h, the efficiency of this hot point was only 0.53, which is slightly lower than the average trend shown in Fig. 1.56. In addition, silver casting is much more expensive than copper or aluminum casting.

The major problem regarding ice melting is the intrusion of components that cannot be molten, such as dust or rock particles. Mineral inclusions and dust are always present in glacial ice, and their size and content depend on the site's location (Vallelonga and Svensson 2014). In particle-laden ice, the efficiency of hot points dramatically drops. Thus far, no satisfactory solution has been demonstrated for melting dusty ice, because dust concentrates underneath the melting tip and builds up an impenetrable layer. However, one of the possible solutions for drilling in slightly dirty ice is to use a melting tip with a parabolic or conical shape, which would push dust aside from the tip.

1.3 Freezing-in Hot-Point Drills (Philberth Probes)

Freezing-in hot-point drills are able to move towards the glacier base while the melted water refreezes behind the probe. One of the first freezing-in hot-point drills, subglacial wireless autonomous station, was supposed to use nuclear energy for melting through the ice sheet, but it was never realized owing to technical challenges and environmental considerations. In the middle of the 20th century, brothers Bernhard and Karl Philberth, which were independent German physicists, engineers, and Catholic priests, invented the freezing-in hot point to study the temperature distribution in ice sheets. The most outstanding characteristic of these hot-point drills is that the wires used for the transmission of electrical power to them and the signals from them pay out of the advancing drill and became fixed in the refreezing meltwater above it. Recently, this type of thermal drills has often been called as Philberth probe. Some drilling parameters of freezing-in hot-point drills are shown in Table 1.2.

1.3.1 Subglacial Nuclear Autonomous Station

In 1963, I.A. Zotikov and A.P. Kapitsa proposed a subglacial autonomous station with a small nuclear power plant on board to study the Antarctic subglacial environment (Zotikov 2006). The nuclear power plant was to be located in the lower part of a hermetically closed 0.9-m-diameter container of the station, which would also include various instruments and equipment (Fig. 1.57). The nuclear power plant would have to produce enough energy to melt to the bottom of the ice sheet. No hole was needed because the station was supposed to descend into a water cavern melted by the unit, which would refreeze above it. Communication with the device from the surface of the ice sheet would be maintained via wireless transmissions.

The Atomic Energy Institute of the USSR Academy of Sciences approved the project and agreed to provide a small nuclear energy reactor ~ 0.5 m in diameter and capable of producing 0.5–2 MW of heat and 100 kW of eclectic power. According to estimations, this power would provide an average rate of penetration slightly more than 1 m/h. This means that the bottom of the central part of the Antarctic Ice Sheet could be reached in less than four months.

This project was never realized owing to issues with the classified status of the reactor. At the early stages of the project, the authors had no idea on how to bring the station back to the surface, and the unit could not be left at the bottom of the ice sheet because the Antarctic Treaty signed in 1959 prohibited nuclear waste disposal. The problem was solved in 1975 when the authors suggested the release of ballast at the bottom of the ice sheet and use buoyant forces to recover the station. To move up and melt ice on the way,

Table 1.2 Drilling parameters of freezing-in hot-point drills

Drilling site	Year	Ice temperature at final depth/°C	Power/kW	Hot point diameter/mm	Penetration rate in ice/(m h⁻¹)	Max depth reached/m	References
Camp Century, Greenland	1965	−24[a]	NA	108	NA	∼90	Aamot (1967b)
Camp Century, Greenland	1966	−24.4	3.38–3.68	108	2.72–2.95	∼259	Aamot (1967b)
Jarl-Joset, Greenland	1968	−29.0	3.7	108	2.0	218	Philberth (1976)
Jarl-Joset, Greenland	1968	−30.0	3.7	108	1.9	1005	Philberth (1976)
South Pole Station, Antarctica	1971	−46.3[a]	NA	NA	NA	6	Aamot (1968b)
S2 site, 80 km of Casey Station, Antarctica	1973	−19.4	6.06	129.5	1.08	112	Morton and Lightfoot (1975)
Lab tests in PICO, Nebraska	1979–1980	−25	5.4	127	1.6	NA	Hansen and Kersten (1984)
Neumayer Station, Antarctica	1992–1993	NA	3.4	100	2.9	225	Ulamec et al. (2007)
Summit, Greenland	1993	−32[a]	NA	127	NA	7	Kelty (1995)
Summit, Greenland	1994	−32[a]	4.04	127	1.45	∼120	Kelty (1995)
Dye 2, Greenland	1996	−21[a]	NA	127	NA	∼60	J. Kelty, personal communication
Lab tests, Jilin University	2017–2018	−20	5.0	160	1.8–1.9	10	P. Talalay, personal communication

Note: *NA* not available; [a]author's estimated values

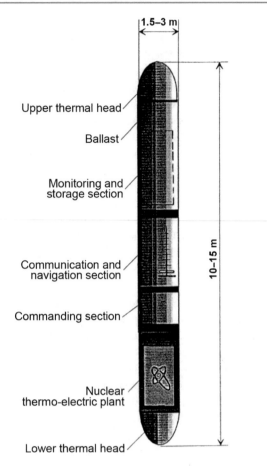

|1.5–3 m|

Upper thermal head

Ballast

Monitoring and
storage section

10–15 m

Communication and
navigation section

Commanding section

Nuclear
thermo-electric plant

Lower thermal head

Fig. 1.57 Subglacial nuclear autonomous station (Zotikov 2006)

the heat generated by the nuclear reactor would have to be moved to the top of the unit. This new type of returnable unit was supposed to stay many years at the bottom of the ice sheet collecting various data and would be powered by a nuclear thermo-electric plant. However, in the 1980s, an additional protocol added to the Antarctica Treaty declared Antarctica as totally nuclear-free zone, which prohibited the use of nuclear power within the continent.

1.3.2 Philberth Probe

In 1956, B. Philberth proposed a plan for storing nuclear waste in the central part of Greenland Ice Sheet (Philberth 1956). Although his plan would offend some modern sensibilities, it has to be considered in the context of its time. According to his plan, radioactive waste would be incorporated into glass or ceramic ingots, and each ingot would be sealed inside a metal sphere roughly 200 mm across. He estimated that all of the nuclear waste produced worldwide throughout the year 2000 could be held inside 30 million of these spheres, which could be spread over a 30-km-wide patch of ice sheet. Those vessels of waste would melt their

way into the ice sheet and remain there for 20–50 thousands of years before the slow flow of the ice brought them to the coastline. There, they would be shed into the ocean in icebergs and sink to the bottom of the sea once the icebergs melted.

To estimate the ice movement, he needed to map the internal temperatures of the ice sheets. This was information that glaciologists of all stripes wanted to obtain, even aside from the issue of storing nuclear waste. However, in the 1960s, this had never been done. In 1962, prior to the drilling of any of the deep boreholes, Karl and Bernhard Philberth were the originators of the idea of using an ice-penetrating thermal probe to measure the temperature inside ice sheets (Philberth 1962). During penetration, the probe would release wires that would freeze inside the ice. At the desired depth, the power supply of the heating elements would be switched of, penetration would stop, and the probe would totally freeze in ice. After reading the temperature, pressure, and other parameters, the power supply would be restarted, the probe would melt out of the ice, and penetration would continue down to the base of the ice sheet.

In 1964, the development of a Philberth thermal probe was undertaken at the CRREL with the assistance of Karl Philberth (Aamot 1967b, c, 1968b). This 108-mm-diameter probe consisted of a hot point for melting penetration, instrumentation for control and measurement functions, two supply coils of conducting material to connect the probe with the surface for the transmission of power and measurement signals, and a reservoir section (Fig. 1.58).

The hot point was made of copper with a cartridge electric heater with resistance of 240 Ω in its center. The ring-shaped outer section, consisting of a mercury steering ring, was partially filled with mercury. When the probe began to "lean", the mercury would move towards the plumb side in the cavity. Thus, the heat flow in the steering ring was directed towards the plumb side and the hot point penetrated with a corresponding lateral movement towards that side until the probe was plumb again. In 1962, the concept of using a mercury steering-ring to stabilize the vertical course of a thermal probe was successfully tested at Jungfraujoch, Switzerland (Philberth 1964).

The pressure-protected cylindrical chamber for the instrumentation was located above the hot point. The instrumentation package consisted of three thermistors for temperature measurements, two silicon-crystal strain gauges on the inside of the sealed chamber to measure hydrostatic pressure, calibration resistors for line-resistance measurements (the increasing line resistance due to the paying out of bare wire was an indicator of probe depth) and insulation-resistance measurements (insulation-fault detection), and a geophone for seismic soundings. A multi-position twelve-stepping relay served for switching between the various circuits for measurements and probe

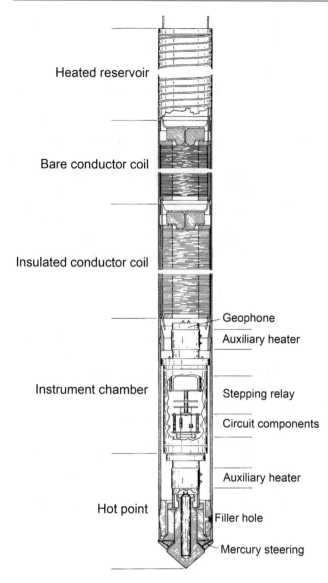

Fig. 1.58 Philberth probe (Aamot 1968a)

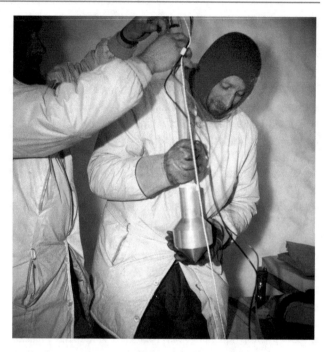

Fig. 1.59 Karl Philberth (right) and Father Hugo Jännichen, a Benedictine monk, priest, and physicist (left) in the dim bottom of a snow pit, readying a thermal probe for drilling into Greenland Ice Sheet in 1968 (These priests' invention… 2015)

operation. Two auxiliary heaters were fixed above and below the instrumentation package.

Two conductors were available for the transmission of power to the probe and for the transmission of information from the built-in instrumentation. Depending on the types of wires used, two versions of the probe were constructed. Probe I had an insulated wire, a bare wire, and an open oil reservoir that was attached to the top (this version is shown in Fig. 1.59). At the start, the reservoir was filled with a silicone oil with a density higher than the density of water to prevent displacement by meltwater. The oil protected the probe from damage by the refreezing meltwater after stopping. The level in the reservoir dropped gradually as the probe advanced and the conductors payed out.

Probe II had two insulated wires and no reservoir. The advantages of the arrangement of Probe I is the reduced coil size of the bare wire. Moreover, the sidewall around the bare-wire coil and of the oil reservoir of Probe I contained a 100-Ω resistor wire for additional heating, which was used for melting the probe out of the ice when restarting and for increased lateral heat transfer during slow penetration. Probe I was 2.92 m long (including the reservoir) and Probe II was 2.55 m long; almost 80% of the probe length consisted of coils.

Each coil contained approximately 3150 m of copper wire, stored in 23 layers of 600 windings each of insulated wire and 29 layers of 485 windings each of bare wire. The diameter of the bare wire of Probe I was 0.9 mm, and those of the insulated wire were 0.95 mm (net) and 1.26 mm (gross). The insulation consisted in Teflon-sealed Kapton tape. The self-heating of the insulated cable coil due to its conductor resistance supplied heat for lateral transfer during penetration and for melting the probe out when restarting. The resistance of the insulated cable (18AWG copper) was 15.1 Ω at 25 °C; the bare-wire line resistance increased with the paid-out length.

The orthocyclic method (Aamot 1969) was selected for winding the coils because it offers the highest possible coil density (packing factor) for a minimum coil size and the best heat transfer. A regular winding pattern also ensures reliable unwinding from inside the coil as the probe advances. The completely wound coil was potted into the probe's housing and the collapsible mandrel was removed. The decision to power the probe with alternating current was based on the observation that leaking currents at possible insulation faults in a water-immersed cable are smaller and grow less under an AC potential than under a DC potential.

The first prototype of Philberth Probe I was built at the USA CRREL and tested at Camp Century, Greenland, in the summer of 1965. An insulation fault destroyed one of the two conductors and contact was lost with the probe at a depth of ~ 90 m (Aamot 1970a).

Probe II was tested in June 1966 at the same site. The probe was operated at 3.38 kW and a corresponding speed of 2.72 m/h during the first day of penetration. For the remaining penetration work, the probe was operated at 3.68 kW with a speed of 2.95 m/h. After four days of penetration, the probe was stopped at a scheduled depth of ~ 259 m below the trench floor. Immediately after power was turned off, readings were taken to observe the cooling rate and the rise of hydrostatic pressure due to the refreezing of meltwater. Seismic soundings were carried out to determine the depth of the probe. In addition, line resistance measurements and the final relaxed hydrostatic pressure (overburden stress) served as depth indicators.

The probe temperature dropped rapidly at first, approaching the final temperature hyperbolically. The readings covered a period of nine days and indicated a final temperature of -24.4 °C. During the last six days, temperature decreased very slowly, meaning that three days was enough for reliable temperature measurements.

During the refreezing of the confined meltwater in the ice surrounding the probe, a high crystallization pressure is developed. An observed peak of ~ 8.8 MPa above the overburden stress was reached 75 min after power was turned off. Then, pressure decreased to a final reading of 2.6 MPa. At 259 m, the overburden stress, computed from the density profile, was 2.32 MPa, indicating that the accuracy of the transducer was insufficient to confirm the probe's depth.

Seismic soundings to determine probe depth were not successful. The transmission line between the probe and the surface picked up strong signals from the electrical fields of the camp's power equipment. The weaker geophone signals disappeared in this noise. Instead, a line resistance measurement to determine the length of the bare wire that was paid out was attempted to measure probe depth. The depth computed on this basis was ~ 244 m.

After nine days of cooling and stabilization, an attempt was made to melt out and restart the probe. The head section was heated via internal heat transfer from the hot point. The reservoir heater heated the upper parts of the probe and the internally stored self-heating cable heated the midsection. Unfortunately, the transferred heat was insufficient to recover the probe (it is most likely that the self-heating of the cable coil was not enough for melting out the midsection) and the main heater finally became short circuited owing to overheating.

In the summer of 1968, two other sets of Philberth probes were used in the Expédition Glaciologique Internationale au Groenland (EGIG) at station Jarl-Joset (Fig. 1.59) (Philberth 1976, 1984). During testing at Camp Century, the constantan resistance element proved unstable and was changed to a Chromax wire that had a uniform resistance increase with temperature and that allowed for easy monitoring of the heater temperature. The probes had sufficiently long wires for penetrating the ice sheet (2500 m), but the breakdown of the main heater stopped Probe I at a depth of 218 m and Probe II at a depth of 1005 m. The cause was probably the moisture content of the insulation, resulting in electrolytic action and then failure.

Probe I was started at a depth of 7 m. Down to a depth of 35 m, the firn absorbed the meltwater. Therefore, the mean coil temperature reached 90 °C for a current of 1.8 A. Later, for a current of 2.7 A (corresponding to a total power of 3.7 kW) and a rate of 2.0 m/h, the temperatures were as follows: less than 80 °C (mean value) in the coil, less than 55 °C in the thermistors, and ~ 300 °C in the Chromax wire in the cartridge heater. Probe I measured a cooling curve over a 5-day period at a depth of 218 m, yielding a temperature of -29.0 °C.

Probe II was started at a depth of 45 m and reached a current of 2.3 A (corresponding to a total power of 3.7 kW) and a velocity of 1.9 m/h. Under these conditions, the temperatures were as follows: less than 80 °C (mean value) in the coils, less than 50 °C in the thermistors, and less than 250 °C in the Chromax wire.

Probe II was stopped intentionally at a depth of 615 m, where the cooling curve for a 6.5-day period was measured, yielding a temperature of -29.3 °C. For approximately 30 h, only the coils were heated; the heating current was increased until the temperature of the coils made it evident that the upper part of the probe had reached the melting point. Penetration continued to a depth of 1005 m. The acquisition of the cooling curve by Probe II was interrupted after 5.2 h because of wire breakage. The temperature was determined to be -30.0 °C.

The depth of the probes was measured via four methods: the inductance of the coil, time integration of velocity, the resistance of the length of the bare wire (Probe I only), and the hydrostatic pressure measured using strain gauges. The first two methods worked continuously. The inclination of Probe I over its entire path and that of Probe II over the second section were always less than 4°. However, in the first section, Probe II reached an inclination of up to 10°.

1.3.3 CRREL Pendulum Probe

In order to solve the probe stabilization problem, Aamot (1968a, b, e, 1970b) proposed the pendulum-steering method, so called because it places the center of support above the center of gravity of the probe. The power levels of the lower circular hot point and the upper annular hot point were regulated so that the slightly underpowered upper hot point supported most of the probe weight (Fig. 1.60). Both hot points, made of copper, were heated via cartridge heating elements. The diameter of the probe was 127 mm, and its nominal length was ~2.5 m. The power requirement for a speed of 6 m/h in ice at −28 °C was approximately 15 kW.

The transmission line used was a coaxial cable consisting of 20AWG conductor with a solid copper center, a wrapped and fused film insulation, and a braided outer conductor having approximately the same resistance. The insulation

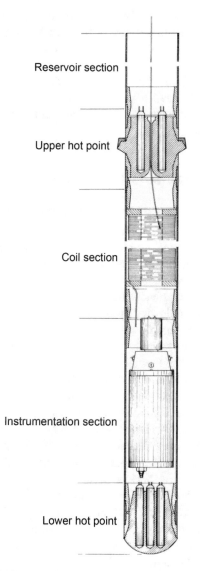

Fig. 1.60 CRREL pendulum probe (Aamot 1968a)

Reservoir section

Upper hot point

Coil section

Instrumentation section

Lower hot point

protected against stresses of 2500 V DC even during water immersion. No cover insulation was used to minimize the cable diameter. The outer conductor resistance of the payed-out cable was used as a measure of probe depth. The cable coil was wound via the orthocyclic method. The length of the coil was ~1.1 m for 3000 m of cable. The housing consisted of filament wound tube structures (glass/epoxy) with built-in heating elements over their whole length.

The hot point was made from solid copper. The probe itself was completely filled with silicone oil to keep water out to protect it from ice action and for electrical insulation. The top of the straight cylindrical reservoir was open to ensure reliable release during restarting of the probe after stopping and freezing in. The oil level in the reservoir dropped as the cable was payed out.

In 1971, a CRREL pendulum probe was tried at South Pole Station, Antarctica by J. Rand (Hansen and Kersten 1984). It failed at a depth of 6 m, probably because the coil was inadvertently overheated by operating at too high a power level in porous firn.

1.3.4 Meltsonde Probe of the Australian Antarctic Division

The Australian Antarctic Division began developing the meltsonde technique in 1972 based on the concept of the CRREL pendulum probe (Morton and Lightfoot 1975). The meltsonde probe consisted of a lower hot point, a pressure chamber with an instrumentation section, a coil made with the cable, an upper heater with a diameter larger than the rest of the probe, and a reservoir section on the top filled with silicone oil to fill the volume of the cable as it was paid out (Fig. 1.61). The diameter of the probe was 129.5 mm at the lower heater and 152 mm at the upper heater. The overall length was 1.9 m.

The heaters were fabricated using sheathed tubular elements, cast in aluminum, and fitted with a duplicate set of elements to provide backup heating in the event of failure of the main elements. The duplicate elements were of a lower power rating than the main elements. The main heater was rated at 5.5 kW in the lower tip and 2 kW for the upper heater. The standby elements were rated at 4 and 1.5 kW, respectively. The upper heater was slightly underpowered compared with the lower heater in order to maintain the pendulum action and a vertical attitude (the center of gravity of the probe was located below the upper heater).

A ribbon heating element was fitted on the outside of the probe over its entire length to free the instrument from the ice that would form following each freeze-in for making measurements. The sidewall heating elements consisted of a bifilar winding of nichrome ribbon over a layer of fiber-reinforced resin, which had been laid previously over

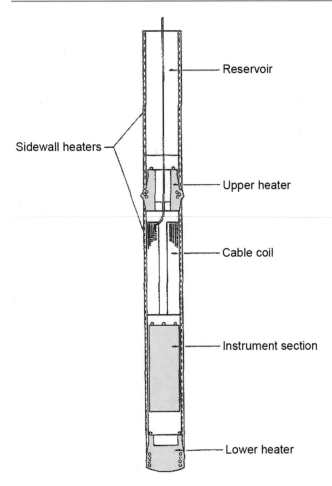

Sidewall heaters —

— Reservoir

— Upper heater

— Cable coil

— Instrument section

— Lower heater

Fig. 1.61 Meltsonde probe of the Australian Antarctic Division (Morton and Lightfoot 1975)

the outside of the probe cylinder to a thickness of 0.8 mm. The ribbon was coated with another layer of resin with a thickness of 0.8 mm. The resin provided insulation and sealed the probe against the entry of water. The power rating of the instrument section of the winding was 969 W, whereas it was 2047 W for the coil section and 1248 W for the reservoir section. The power dissipating within the coil itself reached a maximum of 1400 W with the sidewall heating off.

The coil contained 3050 m of 2.3-mm-diameter coaxial cable with no insulation over the outer braid. The cable was capable of withstanding a DC voltage of 2500 V, even during water immersion. The cable length indicator section was mounted immediately above the coil section. The cable was gripped between a knurled wheel and a pinch wheel. The knurled wheel drove a potentiometer that measured the paid-out cable length.

A twelve-stepping switch in the instrument section allowed for checking the insulation of the cable, reading out the temperature at the lower end of the cable coil, making

loop resistance measurements of the supply cable, reading out the ice temperature, making resistance measurements of the depth potentiometer, reading out the hydrostatic pressure, and switching the lower, upper, and sidewall heaters in various combinations as required.

In 1973, the meltsonde probe was tried at site S2, 80 km south of Casey Station, where the thickness of the ice was estimated to be ~ 1000 m. A hole was drilled in the ice to a depth of 2 m and the sonde was placed in it. In firn, the operating power was limited to 1.9 kW to prevent overheating. In this case, the penetration rate was 0.5 m/h. Then, power was increased, reaching the rated power of 6.06 kW at a depth of 80 m and yielding a penetration rate of 1.08 m/h. At that depth, the temperature of the instrument chamber was 17 °C, and that of the coil was 70 °C.

The probe operated consistently down to a depth of 112 m, where it failed without warning. The instrument was switched to sidewall heating but this also failed, preventing further penetration. The failure of this probe has been attributed to the breakdown of the main heating elements or to inadequate insulation on the stepping switch. It was noticed prior to the failure of the sonde that the temperature of the instrument chamber rose rapidly from 17 to 40 °C. However, the switch was still operative, and temperature data was obtained. The final temperature was −19.4 °C measured after 170 h.

1.3.5 PICO Thermal Probe (Hansen Probe)

In 1979, the Polar Ice Coring Office (PICO), University of Nebraska-Lincoln undertook the development of a freezing-in thermal probe, the design of which differed from that of all its predecessors in that it included a telemetry system to allow for measurements while advancing under the applied power, provided means for making measurements on the meltwater flowing from the tip of the probe through high-pressure sample cells within the probe, and included transducers to measure changes in the orientation of the probe due to the flow of the ice sheet during long freeze-in periods (Hansen and Kersten 1984).

This probe was pendulum-stabilized; its center of gravity was below the 165-mm-diameter upper heater, which controlled the rate of advance. The lower hot point and the body of the probe had a diameter of 127 mm. The overall length was 3.45 m, and the weight of the probe was 150 kg.

The cartridge heaters were hermetically sealed and operated at 50% of their rated power. At a maximum operating voltage of 1325 V DC and a current of 4.07 A, the total heating power of 5.4 kW was divided so that 1.35 kW were dissipated in the upper hot point and 4.05 kW in the lower hot point. This power distribution consisting in sending 1/4 of the power to the upper heater and 3/4 to the lower hot point ensured the proper vertical pendulum steering of the probe.

The lower 2.11-m segment of the probe was a thick-walled, non-magnetic stainless-steel chamber, which was designed to withstand an external pressure of 50 MPa. This segment was hermetically sealed using a U-type Variseal and a pair of O-rings. This chamber contained the two heaters, a rotary selector stepping switch, transducers, the high-pressure sample cells, and the telemetry equipment. The switch was actuated by reversing the polarity of the feeding DC power and allowed for the application of power for any one or a selected combination of the following: powering the heating elements, measurement of the resistance of the heating elements or various thermistors, measurement of the voltage, charging of the battery pack used to energize the transducers for meltwater conductivity measurements, measurement of the DC voltage outputs of the inclinometers, powering of the compass and pressure transducers, measurement of the conductivity of the meltwater as the probe advanced.

The upper 1.32-m segment of the probe contained approximately 3 km of coaxial high-voltage cable tested to withstand 3000 V DC while immersed in water; this segment was open to the ice/water environment. The cable consisted of a solid-center conductor insulated with Tefzel, which is a modified ethylene-tetrafluoroethylene (ETFE) fluoropolymer, covered with a braided silver-plated outer conductor. The center and outer conductors were made from annealed copper. This allowed for a 20% elongation due to movement of the ice before the cable ruptured. The diameter of the central conductor was 0.823 mm, and that over the insulation was 1.245 mm. The overall cable diameter was 1.748 mm. The DC output voltages of the transducers were converted into pulses in the audio frequency range. These variable frequency pulse trains were sequentially transmitted over the coaxial cable to the surface monitoring station and were collected using a portable computer.

The coil section consisted of four individual sections, each of which contained 550 m of coaxial cable and was 29.2 cm long. The main advantage of this sectioned coil segment is the reduction in cable and winding costs. To test the prototype of the PICO probe, a CRREL coil section was used.

To melt the ice around the probe when starting up after a freeze-in period, most of the exterior cylindrical surface of the probe was covered with a nichrome V ribbon heating element sandwiched between fiberglass-reinforced epoxy layers. It provided a power density of 0.3 W/cm^2 on the exterior surface area. The performance of the prototype probe was tested at the PICO laboratory. Tests showed that the proposed technique melted the probe free from ice at a temperature of −40 °C in approximately one hour, with

the interior chamber temperature when melting out being ∼20 °C. The penetration rate in ice at a temperature of −25 °C was 1.6 m/h. Unfortunately, the probe was never used in the field (Kelty 1995).

1.3.6 Probes of the Snow and Ice Research Group

In 1990, the PICO thermal probe was reinvestigated by the Snow and Ice Research Group (SIRG) at the University of Nebraska-Lincoln and modified so that both meltwater conductivity and micro-particle concentration could be continuously monitored (Fig. 1.62) (Kelty 1995). Micro-particle concentration was measured via the scattering of light at approximately 90° using a small (5 mW) laser beam.

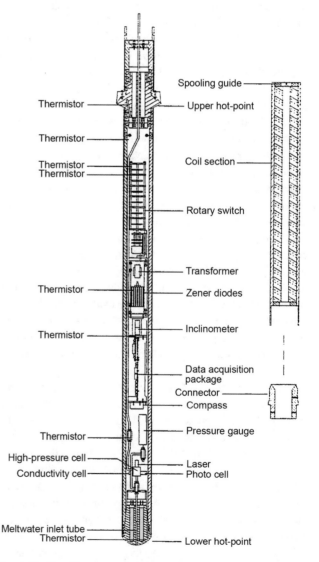

Fig. 1.62 New Hansen thermal probe assembly (Kelty 1995)

Tests of this probe (named the New Hansen thermal probe) were carried out by SIRG in 1990 in an ice well in Nebraska, which included completely freezing the probe in the well for several days and then assessing the melt-out characteristics of the probe. Many problems with the thermal probe were detected during these initial tests, and the appropriate modifications were made to the probe design. Tests showed that the probe penetrated into the ice at a rate of 1.37 m/h with a total heating power of 3.8 kW.

This thermal probe became an official part of the Greenland Ice Core Project (GRIP) and was taken to Summit, Greenland in 1993. During the first field trial, the probe failed at a depth of 7 m when the lower hot point was overheated and a cartridge heater became shorted to the case.

Seventeen more well tests were performed before the next field deployment. In 1994, the probe was launched at Summit again. It was decided to drill through the firn using an additional power line connected directly to the heaters. The probe was attached to the winch with the cable left slack as the probe melted its way down. The winch provided a means of hoisting the probe back out of the firn and of alternatively powering the probe using the connection that bypassed the coil section. Thus, the coil section was not heated during this running time in the firn. The average penetration rate in firn was ~1 m/h.

At a depth of 31.95 m, the probe became stuck and wall heating was used to melt it out of the ice. Then, the probe got stuck 15 more times before reaching the firn–ice transition layer. Each time, wall heating was used to free the probe, after which the probe advanced 1–2 m and froze-in again. At a depth of 59.5 m, the probe was released and the bypass cable was intentionally broken from the probe. The probe could not be pulled up for checks before its release because of ice bridging and the partially closing hole. The probe was powered through the coaxial cable, thus heating the coil section. However, it could not penetrate further. In addition, the coaxial cable failed. Luckily, the cable did not freeze in the hole. A total of 251 m of the coaxial cable were removed and the power supply of the probe was restarted. Wall heat was transferred to the probe and it was successfully melted out.

After the probe was freed, penetration was continued, but the tilt angle increased to a maximum of 9.04° over a period of 11 h. As the tilt angle increased, the probe froze-in several times after various efforts to correct the tilting failed. The main persistent problem was that there was insufficient heat in the sidewalls to keep the probe from getting stuck. This was due to one of the sidewall heaters having been damaged in transit, so the probe had only two thirds of the intended sidewall heat. However, the sidewall heater problem was twofold. With only two sidewall heaters, the maximum voltage that could be applied to the probe's hot point heaters with the two wall heaters powered was 1200 V DC. This

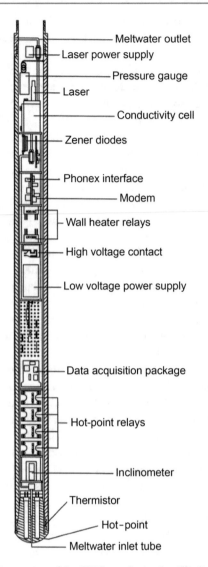

Fig. 1.63 Lower part of the SIRG steering probe (*Credit* J. Kelty)

provided less than the optimum power required for the thermal probe to penetrate fast enough to avoid getting stuck. Finally, the coaxial power cable failed again at an estimated depth of 120 m. Attempts to pull out some of the cable failed as it had become frozen in the ice. All efforts for restarting the probe were unsuccessful, thus ending the field test.

The final version of the probe, called the SIRG steering probe, had eight independent 750-W heaters identical to those of the Hansen probe and no upper annulus heating ring (Fig. 1.63). Heaters wired in parallel were supplied with AC power and individually switched on and off using solid-state relays, which were automatically controlled by an independent microcontroller board separate from the data acquisition package of the thermal probe. The inclinometer outputs were converted to angles and the total probe tilt was calculated. This tilt was checked against the maximum allowed tilt,

Fig. 1.64 Upper part of the SIRG steering probe during preparation for testing (*Photo* J. Kelty)

which was selected by an operator. If the maximum tilt was exceeded, the high-side angle calculated from the inclinometers was used to decide which of the probe heaters needed to be turned on more than the others to compensate for the tilt. Tests in the Lincoln ice-well facility showed that, during normal running, only one or two heaters needed to be changed, whereas the remaining ones were kept at a 70% power level for the probe to remain vertical (Fig. 1.64).

That probe was tested at Dye 2, Greenland, deployed on July 16, 1996, and failed at a depth of 60 m owing to a short in the coil (J. Kelty, personal communication, 2016). The failure of this probe was somewhat similar to what happened in 1994. The 20-year-old CRREL wire had been tested at 5000 V DC before and after the coil section was constructed, but it failed in the field at 1200 V AC.

1.3.7 Sonde Under Shelf Ice (SUSI)

In 1990s, researchers at the Alfred Wegener Institute for Polar and Marine Research (AWI) constructed a new probe, called the sonde under shelf ice (SUSI), for the investigation of the environment under Antarctic ice shelves (Ulamec et al. 2007). This sonde used a simple stabilization mechanism regulated via wire tension that was based on information of the temperature of the thermal head and the tilt of the sonde to control the friction clutch of the sonde's tether/cable pay-out mechanism. The SUSI-II sonde was successfully tested on the Rettenbach glacier (Sölden, Austria) in 1990 and reached a depth of 60 m. The sonde could be retrieved from the glacier because the melted hole did not refreeze within 8 h.

Two-years later, during winter in 1992–1993, this sonde was able to penetrate a 225-m-thick Antarctic ice shelf near Neumayer station at the Ekstrom Ice Shelf in 75 h with an average penetration rate of 2.9 m/h. The sonde accessed the ocean cavity below the ice sheet. The specifications of SUSI-II were as follows: length of 2.25 m, diameter of 100 mm, power of the thermal head of 3.4 kW, and supply power of 600 V/8 A.

A newer version of this sonde, named SUSI-III, was built for penetrating ice with a maximal thickness of 800 m (Tibcken and Dimmler 1997). The SUSI-III drilling set consisted of four subsystems: (1) a melting sonde; (2) electronics for control from the surface; (3) a 12-V power supply for the electronics control unit; and (4) a DC generator for the thermal drill head. In turn, the sonde consisted of three sections (Fig. 1.65): (1) a thermal head with protective housing; (2) an electronic unit with the control system, the payload, a trigger, and a friction winch; and (3) coil with wires.

The sonde was suspended by an 800-m-long and 1-mm-diameter Kevlar rope with a breaking strength of 135 kg, released from the winch in the sonde. Power supply and signals were transferred through two 900-m-long Teflon-insulated single copper wires with a cross-section of 1 mm^2. Wires could be released from the coil with a slight pull. The friction winch with the rope was located below the coiled wire, and the rope was passed through the inner hole of the coiled wire so that the wires twisted bifilarly around the rope when the sonde descended. As the rope and wires were frozen in the melted borehole, the rope was able to carry the whole weight of the sonde (the starting weight was 75 kg). The winch was braked in the de-energized state. During drilling, the winch was controlled by a microprocessor to adjust the feeding speed. The inclination and thermal head temperature were used as control gauges to reach an optimum melting process.

Fig. 1.65 Sonde under ice shelf
SUSI-III (Tibcken and Dimmler
1997)

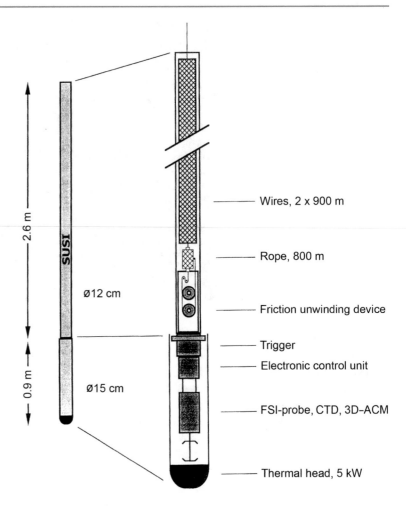

Using the trigger, the thermal head with the protective housing could be released when the ice sheet bottom was reached so that the sensors could be released in the water. The 150-mm-diameter thermal head was supplied DC power from a specially designed generator with an output power of up to 12 kW, driven by a commercial diesel engine. This ready-to-install system weighed 500 kg. However, it was possible to separate it in two units, each with a weight of $\sim$250 kg. Voltage could be adjusted continuously with a remote control between 350 and 1000 V by changing the engine's rotation speed.

At the bottom part of the sonde, a commercial probe from FSI Technologies Inc. was installed, with sensors for temperature, pressure, salt content, flow direction, and speed. The surface microcontroller controlled the winch, monitored the communication with the FSI probe and the operating parameters, and modulated the data using frequency-shift keying. The system worked in two operating modes: melting and measuring. Switching between the melting and measuring modes was performed by reversing the supply voltage. For melting, up to -1000 V/12 A were supplied, whereas $+300$ V/0.1 A were supplied for measuring.

During drilling, all parameters were displayed and controlled continuously by the operator. As soon as the sonde reached the bottom of the ice shelf and the FSI-probe was hanging freely in the water, the sonde was powered by the 12-V supply. In the measuring mode, the system was fully automatic, and the FSI probe was only switched on at certain intervals (as planned, 5 min running every hour). The data were transferred to the ice surface and then processed for satellite transmission via the Argos system during at least one year.

The surface electronics control unit was powered by the 12-V supply. Solar cells and a wind generator, each buffered with lead-acid batteries, served as voltage sources. In the melting mode, the starter battery of the DC generator could also be used. Plans were made to add zinc-air batteries to the power supply circuit so that a power would be supplied without failures during the dark season, or if the wind generator failed.

In January–February 1995, the SUSI-III sonde was planned to be tested at Foundation Ice Stream in Antarctica, between Berkner Island and Henry Ice Rice, where the estimated ice thickness was 700 m. Unfortunately, during the preparation of the device for drilling, no data were received from the FSI probe. It also did not respond to software commands and could not be handled by the control program. The cause of the error could not be determined on site. Changing the control program was not possible within the given time frame with the available equipment. Because the inclination sensor of the FSI probe was used to control drilling, the sonde could not be started without the payload and the drilling tests were terminated. Further testing of the SUSI probe and its development were cancelled. However, this probe design was used for the development of a melting probe prototype for extraterrestrial investigations by the German Aerospace Center (DLR) in Köln (see Sect. 1.4.3).

1.3.8 Recoverable Autonomous Sonde (RECAS)

The main function of the RECoverable Autonomous Sonde (RECAS) is the exploration of subglacial lakes while causing minimal chemical and microbial contamination (Talalay et al. 2013, 2014). A unique feature of this sonde is that it was designed to be recovered by using an upper melting tip and a cable recoiling mechanism. Two versions of the RECAS design are being considered: one with 200 m of coiled cable (RECAS-200, to be used as a prototype) and a second one with a 2500-m-long cable (RECAS-2500).

This drill is equipped with two 160-mm-diameter electrically powered melting tips located at the upper and lower ends of the sonde (Fig. 1.66). The copper cone-shaped (38°) tips have height of ∼ 200 mm. Both of them include 16 cartridge heaters (6 mm in diameter). The total power of each thermal head is 5 kW. The upper melting tip has a small central hole for the cable to slide in and out of the RECAS. To protect against refreezing, the exterior cylindrical surface of the sonde is heated by an element with a power of 0.7–1.2 W/cm^2 of the housing area; for a 6.8-m long sonde, the power required for the side heating of RECAS-200 at −30 °C is estimated to be 2.2 kW.

The driven section in the upper part of the probe contains three gear motors. One gear motor drives the pulley system with a controlled speed of 0–4 m/h and a maximum load of 8 kN; the two others are used to rotate the cable coil and level winder. To minimize the size of the cable, power is supplied at 2500 V AC and converted to the required voltage according to component specifications. The 6-mm-diameter cable used consists of two 2-mm^2 lines used for power and signal communication. A pressure chamber with transformers and the control system occupies the middle part of the drill. The pressure chamber is instrumented with an inclinometer. Coded data from the RECAS are transmitted to a computer on the surface.

Two air-filled sample bottles (200 mL), sensors (pressure, temperature, water conductivity, and pH), and a video

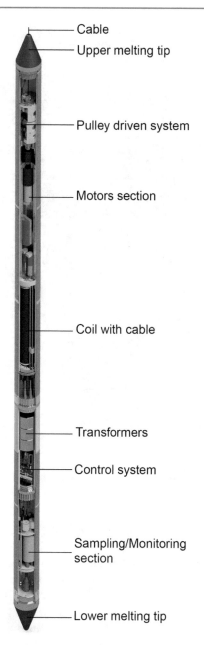

Cable
Upper melting tip
Pulley driven system
Motors section
Coil with cable
Transformers
Control system
Sampling/Monitoring section
Lower melting tip

Fig. 1.66 Recoverable autonomous sonde: RECAS

camera are installed in the lower part of the sonde. The bottle valves are actuated with magnetically coupled electric motors that open and close them on demand. The samples are maintained at in situ hydrostatic pressure, allowing for a quantitative analysis of dissolved gases. Another option is to add a laser spectrometer into this section for in situ gas analyses of the melted water. For this purpose, the OF-CEAS spectrometer system designed by the Interdisciplinary Laboratory of Physics, Grenoble, France, is being considered (Grilli et al. 2014).

The RECAS will operate in a borehole isolated from the glacier surface and atmosphere. All downhole RECAS components will be sterilized using a combination of

Fig. 1.67 RECAS's lower melting tip; a dust trap is fixed to the tip's top (*Photo* Y. Li)

chemical wash, hydrogen peroxide vapor, and ultraviolet sterilization prior to deployment. On the way down, the lower melting tip is switched on. Just as for other Philberth probes, the melted water is not recovered from the hole and it refreezes behind the sonde. The power and signal cable is released from the coil inside the sonde. The coil motor slows down the paying out of the cable in such way that the support point is above the center of gravity of the RECAS. Because of this, the sonde is inherently stable, and its attitude is plumb at all times.

When the sonde enters the subglacial lake, it samples the water and examines water parameters. After the completion of the sampling and monitoring operations, the coil motor is activated and the top melting tip is powered. The recovery of the sonde to the surface begins by spooling the cable. Finally, the sonde reaches the surface with subglacial water samples ready for servicing.

The sounding of a relatively shallow subglacial lake can be conducted during an Antarctic summer; however, the exploration of a deeply buried subglacial lake requires

4–5 months. In the latter case, the research personnel leave the site after deployment and the sonde operates as a fully autonomous system. Coded data from the RECAS are recorded and then automatically sent to a remote server for inspection. The RECAS power system has to operate continuously without service or refueling for several months and includes two automatically controlled 16-kW diesel engines.

Subassembly tests were carried out in 2018. Melting tip tests (Fig. 1.67) in an ice well with a temperature of −20 °C yielded a penetration rate of 1.77–1.88 m/h at an electric power of 5 kW and a load of 24–90 N. Field tests of the RECAS-200 near Zhongshan station in Antarctica are planned for the 2019–2020 season.

1.3.9 Summary

The first concept for a freezing-in hot-point drill consisted in the use of a small in-board nuclear reactor for supplying power to the thermal heads and downhole equipment. However, environmental considerations prohibited the realization of such projects. All realized hot points delivered electric power to the downhole unit from the surface. The traditional power supply design consists in using a cable, paid out from a coil stored in the aft of the melting probe. Delivering heating power via a cable to great depths is challenging owing to both mechanical and electrical constraints, but simplifies communication and stabilization issues.

Freezing-in hot-point drills require a means of stabilization if the required plumb attitude is to be maintained. While drilling, they stand on their tip—the hot point. The slightest deviation from the vertical results in an increasing tendency of the probe to "lean" and "topple over". This instability must be counteracted, unless it can be eliminated (Philberth 1966). The following practical methods for drill stabilization have been considered:

(1) Philberth (1976) proposed a mercury steering principle, which has proved to be effective. As the heat transfer medium within the hot point, mercury directs the penetration of the probe towards the plumb line when a deviation develops. There is a continuous correcting action to counter the leaning tendency of the probe.

(2) The pendulum steering concept, proposed by Aamot (1967a), eliminates the aforementioned instability by adding another hot point at the upper part of the probe and placing the support point above the center of gravity, i.e., the probe "hangs" plumb at all times and freely seeks the gravity vector as a means of steering.

(3) Aamot (1968c) suggested a method of hot-point stabilization consisting in adding a light upper end (a buoyancy chamber) to the probe, whose upward force keeps the drill erect.

(4) Kelty (1995) suggested individually controlling cartridge heaters in the lower hot point (the thermal probe had eight cartridge heaters spaced radially around the hot point) to steer the probe vertically downwards.

(5) The SUSI thermal probe included a mylar tether that bore the weight of the probe and was used for steering in such way that the weight of the drill was partially compensated by the rope's tension and was partially transmitted to the bottom (Tibcken and Dimmler 1997). Talalay et al. (2014) suggested slowing down the release of the carrying cable using a downhole motor-gear drive.

Typically, high voltages (>1000 V DC) are used for the power supply to minimize losses in the cable. Thus, two wires, of which only one has to be insulated, suffice for power and telemetry. Clearly, the thickness of the cable, along with the available total volume for the storage of coils, is the limiting factor of the maximum possible penetration depth.

In comparison with the CRREL/Australian Antarctic Division/PICO pendulum probes, the Philibert probe is simpler in concept and does not require an upper heater. The upper heater of a pendulum probe should support most of the probe weight. It requires complicated automatic control, which regulates the power to the upper heater depending on the hole size, which in turn is influenced by the heat transfer from the lower heater of the probe. There is no documented reason found as to why the diameter of the upper coil section in pendulum probes is not the same as the upper heater diameter.

One of the main sensitive items of freezing-in probes is the cable, because the failure of its isolation was often the reason behind probes getting stuck. Another reason for this was insufficient lateral heating when the refreezing of ice laterally and behind the probe was high compared with the melting velocity. Although this is not a big issue in warm ice, it is critical in cold ice, in which the heating elements should be distributed along the length of the probe and heating should be careful controlled.

All freezing-in probes are considered to be non-recoverable except the RECAS returnable probe proposed by Talalay et al. (2014). This probe has electrically heated hot points in its lower and upper ends. The tethered Kevlar-armored cable is spooled on a motor-driven winch that allows for pulling the RECAS to the surface when the research mission is complete.

1.4 Thermal Probes for Extraterrestrial Investigations

Ideas of using hot points for accessing the subsurface layers of planetary ice sheets arose in connection with the proposed Mars and Europa missions (Paige et al. 1993; Zimmerman et al. 2001; Ulamec et al. 2007). Thermal probes are also suggested to use for collecting biological samples from cryogenic ice at the surface of icy moons in the Solar System (Davis 2017). Some theoretical and experimental work has been done to understand the behavior of such probes under extraterrestrial conditions—very low temperatures, vacuum, and low gravity (Di Pippo et al. 1999; Biele et al. 2002; Kömle et al. 2002; Lorenz 2012; Horne 2018; Schüller and Kowalski 2019). Table 1.3 summarizes parameters of the icy surfaces of Mars, Europa, and Earth. It is clear that each environment requires different technical solutions for melting probes. Dust content, salinity, temperature, and radiation vary considerably among these three examples. Whereas on Earth a heated probe deployed into an ice sheet always causes melting with subsequent refreezing, the behavior of such a probe in a low-pressure environment is quite different (Kaufmann et al. 2009).

None of the existing melting probes is nearly mature enough for planetary applications without considerable technical development. Critical parameters for any space mission are weight, overall dimensions, and power supply. Missions to explore the surface and subsurface of icy satellites can use a wide variety of schemes and technologies (Table 1.4), which are discussed below.

Many of the components of thermal probes for extraterrestrial investigations were tested in Earth's glaciers and ice sheets. Moreover, some of them were involved in different polar projects to study subglacial environments or deliver sensors into ice bodies.

Table 1.3 Relevant parameters for hot-point's application on Earth, Mars and Europa (Ulamec et al. 2007)

Parameters	Earth (polar)	Mars (polar)	Europa
Temperature/°C	−60...0	−68...−63	~ -170
Gravity/(m s^{-2})	9.8	3.7	1.3
Atm. Pressure/mbar	1000	6.7	0
Radiation/(rad day^{-1})	~ 0	$\sim 10^2$	4×10^6
Composition of ice	H_2O, some air	H_2O, CO_2, dust	H_2O, possibly others (salt, CO_2, CH_4)

Table 1.4 Hot points for extraterrestrial investigations

Thermal probe	Years	Power/kW	Hot-point diameter/mm	Penetration rate in ice/ (m h^{-1})	Max depth reached/m	References
Cryobot	1998–2002	1.0	120	0.35–0.6	23	Zimmerman et al. (2001)
SIPR	2002–2006	0.25	75	1.0	50	Bentley et al. (2009)
IceMole	Since 2009	2.4	150 × 150	1.0	25	Dachwald et al. (2014)
VALKYRIE	2010–2016	5.0	250	0.9	30.5	Stone et al. (2014, 2018)
SPINDLE	Since 2015	50	400	NA	NA	Stone et al. (2018)
Ice Diver	Since 2011	2.15	65/85[a]	2.4	400	Winebrenner et al. (2013, 2016)
IceShuttle Teredo	Since 2015	3.5	180[b]	NA	NA	Wirtz and Hildebrandt (2016)

Note: *NA* not available; [a]Diameters of the thermal head and the upper thermal flange; [b]Author's estimation

1.4.1 Cryobot

The design of a fully autonomous robotic melting probe Cryobot was started in the Jet Propulsion Laboratory (JPL), California Institute of Technology, Pasadena under NASA's research program in 1998 (Zimmerman et al. 2001, 2002). This system was designed to penetrate through Mars Polar Cap and Europa's ice pack, which are supposed to be 3–10 km thick. This probe can only move down. Another opportunity to test this technology was to explore the lakes buried deep under the Antarctic ice sheet (Lane et al. 2001). In Earth-based applications, manual mode was considered, with an operator monitoring the state of the probe in real time. In these circumstances, the vehicle would be placed in a deep, hot-water predrilled hole that stops a few hundred meters above the ice roof of subglacial lake.

Cryobot was allotted with integrated radioisotope thermoelectric generators (RTGs), providing 1 kW of thermal direct-melting energy. Fluid thermal modelling revealed that, with 1 kW of power, a melting rate of 0.3 m/h can be realized in very cold (−170 °C) ice; the water jacket maintained around the probe is 1–2 mm in width and the melt plume would not refreeze for 1.25 m behind the vehicle. Another option to convey power and control signals down and data up to a surface station is to use a tether in the same way Philberth probes do (this version is shown in Fig. 1.68). In such cases, the probe would be powered by a solar photovoltaic system at the surface.

The prototype version of Cryobot was proposed to be from 1.0 to 1.25 m long and 120 mm in diameter and to have a weight of ∼40 kg. The actual flight version of the probe was planned to be between 0.8 and 1 m in length and weigh 20–25 kg. The probe was divided into following

sections: nose, pump, instruments, electronics, and tether. Cryobot used two sources to melt ice at the bottom: one was a traditional spheroid solid penetration tip (passive heater) and another one was a hot-water jet nozzle at the central part of the tip (active heater). The active heater (the water jet) formed a melt zone with a parabolic shape in front of the Cryobot probe and allowed for a faster penetration in dense ice. The velocity of the water jet was 10–20 m/s at a flow rate of 1–1.5 L/min. The temperature of jetted water was 20–25 °C.

The solid tip contained four quadrants with heat-conductive copper inserts. These quadrants were split with ceramic fins into sections to prevent heat transfer between them during drilling and steering. The tip was covered with a copper nose. Heading corrections were executed by differentially heating one side of the thermal tip while turning off the opposing heaters.

The sonde was equipped with temperature/pressure sensors (ambient and inner), a tether odometer, an encoder, a two-axis inclinometer, a water-cavity capacitive float sensor (for detecting the case when the ice was broken up and fractured and the meltwater was lost to the surrounding ice cavity), a pressure sensor in the water pumping circuit, and an acoustic sonar at the tip for acquiring obstacle information. Cryobot had a window in the pressure vessel that provided an equivalent sampling interface for examining the ice walls of the borehole and the meltwater as the vehicle descended. The two scientific instruments selected for this were a 100-mm camera with flash and a UV laser spectrometer.

The tether was used to supply power and serve as a data and command communication cable. It was spooled out of the probe during its descent into the ice. In the first

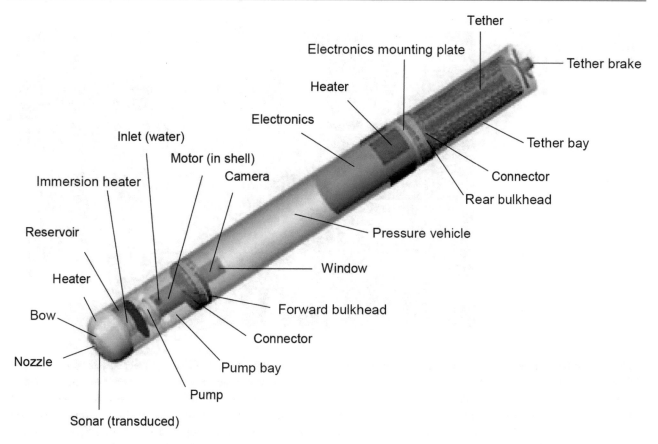

Fig. 1.68 Overview of the tethered version of Cryobot (Zimmerman et al. 2001)

prototype, this section was designed to accommodate a 2.5-mm diameter tether, which was 150 m long and consisted of two twisted-pair cube conductors with enamel and Teflon insulation, one or more optical fibers, and a Kevlar strength member (Bentley et al. 2009).

In September 2000, Cryobot was tested by using it to melt through a 5-m column of clear ice with a temperature close to 0 °C (Fig. 1.69). Testing was performed without either analytical instruments or an internally deployable tether. The average rate of penetration was 0.43 m/h at 0.42 kW of consumed power. At that power, the passive melting rate was 0.35 m/h, whereas the active melting rate was as high as 0.6 m/h. Two corrections of the sonde's attitude were carried out. The first 3° off-vertical drift was corrected by inclining the sonde $\sim 10°$ in the opposite direction. The vertical travel during this correction was 0.28 m. The second time, the sonde position was corrected from a 7° overshoot back to vertical during a vertical penetration of only 8 cm.

In the season of 2000–2001, Cryobot, lacking the full suite of instruments and the compact tether, was tested on Ice Stream C, Antarctica, using a passive drilling prototype in the upper cold firn layer, which had a temperature of −26 °C (French et al. 2001). It was found that, at low drilling speeds, the passive method became very inefficient because of the refreezing of the water percolating down into the firn.

In October 2001, NASA teamed with the Norwegian Polar Institute and the Norwegian Space Center and tested a simplified Cryobot at the Longyearbreen glacier in Svalbad (Bentley et al. 2009). After its initial deployment from a light tripod, the probe penetrated to a depth of 23 m. With water jetting enabled, the probe descended at a rate of 0.6–0.7 m/h using 750 W of power. The probe successfully penetrated multiple layers of dust and debris. The water above the probe remained unfrozen at this relatively warm location, allowing Cryobot to be lowered by its external tether.

Additional research efforts were undertaken to evaluate an alternative design of Cryobot in which the full thermal power output was exploited by incorporating the reactor directly into the body of the Cryobot probe (Elliott and Carsey 2004). The reactor would use a thermal-to-electric power conversion system to provide electricity to the surface elements to support scientific measurements, telecommunications, and survival heating. The probe under this preliminary design concept was ~ 3 m in length and 0.5 m in diameter. It was designed to use heat pipes to transfer the thermal output from the reactor to heat the water jets

Fig. 1.69 Cryobot prototype:
a ready for testing; **b** melting
through a 5-m ice column
(Zimmerman et al. 2001)

(a)   **(b)**

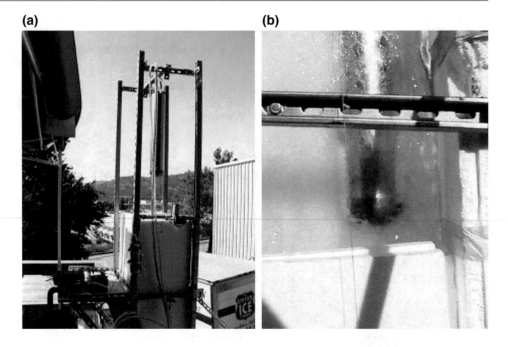

expelled through the nose. A heated sampling probe was
included in the nose of Cryobot to obtain pristine meltwater
samples slightly ahead of the main body. Water samples
would be either transferred to the instrumentation within
Cryobot itself or pumped via a tubing system up to analysis
equipment on the lander.

Further Cryobot development was abandoned in favor of
the open-hole SIPR probe, which was seen as more compatible with a near-term Mars polar mission (see the next
section).

1.4.2 Subsurface Ice Probe (SIPR)

Since 2002, the JPL has been developing an open-hole
thermal drilling system customized for the Martian polar
layered deposits (Bentley et al. 2009). On Earth, it is generally accepted that drilling safely in open-holes (without
fluid) can be carried out to depths of 300 m to 350 m
because the borehole closes due to the overburden ice
pressure that is not compensated for by the pressure inside
the borehole (Talalay et al. 2015). Problems associated with
borehole closure are over-pulling of the drill and an increase
in the probability of the drill getting stuck. However, models
of the Martian North Polar Cap indicate that the reduced
gravity and the low temperatures of the ice at high depths
would allow for the drilling of open holes through the entire
depth of ~ 3 km of the ice cap (Cardell et al. 2004).
A 75-mm-diameter drill with an applied power of 200–
400 W is expected to descend at speeds of 0.20–0.45 m/h in
ice with a temperature of $-110\ ^{\circ}$C.

The Subsurface Ice PRobe (SIPR) consisted primarily of
a downhole heating element and a pump to send meltwater
to the surface (Fig. 1.70). The nose was a heated copper disk
that melted the ice ahead of the drill. A small forepump
delivered water to the top of a reservoir, where air bubbles
were allowed to separate and the occasional particles larger
than 15 μm were filtered out. This water was then gravity fed
into the main pump, which pressurized it for delivery to the
surface.

Two different SIPR probes have been constructed. In the
first version of the probe, the main body, including the
heating elements, the pump, and the support structure, was
submerged. In the second version of the probe, only the
heating element was submerged.

Supporting the melting and pumping operation was a
tether that provided power and command information and
also sent back meltwater and data. The power for the heater
and pump was carried through shielded copper wires. A
braided nichrome heater in the tether, divided into zones,
kept the meltwater liquid all the way to the surface. On the
surface were the tether reel, electronic controls, and the
chemical analysis instruments that sampled the meltwater
before it was discarded. A manifold distributed the meltwater to the various instruments and selected samples for
laboratory analysis.

To ensure a vertical penetration, the SIPR operated as a
plumb bob, using feedback to touch the base of the hole
lightly while another feedback system, using down-hole
water level sensors, adjusted the pumping rates to match the
rate of melting. Additional heaters allowed the drill to be
brought back up after the conclusion of the descent.

Fig. 1.70 Elements of the SIPR probe; diagram by M. Hecht (Bentley et al. 2009)

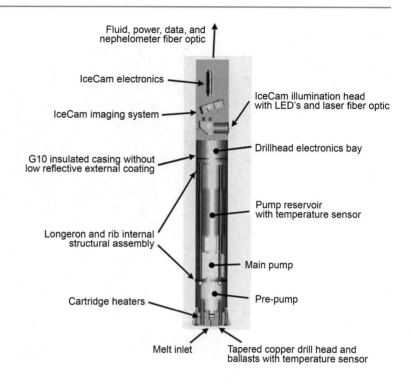

A preliminary version of the SIPR was field tested in the Athabasca Glacier, Columbia Icefield, Canada, in October 2003, where it drilled to a depth of 20 m in approximately 24 h (Fig. 1.71a). In that version, the main body, including the heating elements, the pump, and the support structure, were submerged. A fully integrated drill was subsequently deployed in Greenland to a depth of 50 m in July 2006 (Fig. 1.71b). In addition to having flight-compatible packaging and thermal control, the drill was autonomously deployed with an aerogel-insulated tether using a load cell in a force-feedback control system. It was instrumented with downhole cameras and sensors, and meltwater was characterized in a field-portable isotopic water analyzer. The drill melted an 80-mm-diameter hole and recovered water through a Teflon tube with a diameter of 1.6 mm and a length of 100 m. Drill power was selectable up to 250 W, corresponding to penetration rates of up to 1 m/h.

1.4.3 Modified SUSI Probe

To confirm the feasibility of using melting probes under extraterrestrial conditions, a number of experiments have been performed at the German Aerospace Center (Deutsches Zentrum für Luft- und Raumfahrt, DLR) in Cologne, Germany in a cold lab as well as in a vacuum chamber (Treffer et al. 2006; Ulamec et al. 2005, 2007) using a modified SUSI melting probe (see Sect. 1.3.7). The main part of the probe consisted of a copper hemisphere melting head (115 mm in diameter and 75 mm height) containing three 230-V heating

elements with a power of 200 W each. The heating elements could be operated independently from each other but, during tests, the full heating power of ~0.6 kW was employed. Permanent monitoring of the temperature of the thermal head at a single point was enabled by a PT100 thermo-resistor.

The tether was not stored inside the probe, but rather connected in a stretched condition to a motor via a strain gauge and two pulleys (Fig. 1.72). The strain gauge was used as a control sensor ensuring that the probe always exerted its full weight on the ice sample during the experiments and was not kept back by the tether.

The average rate of penetration achieved in a vacuum was 0.087 m/h above the predicted rate for ice sublimation (0.064 m/h), but much less than the calculated rate for melting (0.295 m/h). During low-temperature vacuum experiments, the melting borehole froze again very rapidly, and thus the probe's hot point was already in an almost-closed cavity shortly after penetration; the probe got stuck at its unheated aluminum body. To avoid such situations from happening, the body of the probe must also be supplied with a sufficient amount of heat (Biele et al. 2011).

1.4.4 Melting Probe of the Space Research Institute

Several hot-points aimed to explore the icy layers in the Solar System were designed at the Space Research Institute (Institut für Weltraumforschung, IWF), Graz, Austria. Initial

Fig. 1.71 SIPR probe:
a beginning its descent into
Athabasca Glacier, October 2003
(Cardell et al. 2004);
b descending into the Greenland
ice sheet, July 2006 (Bentley
et al. 2009)

(a)

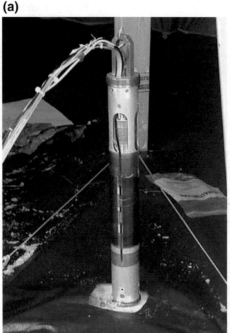

(b)

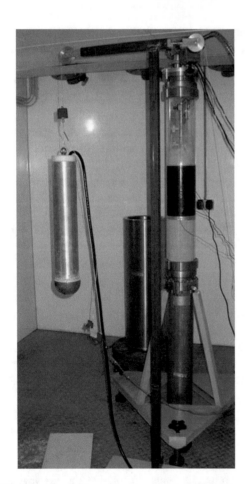

Fig. 1.72 Setup for melting probe experiments at DLR (Ulamec et al.
2007)

tests with a 40-mm-diameter hot point consisting of two half
spheres of brass screwed together to form a full sphere were
performed in a cylindrical vacuum recipient (Kömle et al.
2004). Between the two half spheres, a circular heater foil
using Kapton as a substrate and isolation material were
inserted. The average penetration rate of this hot point
heated with 25 W in compact snow that was cooled down to
a temperature of approximately −100 °C was 0.048 m/h
under vacuum conditions, an order of magnitude less than
the rate obtained for a similar setup but under air pressure
(0.45 m/h).

Later on, simple prototypes of thermal probes with an
ogive-shaped heating tip and a flat blunt tip were tested
(Kaufmann et al. 2009). It was assumed that the hot point
with the ogive-shaped tip would make it possible to move
through dusty ice layers and would help in shifting the
center of gravity close to the front end. The tip was made of
brass and had a total height of 65 mm (Fig. 1.73). Its largest
outer diameter was 63.5 mm and it was fixed to a
stainless-steel tube with the same outer diameter. Three
heaters were placed inside the tip: a ring-shaped heating foil
at the bottom of the cutout and two rectangular heating foils
electrically connected in series, fixed to the inner sidewall.

The ogive-shaped melting tip was tested at the DLR
Planetary Environments Simulation Chamber at Cologne in
a vacuum with solid ice at a temperature of −15 °C. The
total heating power was minimized to ∼80 W. During the
whole test, the melting probe was only able to penetrate the
ice to a depth of ∼15 cm. The melting probe slowed down
from a speed of 1.5 cm/h at the beginning down to 0.3 cm/h

(a)

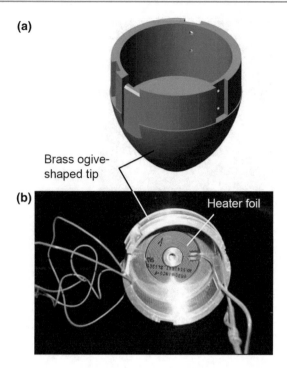

Brass ogive-shaped tip

(b)

Heater foil

Fig. 1.73 Ogive melting tip: **a** 3D model; **b** top view, which shows the two heating circuits (Kaufmann et al. 2009)

at the end of the test. The melted borehole had a crater shape with a maximum diameter of ~ 100 mm, which is approximately one and a half times the diameter of the probe. This means that a considerable amount of heat was transferred from the tip to the sides instead of to the front, and more energy was dissipated laterally than downwards.

The hot point with a flat blunt heating tip had the same outer diameter of 63.5 mm. It was tested under vacuum

conditions in a smaller chamber at the IWF in Graz. Although the ice was significantly colder than during the Cologne test ($-57\,°C$), the penetration rate reached 1.1 cm/h. The probe marginally tilted into the ice while melting and reached a penetration depth of ~ 10 cm.

1.4.5 Melting Probe of Hong Kong Polytechnic University and the Space Research Institute

An original version of a melting probe prototype was designed by Hong Kong Polytechnic University in cooperation with the IWF within the Europa Jupiter System Mission (EJSM) project, which planned to place scientific instruments on the surface of Jupiter's moon Europa in the late 2020s (Weiss et al. 2008, 2011). In the same way as a Philberth probe, the vehicle included a cable magazine with a passive cable-release system that paid out the tether while the probe sank into the ice. The combined drill head, containing melting and drilling mechanisms, was designed to be able to penetrate through debris-laden ice. To provide an antitorque reaction, the probe had a rectangular cross-section and did not rotate around its main axis when torque was applied.

During the first stage, the performance of two different copper thermal head shapes was compared (Fig. 1.74). In both cases, the head was slightly bigger than the body (+1 mm) in order to ensure that the melted channel was big enough for the body to pass. A heating cartridge with a power of 75 W was integrated in the middle of each head, and each system had a total weight of 640 g (without the

Fig. 1.74 Two different thermal heads: **a** conical and **b** spherical; ① probe body, ② melting head, ③ heating cartridge (Weiss et al. 2008)

(a)

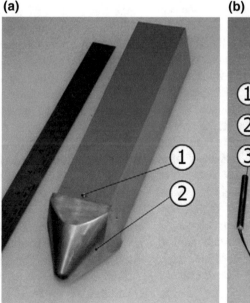

(b)

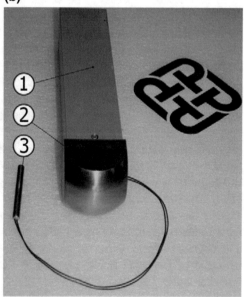

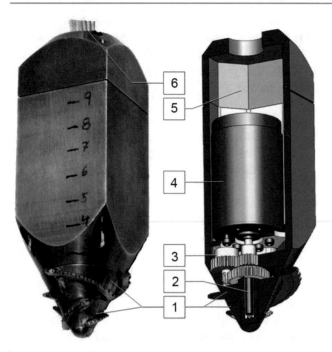

Pure ice (3 cm)

Sand (1 cm)

Pure ice

Fig. 1.75 Hybrid drilling–melting drill head: (1) rotary blades, (2) cartridge heater (four others in the corners are not shown), (3) gear, (4) motor, (5) tether compartment, (6) cable (Weiss et al. 2011)

Fig. 1.76 Penetration test through fine sand (grain size <1 mm) (Weiss et al. 2011)

cable). Tests were carried out with a simple probe architecture under atmospheric conditions and at a temperature of 21 °C. The probes did not include any lateral heating mechanism. Each of the two melting heads was tested separately to a maximum depth of 22 cm. In most cases, the probe started to incline. A penetration rate of ∼0.23 m/h was attained with both head types. Therefore, it was assumed that the shape of the head has only little or no influence on penetration rate.

To penetrate through the impurities (e.g., regolith and rock) that can be expected to be within planetary ice, the hybrid drilling–melting drill head was designed based on the assumption that particles are moved backwards by the blades (Fig. 1.75). The head included two separate heating circuits: the "hot nose" in the front of the probe (35 W) and four heaters in the edges of the body (5 W each). The purpose of the hot nose was to stabilize the probe during the process and to heat the drill blades. The four edge heaters served to keep open the remaining profile of the borehole and to melt down (or sublimate) the drilling cuttings that were transported to the rear by the blades. The drilling mechanism consisted of two counter-rotating wheels on which three steel blades were fitted. The wheels were driven by a 15-W motor. The application of counter-rotating propellers was chosen to further reduce the drilling torque, but the probe still required a deployment system on the surface to eliminate the torque produced by the small frontal propeller before the second propeller reached the ice.

A drill head size of 40 mm × 40 mm was used, which was equal to the size of the probe. The probe weighed 760 g excluding the cable at the rear of the motor. The prototype was first tested in clear ice and then in ice with a layer of sand of different grain sizes under normal atmospheric pressure (Fig. 1.76). To stabilize the vehicle while penetrating, a simple support structure was used. The experiments were done using either the drilling and melting systems together, or only the melting one. After penetration of the first 2 cm, the rate attained with the melting-only mode was quite stable at ∼0.36 m/h, whereas that attained with the melting–drilling mode was rather unstable and suddenly changed in the range of 0.25–1.04 m/h. The results for penetration into the sand layers were similar. The drilling system showed higher speeds than the melting one for the first 2 cm and then slowed down significantly.

A melting probe prototype refurbished from an earlier 40 mm × 40 mm square model was tested in a vacuum in the IWF lab (Kömle et al. 2018a) (Fig. 1.77). The experimental setup consisted of a cylindrical vacuum chamber with a diameter of 40 cm and a total height of 80 cm, composed of two segments with a height of 40 cm each (Fig. 1.78). The cooling system consisted of a copper cold shroud covering the inner mantle of the lower segment and a cold plate positioned at the bottom of the chamber. The cooling machine could adjust temperature down to −70 °C.

Four experiments were performed under largely identical conditions, except for the environmental gas pressure. The

Fig. 1.77 Melting probe used for tests in a vacuum (length: 12 cm, cross section: 40 mm × 40 mm) (Kömle et al. 2018a)

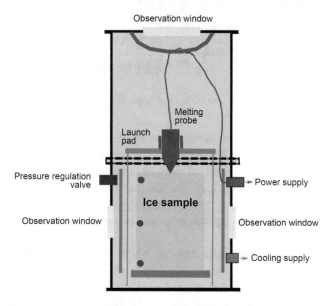

Fig. 1.78 Experimental setup for the melting probe tests; the blue circles mark the positions of Pt100 temperature sensors (Kömle et al. 2018a)

experiment, the gas pressure in the chamber was reduced to ∼20 Pa, the smallest value that could be reached with the available pumping system. This pressure is in the sublimation regime for a free icy surface, and the behavior of the melting probe upon heating should resemble its performance on atmosphere-less bodies, such as Europa or Enceladus. The average melting rate attained with this setup was only 0.083 m/h. Finally, for the purposes of comparison and calibration, another experiment was performed, where the environmental gas pressure was the standard surface pressure of Earth (10^5 Pa), and the probe could drill at a rate of 0.3 m/h. Thus, experiments clearly indicated that the penetration rate of a melting probe is lower by a factor of 2–3 under the pressure of Mars than what it would be under terrestrial ambient pressure.

The next series of trials used a 60-mm-diameter circular melting probe with a hemisphere heated tip (Kömle et al. 2018b). The cable to be paid out during penetration was initially stored in coils inside the probe. The experiments were extended in several aspects: (1) dust layers were embedded inside the samples in order to study their influence on probe penetration, (2) a surface cover of granular CO_2-ice was added in order to study the performance of the probe in the presence of dry ice, and (3) the performance of the melting probe in a highly porous ice layer (fresh bonded snow) was studied. All melting probe tests were performed at an environmental pressure of 800–900 Pa. While in heating mode, the heaters were operated with a heater power ranging between 85 and 115 W.

The experiments demonstrated that melting probes can be successful even in the presence of embedded sand and dust layers as long these non-volatile layers are thin (in the order of centimeters). The presence of a porous CO_2-ice cover on the surface does not seriously hamper the penetration of melting probes because CO_2-ice is much more volatile than water ice, and probes pass through such layers very fast when heated with the same power. In fresh bonded snow, which has very high porosity, probes penetrate correspondingly faster than in compact ice.

1.4.6 IceMole

The IceMole is a maneuverable subsurface ice probe for the clean in situ analysis and sampling of glacial ice and subglacial materials at the icy moons of the Solar System (Dachwald et al. 2014). It has been being developed and constructed at the FH Aachen University of Applied Sciences, Germany since 2009. Its design is based on the concept of combining melting and mechanical propulsion with a melting head that features a rotating hollow ice screw with an ID and an OD of 15 and 18 mm, respectively (Fig. 1.79).

same power supply was used for all experiments (108 W, corresponding to a voltage of 24 V and a current of 4.5 A). Before the start of each melting test, the vacuum chamber was evacuated and pre-cooled to a set temperature of −40 °C for approximately one hour.

In the first two experiments, the chamber pressure was kept in the ranges of 800–900 and 500–600 Pa, which could both occur on Mars, and the average penetration rates attained were 0.17 and 0.10 m/h, respectively. In the third

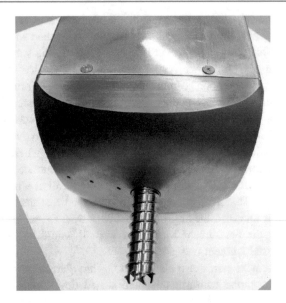

Fig. 1.79 IceMole 1's melting head with a hollow ice screw (Dachwald et al. 2014)

The screw is driven by a 25-W servo-controlled electric motor and a gear system. This continuously rotating ice screw generates a driving force larger than 1 kN by pressing the melting head against the ice. This enhances conductive heat transfer into the ice and aids in steering in the desired direction, including the probe's upward motions against gravity. However, this is only possible in solid ice; cavities exceeding the length of the screw will stop the IceMole's upward motion. Moreover, surface firn layers cannot be penetrated. However, the probe has an enhanced ability for penetrating layers of debris-laden ice.

The IceMole body has a rectangular tube shape with a 150 mm × 150 mm cross-section. The rectangular shape of the IceMole counteracts the torque of the rotating ice screw. The IceMole can change the melting direction via differential heating of the melting head and optional sidewall heaters. The probe is powered with 230 V at 50 Hz (internally, a voltage of 24 V DC is provided for the payload systems). The electric power is generated by a surface generator and transmitted via a three-conductor power cable. The accumulated meltwater can be pumped out of the melting channel to prevent refreezing or, if this is not done, cable may be uncoiled from the probe.

IceMole 1

The first IceMole 1 prototype had a complanate melting head (as shown in Fig. 1.79) with a power of 3.2 kW (0.8 kW 4 = 3.2 kW), distributed over four heating zones. The sidewalls were not heated. This 0.87-m-long probe was originally designed to collect ice samples. Its hollow ice screw was thermally isolated from the melting head to allow

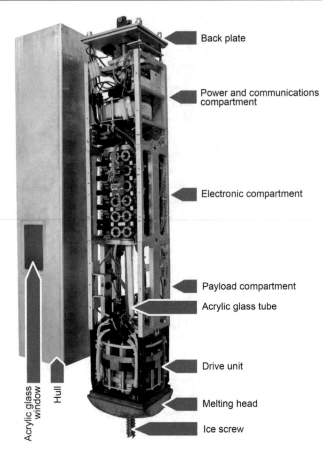

Fig. 1.80 IceMole 1 assembly view (Dachwald et al. 2014)

for the collection of unmelted ice through the interior of the screw. Further up the ice screw, the crushed ice core could be visually inspected or imaged within an acrylic tube (Fig. 1.80). Additionally, the hull had an acrylic window to allow for imaging of the melting channel. A simple off-the-shelf camera was installed in the payload bay for testing purposes.

In September 2010, IceMole 1 was used in three penetration tests in Morteratschgletscher, Switzerland (Fig. 1.81), where it demonstrated its ability to melt upwards at an angle of 45° over a distance of ~1.5 m, horizontally over ~5 m, and downwards at an angle of 45° over ~3 m. During the last test, a 4-cm-thick dust layer was penetrated and the hole channel was curved with a radius of ~10 m. The penetration velocity in each of the tests was ~0.3 m/h.

IceMole 2

A second probe IceMole 2, was developed between 2010 and 2012 (Fig. 1.82). It had a more sophisticated heater control system, with 12 separately controllable heating cartridges in a parabolically shaped melting head (12 × 0.2 kW = 2.4 kW) and eight sidewall heaters (8 × 0.3 kW = 2.4 kW). Both updates were intended to improve the

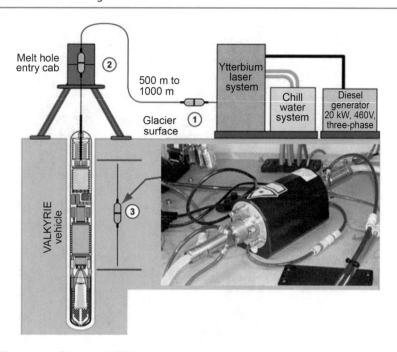

Fig. 1.85   VALKYRIE field test setup (Stone et al. 2014)

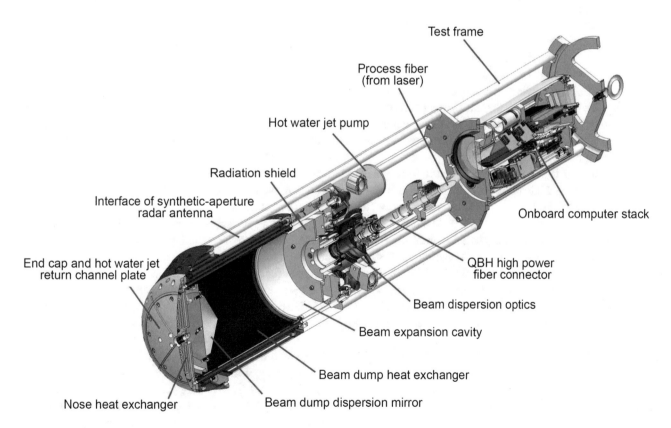

Fig. 1.86   Solid model of the initial test prototype for VALKYRIE (Stone et al. 2014)

possibly avoid obstacles in the ice, a novel end-fire synthetic aperture radar was incorporated into the probe design.

VALKYRIE carried a scientific payload consisting of a fluorescence-based multi-channel flow cytometer and water and filter sample collection systems. Autonomous

Fig. 1.87 Assembled
VALKYRIE prototype beam
dump and melt head (*Credit*
Stone Aerospace 2013)

sample-collection algorithms were developed, which could evaluate real-time multichannel flow cytometer data to determine the optimal triggering and timing of sampling routines. It was planned that, upon reaching the subglacial lake, the probe would operate as autonomous underwater vehicle and, after sampling operations were completed, it would be able to turn 180° so that the thermal head would be at the top of the vehicle. Then, a buoyancy engine or a drop-mass release would activate and cause the vehicle to rise up through the ice cap.

The probe was tested over the course of two month-long field seasons in the Matanuska Glacier, Alaska in 2014 and 2015 (Fig. 1.88). The length of the test vehicle was 1.5 m for the 2014 field tests and it was lengthened (with the addition of the scientific payload) to 2.5 m in 2015. The total weight of test system was ∼1800 kg. During the first tests, in 2014, the probe descended to a depth of 30.52 m at an average penetration rate of 0.9 m/h (when driven by a 5-kW laser) using active jetting (Stone et al. 2018). Using the turning jets, the vehicle was able to achieve a deviation angle from the vertical of up to 7° and then was brought back to the vertical, thus proving the ability of the vehicle to divert in the presence of obstacles.

In the second field season, the scientific payload was integrated into the vehicle. The sample collection algorithms were tested and the probe collected filter and water samples for subsequent laboratory analysis. The radar system was tested independently and demonstrated its ability to detect and model for obstacle avoidance behaviors for a 1-m object 80 m in advance.

1.4.8 SPINDLE

Recently, Stone Aerospace started to modify the VALK-YRIE probe within the Sub-glacial Polar Ice Navigation, Descent and Lake Exploration (SPINDLE) project aimed to develop an autonomous two-stage probe consisting of a robotic ice-penetrating carrier vehicle—Cryobot—and a marsupial, hovering autonomous underwater vehicle (AUV) (Stone et al. 2018). The Cryobot probe was to descend through an ice body into a sub-ice aqueous environment and deploy the AUV to conduct reconnaissance, life searching, and sample collection operations. The AUV was to return to and auto-dock with Cryobot after the conclusion of the mission for subsequent data uplink and to return the samples to the surface. The Cryobot probe was designed for penetrating through terrestrial ice sheets and the AUV was designed for persistent exploration and scientific presence in Antarctic subglacial lakes. If terrestrial tests are successful, SPINDLE will be available for a flagship-class mission to either the shallow lakes of Jupiter's moon Europa, the sub-surface ocean of Ganymede, or the geyser sources on both Europa and Enceladus (Stone et al. 2016).

The principal module elements of the Cryobot prototype included (Fig. 1.89): forward closed-cycle hot-water jets, the AUV cargo bay, a guidance, navigation, and control electronics pod, jet pumps, a thermal power source (laser or high voltage) and back-power conversion systems, a power and communications tether spooler, an elevation control servo spooler, and aft hot-water jets. The vehicle measured 40 cm in diameter and was 8 m long.

Fig. 1.88 The VALKYRIE 2014 Expedition. *Credit* Stone Aerospace (VALKYRIE 2015 Matanuksa Glacier, Alaska, 2015)

Fig. 1.89 General concept of the SPINDLE two-stage, bi-directional Cryobot (Stone et al. 2018)

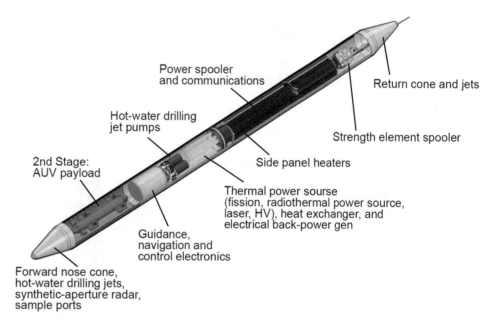

Cryobot uses two servo-controlled tether spoolers: a dedicated strong spooler for descent and ascent and a dedicated power/communications spooler. The initial prototype was designed with the cables needed for a 2500-m-deep Antarctic subglacial exploration mission. For the high-voltage transmission of 50 kW of power, the use of a 4-mm-diameter cable with four conductors insulated with Kapton was proposed. The center of the cross section was occupied by an armored K-tube fiber optic link. The strong spooler used a compression- and shear-insensitive Spectra 4-mm-diameter recovery line.

SPINDLE uses a closed-loop heater system that is in thermal communication with an open-loop hot-water drill via a heat exchanger (Fig. 1.90a). The primary heater loop uses a process fluid with a depressed freezing point so that the vehicle can restart even after being frozen in the ice. Primary loop circulation is accomplished using a high-volume, low-pressure centrifugal pump. The process fluid transits through the high-voltage heater core and into the primary side of the heat exchanger. Meltwater enters inlet ports aft of the nose cone and is pumped through the secondary side of the heat exchanger by a series of

(a)

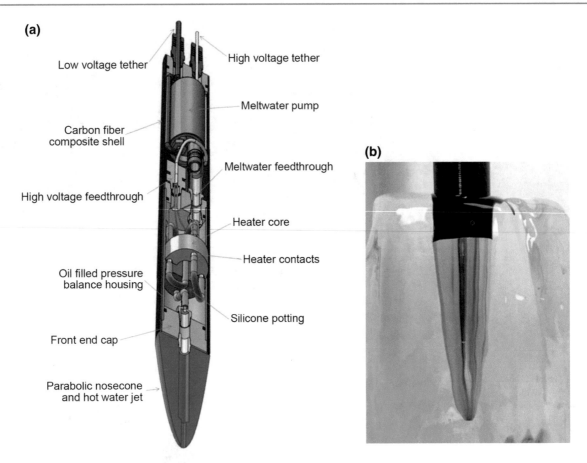

Fig. 1.90 SPINDLE melting probe: **a** 3D model of a prototype with a high-voltage heater; **b** initial testing at Stone Aerospace, August 2016 (Stone et al. 2018)

high-pressure, high-volume diaphragm pumps. After the water travels through the heat exchanger, it is ejected from the vehicle in a series of jets that can be controlled via a series of solenoid valves. These principles were tested during initial tests at Stone Aerospace in August of 2016 (Fig. 1.90b).

1.4.9 Ice Diver

A thermal probe Ice Diver was developed in the Applied Physics Laboratory, University of Washington (APL-UW) with the main perspective task of exploring Earth-analogs, such as Mars, Europa, and Enceladus (Winebrenner et al. 2013). Ice Diver included the following units (Fig. 1.91): a paraboloidal thermal head, a pressure chamber with electronics and the payload, a coil with an insulated wire spool, and an upper heating flange and coil with an un-insulated wire spool. This probe used Aamot's pendulum steering (see Sect. 1.3.3), in which the verticality of the drilling-melting process is achieved via braking with the upper thermal flange. The maximal diameter of the thermal head and the

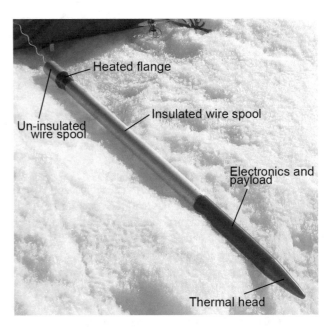

Fig. 1.91 Ice Diver at southwestern Greenland, June 2013 (*Credit* D. P. Winebrenner)

body was 65 mm, whereas the diameter of the upper thermal flange was 85 mm.

Two wires were spooled out from two distinct reels as the probe descended and were connected to a generator and telemetry transceiver on the ice surface. The water above the probe refroze, so transit was only downwards. One wire (20AWG, 0.81 mm in diameter) was un-insulated and was spooled in the chamber above the thermal flange. The wire in the larger section between the flange and the electronics section had the same conductor diameter but was insulated with Teflon.

The thermal head and the thermal flange contained 27 electrical Watlow cartridge heaters filled with dielectric oil, which allowed for using them above their rated voltages. Covered with heat-transfer paste, nine heaters were inserted into precision-machined mounting holes of the cylindrical bank. One bank was placed in the front half of the thermal head, another in the back half, and the last one in the flange (D.P. Winebrenner, personal communication 2016). The banks were connected in parallel across the voltage source but each had a separate commandable thermostatic control. When all three banks were "on", each one got one third of the total power supplied to the probe. The increments in the frontal area of each separately heated component were also one third of the total power.

Analog thermistors were used to sense the temperatures of the thermal head and the rear pendulum flange, together with thermostatic circuitry to maintain specified temperatures in each location, typically ~ 20 °C in the front thermal head and 15 °C in the back. Temperatures were controlled by switching the power to the banks of the heaters in each location "on" and "off" within a fixed time interval (0.2 s).

Two different power-supply systems were developed. In early tests, the output AC voltage of the generator was stepped up to 480 V with a transformer and rectified to DC to power the probe. The second power system was a complete revision and used high-voltage DC power supplies. In both cases, all power conversions were done on the surface.

Initially, probes were tested in 1-m-thick ice blocks in an APL-UW cold room, and then at a research camp on sea ice in Beaufort Sea in April 2011. Three prototype probes melted through 2 m of saline sea ice at temperatures of ~ -15 °C with a penetration rate of 2.0–2.7 m/h using ~ 1.25 kW of electrical power. In early June in 2011, the probe was tested in Easton Glacier, on the flank of Mt. Baker in the North Cascades of Washington, USA. The probe was run for 3 h, during which it traveled through 7 m of seasonal snow and 3 m of glacial ice. The test was ended before the glacial ice could freeze back behind the probe to avoid losing it.

In July 2013, the probe, now with a 700-m cable on board, was tested on the southwestern flank of the Greenland Ice Sheet at an ice temperature of −15 °C at a depth of 10 m.

The ice thickness at the site was ~ 980 m. The probe was operated at 1050 V/2.15 kW and turned off at a depth of ~ 80 m because the camp was about to be pulled out (Winebrenner et al. 2016). During the descent, the depth was estimated by briefly stopping the probe for a resistance measurement of the un-insulated wire.

In 2014, tests were continued at the same site with a revised probe having 24AWG wires with a diameter of 0.51 mm. The probe operated at 2000 V/4.5 kW during 7 h before an electrical fault in a heater occurred. Then, it was decided to go back to the probe that had frozen in the year before. The probe was powered with 1050 V and was revived. Penetration was continued and then stopped at a depth of 400 ± 50 m, again because logistics forced an end to the camp. The average rate of penetration was 2.4 m/h. Because tilt measurements indicated a very nearly vertical travel, the length of wire spooled out was assumed to be equal to the depth of the probe. At the end of the run, time-domain reflectometry measurements on one of the wires were carried out to estimate the depth independently (the result indicated a depth of 436 m).

Experiments designed to determine the potential for englacial materials to be dragged out by the Ice Diver during melt operations were carried out on a 2.0 m × 0.4 m 0.38 m ice block inoculated with a model laboratory microbe (E. coli) and fluorescent beads (1 μm in diameter) (Schuler et al. 2018). Melt samples were collected through a port in the nose of the Ice Diver over the length of the ice block, namely before, during, and after passing through the microbial layer. Then, the total E. coli cell and fluorescent bead concentrations were determined at different depth. The results showed that the concentrations of E. coli and beads spiked as the probe entered the microbial layer, followed by a ~ 91% reduction as the nose of the probe exited the layer. A second spike was observed when the tail of the Ice Diver (the widest portion of the probe) entered the microbial layer. Finally, a return to background cell concentrations was observed as the tail exited the doped layer. Thus, microbial cells were indeed dragged deeper into the ice by the melting probes.

Current APL-UW researches are focused on the development of a recoverable thermal probe using (1) ethanol to prevent the refreezing of the melted water; (2) cable heating and its deployment from the ice surface; and (3) upward melting to travel back up through the partially refrozen channel.

1.4.10 IceShuttle Teredo

Recently, the Robotics Innovation Center, Bremen, Germany designed an ice-penetrating robot, called IceShuttle Teredo, (Fig. 1.92) which is capable of delivering an AUV

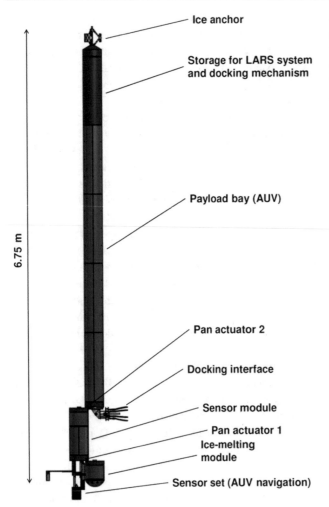

Fig. 1.92 IceShuttle Teredo system overview (Wirtz and Hildebrandt 2016)

through an extraterrestrial ice body into the ocean beneath (Wirtz and Hildebrandt 2016). The ice-melting module at the lower end of the vehicle contains the electronics for the thermal control of the melting tip as well as the mechatronic components for the unfolding mechanism. A sensor module contains a set of sensors to support the navigation of the exploration vehicle from a fixed position under the ice. Because the sensors need to be fully deployed into the water to be functional, the sensor module is simply divided into "wet" and "dry" parts. Whereas the wet part mostly contains the sensors, including the mechanical part of their deployment system as well as a part of the ice shuttle's electrical and communication backbone, the dry pressure housing contains the electromechanics and control units for sensor deployment and the general unfolding mechanism, as well as the electronics for the navigation sensors and other peripheral electronics.

The largest module of IceShuttle Teredo is the payload bay, which contains multiple different subsystems. In addition to the 210-mm-diameter and 4-m-long AUV, it contains a set of freely deployable acoustic beacons for navigation and a docking interface, followed by a mechatronic system for launch and recovery as well as the electric backbone of the system. The payload area was decided to be a wet/flooded compartment.

The rear end of IceShuttle Teredo also consists of a pressure housing including further electronics and mechatronic components and space for an on-board computer. Another component of the rear end is a mechanical system, namely an ice anchor, to fix the IceShuttle at its final position within the borehole at the ice/ocean interface.

In the current state of IceShuttle Teredo, the ice-melting head consists of a trial version that was used for the first functional tests and prior experiments. This head is equipped with six cartridge heaters that together represent 3.6 kW of power. This IceShuttle Teredo prototype was constructed (Fig. 1.93) and the first experiments with a thermal drill head, an ice anchoring system, and the aforementioned docking concept were carried out. An important component missing in the current design that would be necessary for automatic and independent drilling deeper into the ice is a cable-reeling system. The IceShuttle Teredo prototype was aimed to reach only a few tens or a hundred meters while keeping the borehole open using thermal energy, chemicals, or mechanical systems, such as reamers.

1.4.11   Summary

The exploration of subsurface extraterrestrial ice environments on Mars, Europa, Enceladus, and Titan has gained increasing attention in recent years. The main finding from experimental studies was that the penetration performance of classical melting probes under low pressure conditions can be much less favorable than under Earth's atmospheric surface pressure. This is mainly caused by the absence or only intermittent presence of a liquid phase, which can cause poor thermal contact between the probe's hot tip and the surrounding ice. Nevertheless, the interest in the scientific and engineering community to develop Cryobots for application in planetary missions is still high.

A number of mission concepts have been developed to varying degrees to explore these subsurface regions. Although a lot of effort and quite large fundings were invested to develop novel thermal probes for extraterrestrial investigations, the maximal achieved depths during terrestrial tests were quite shallow. The reason for this is that the task is very challenging and lots of research should be done in the future to build a reliable vehicle.

Self-contained active melting, in which heat is exchanged with the ice through the turbulence of a jet, has the advantages of being easy to direct downwards and that heat is

Fig. 1.93 Assembled prototype of IceShuttle Teredo, designed to transport a torpedo-class AUV through sea ice (Wirtz and Hildebrandt 2016)

transferred via vortex flows. This active jet streaming technique also works in debris-laden ice because it stirs up the dust and particulates and holds them in suspension. However, this method does not work in permeable firn because the water for recycling is lost. It is more complex than passive heating and requires pumping, which generates electrical and mechanical noise. In addition, the borehole walls melted by the jetting water are more uneven, whereas a solid nose would produce a smoother borehole.

The open-hole SIPR probe required quite a large amount of additional power to pump out the melted water to the surface and to heat the long delivery hose to prevent meltwater from refreezing in it. In addition, the continuous mixing property of the melted water flow in the delivery hose made it difficult to accurately characterize the vertical resolution.

A few melting probes had a rectangular-shaped body that could compensate for the rotational torque in quadratic boreholes. Using a rectangular shape for the probe's body is also advantageous for integrating the payload inside it. This is generally easier to accomplish with rectangular-shaped bodies owing to the design of most electronic parts. On the other hand, by using a rectangular cross-section, the probe's housing perimeter increases by 13% given the same volume and length.

The novel IceMole probe has high maneuverability (the radius of the melted channel is 8–11 m) owing to the differential heating of the melting head and optional sidewall heaters and is able to "dig" inclined, horizontal, and even upward holes. However, the problem of the removal/refreezing of the melted water at deep ice has not been solved.

The VALKYRIE probe uses a unique power source: a high-power laser transmits laser light to the vehicle via glass fiber. High-power laser systems are developing rapidly, and lasers of this class capable of producing up to 120 kW will be available for designing higher-power probes in the near future. To avoid power losses, the coil used with the fiber must be of a large diameter, and hence the probe should be equipped with a power-consuming big-sized thermal head. The performance of frozen fiber is still unknown. Return missions would need double the storage of fiber inside the probe.

The APL-UW Ice Diver probe has the classical structure of freezing-in probes and demonstrated logistically light access beneath hundreds of meters of ice. Even though it was declared that precision machining of the banks with thermal cartridges ensured good heat transfer and prevented heater failure due to overheating, an electrical fault occurred in the heater after 7 h of operation in the field. Neither samples nor the Ice Diver probe itself could be recovered.

The design of working prototypes of the SPINDLE and IceShuttle Teredo probes are still in progress. A number of aspects still remain to be improved and not all components could be finished so that the systems could show their feasibility in laboratory experiments.

Most of the proposed thermal probes have been based on established radioisotope generator technology, which acts as a thermal energy source through a steady process of particle decay. RTGs are good options when the electric power requirements are relatively low. For more demanding missions, compact fission power systems or Topaz-class reactors operating in the mid-power range (20–100 kW) have been proposed (Elliott and Carsey 2004; Lorenz 2012; Stone et al. 2014). However, the use of nuclear power sources on terrestrial glaciers is problematic owing to environmental issues and has been completely banned in the Antarctic area.

References

Aamot HWC (1967a) Pendulum steering for thermal probes in glaciers. USA CRREL Special Report 116

Aamot HWC (1967b) The Philberth probe for investigating polar ice caps. USA CRREL Special Report 119

Aamot HWC (1967c) Heat transfer and performance analysis of a thermal probe for glaciers. USA CRREL Technical Report 194

Aamot HWC (1968a) Instrumented probes for deep glacial investigations. USA CRREL Technical Report 210

Aamot HWC (1968b) Instrumented probes for deep glacial investigations. J Glaciol 7(50):321–328

Aamot HWC (1968c) A buoyancy-stabilized hot-point drill for glacier studies. USA CRREL Technical Report 215

Aamot HWC (1968d) A buoyancy-stabilized hot-point drill for glacier studies. J Glaciol 7(51):493–498

Aamot HWC (1968e) Pendulum steered thermal probe. US Patent 3,390,729

Aamot HWC (1969) Winding long, slender coils by the orthocyclic method. USA CRREL Special Report 128

Aamot HWC (1970a) Self-contained thermal probes for remote measurements within an ice sheet. In: International symposium on antarctic glaciological exploration (ISAGE), IASH Publ. 86, pp 63–68

Aamot HWC (1970b) Development of a vertically stabilized thermal probe for studies in and below ice sheets. Trans ASME, J Eng Ind 92(2):263–268

Antarctica: ANDRILL (2005) The scientific method. Retrieved 2 Feb 2017 from https://www.flickr.com/photos/orebody/sets/72157622811009641

Bazhev AB, Zagorodnov VS, Rototaeva OV (1988) Burovye raboty v oblasti pitaniya lednika Garabashi na El'bruse [Drilling operations in the ice-feeding region of Garabashi Glacier at Elbrus]. Akademiya nauk SSSR. Institut geografii. Materialy gliatsiologicheskikh issledovanii [Academy of Sciences of the USSR. Institute of Geography. Data of Glaciological Studies] 64, pp 11–12 (in Russian)

Benson T, Cherwinka J, Duvernois M et al (2014) IceCube enhanced hot water drill functional description. Ann Glaciol 55(68):105–114

Bentley CR, Koci BR, Augustin LJ-M et al (2009) Ice drilling and coring. In: Bar-Cohen Y, Zacny K (eds) Drilling in extreme environments. Penetration and sampling on earth and other planets. WILEY-VCH Verlag GmbH & Co., KGaA, Weinheim, pp 221–308

Biele J, Ulamec S, Garry J et al (2002) Melting probes at Lake Vostok and Europa. In: Proceeding of the First European Workshop on Exo/Astrobiology, 16–19 September, 2002. Graz, Austria, ESA SP-518, pp 253–260

Biele J, Ulamec S, Hilchenbach M et al (2011) In situ analysis of Europa ices by short-range melting probes. Adv Space Res 48:755–763

Calciati M (1945) Le perforazioni eseguite del ghiacciaio d'Hosand. Bolletino del Comitato Glaciologico Italiano 23:19–28

Cardell G, Hecht MH, Carsey FD et al (2004) The subsurface ice probe (SIPR): a low-power thermal probe for the martian polar layered deposits. In: 35th Lunar and planetary science conference, 15–19 March 2004. League City, Texas, USA, Abstract no 2041

Clarke GKC (1987) A short history of scientific investigations on glaciers. J Glaciol, Spec Issue, 4–24

Classen DF (1970) Thermal drilling and deep ice-temperature measurements on the Fox glacier, Yukon. A thesis submitted in partial fulfillment of the requirements for the degree of Master of Science. Department of Geophysics. The University of British Columbia, Vancouver, Canada, April 1970, 65 p

Classen DF (1977) Temperature profiles for the Barnes ice cap surge zone. J Glaciol 18(80):391–405

Classen DF, Clarke GKC (1972) Thermal drilling and deep ice temperature measurements on the Rusty Glacier. In: Icefield ranges research project. Scientific results, vol 3. In: Bushnell VC, Ragle RH (eds) American Geographical Society, New York and Arctic Institute of North America, Montreal, Canada, pp 103–116

Dachwald B, Mikucki J, Tulaczyk S et al (2014) IceMole: a maneuverable probe for clean in situ analysis and sampling of subsurface ice and subglacial aquatic ecosystems. Ann Glaciol 55 (65):14–22

Dachwald B, Kowalski J, Baader F et al (2016) Enceladus explorer: next steps in the development and testing of a steerable subsurface ice probe for autonomous operation. In: Proceeding of the conference Enceladus and the Icy Moons of Saturn, 26–29 July, 2016. Boulder, USA, Abstract no 3031

Datta PM (1980) Design and fabrication of hot point probe for Zemu glacier operation. Geol Surv India Spec Publ 4:296–301

Davis A (2017) A prototype ice-melting probe for collecting biological samples from cryogenic ice at low pressure. Astrobiology 17 (8):709–720

Di Pippo S, Mugnuolo R, Vielmo P et al (1999) The exploitation of Europa ice and water basins: an assessment on required technological developments, on system design approaches and on relevant expected benefits to space- and earth-based activities. Planet Space Sci 47:921–933

Dubrovin LI (1960) Rassol v shel'fovom lednike Lazareva [Brine in the Lazarev Ice Shelf]. Informatsionny Byulleten' Sovetkoj Antarkticheskoj Ekspeditsii [Soviet Antarctic Expedition Information Bulletin] 22:15–16 (in Russian)

Dutta RK (1980) Drilling in Zemu glacier, North Sikkem. A study of methodology and problems. Geol Surv India Spec Publ 4:229–233

Ekman SR (1961) Notes on glaciological activities in Kebnekajse, Sweden. Thermal drilling in Insfallglaciären, Kebnekajse. Geografiska Annaler 43(3/4):422

Elliott JO, Carsey FD (2004) Deep subsurface exploration of planetary ice enabled by nuclear power. In: Proceeding of 2004 IEEE Aerospace Conference, 6–13 March 2004, vol 5. Big Sky, Montana, USA, pp 2978–2987

Aviation Epps (2009) Coporate Brochure. DeKalb-Peachtree Airport, Atlanta, USA

French L, Anderson FS, Carsey F et al (2001) Cryobots: an answer to subsurface mobility in planetary icy environments. In: Proceeding of the 6th international symposium on artificial intelligence and robotics & automation in space: i-SAIRAS 2001, 18–22 June, 2001. Canadian Space Agency, St-Hubert, Quebec, Canada, 8 p

Gerrard JAF, Perutz MF, Roch A (1952) Measurements of the velocity distribution along a vertical line through a glacier. Proc R Soc Ser A 213(115):546–558

Gillet F (1975) Steam, hot-water and electrical thermal drills for temperate glaciers. J Glaciol 14(70):171–179

Gillet F, Rado C, Maree G et al (1984). "Climatopic" thermal probe. In: Holdsworth G, Kuivinen KC, Rand JH (eds) Proceeding of the second international workshop/symposim on ice drilling technology, 30–31 August 1982. Calgary, Alberta, Canada. USA CRREL Special Report 84–34, pp. 95–99

Golubev GN, Sukhanov LA, Khromov RS (1976) Beskernovoye termoelektroburenie i ego primenenie dlya izuchenia stroeniya lednika Jankuat [Full-diameter thermal drilling and its using for structure investigations of Jankuat Glacier]. Akademiya nauk SSSR. Institut geografii. Materialy gliatsiologicheskikh issledovanii [Academy of Sciences of the USSR. Institute of Geography. Data of Glaciological Studies] 28, pp 96–104 (in Russian)

Grilli R, Marrocco N, Desbois T et al (2014) SUBGLACIOR: An optical analyzer embedded in an Antarctic ice probe for exploring the past climate. Rev Sci Instrum 85:111301

Grześ M (1980) Non-cored hot point drills on Hans Glacier (Spitsbergen), method and first results. Pol Polar Res 1(2–3):75–85

Haehnel RB, Knuth MA (2011) Potable water supply feasibility study for summit station, Greenland. USA CRREL Report ERDC/CRREL TR-11-4

Hansen BL, Kersten L (1984) An in-situ sampling thermal probe. In: Holdsworth G, Kuivinen KC, Rand JH (eds) Proceeding of the second international workshop/symposium on ice drilling technology, 30–31 August, 1982. Calgary, Alberta, Canada. USA CRREL Special Report 84–34, pp 119–122

Harrison WD (1975) Temperature measurements in a temperate glacier. J Glaciol 14(70):23–30

Harrison WD, Kamb B (1976) Drilling to observe subglacial conditions and sliding motion. In: Splettstoesser JF (ed.) Ice-core drilling: proceeding of the symposium, 28–30 August, 1974. University of Nebraska, Lincoln, USA. University of Nebraska Press, Lincoln, pp 37–43

Hayes D (1995) The lost squadron … Found! Georgia Tech., Spring, pp 40–47

Hodge SM (1974) Variations in the sliding of a temperate glacier. J Glaciol 13(69):349–369

Hodge SM (1976) Direct measurement of a pilot basal water pressures: study. J Glaciol 16(74):205–218

Hodge SM (1979) Direct measurement of basal water pressures: progress and problems. J Glaciol 23(89):309–319

Hooke RLeB (1976). University of Minnesota ice drill. In: Splettstoesser JF (ed.) Ice-core drilling: proceeding of the symposium, 28–30 August, 1974, University of Nebraska, Lincoln, USA. Univ. of Nebraska Press, Lincoln, pp 47–57

Hooke RLeB, Alexander Jr EC (1980) Temperature profiles in the Barnes Ice Cap, Baffin Island, Canada, and heat flux from the subglacial terrane. Can J Earth Sci 17:1174–1188

Hooke RLeB, Hanson B (1986). Borehole deformation experiments, Barnes Ice Cap, Canada. Cold Reg Sci Tech 12 261–276

Horne MF (2018) Thermal probe design for Europa sample acquisition. Acta Astronaut 142:29–36

Hotpoint (n.d.). Retrieved 19 July 2016 from http://www.hotpoint.com/

Ignatov VS (1960a) Izuchenie stroenia snezhno-firnovoi tolshchi v Antarktide termicheskim sposobom [Study of the structure of snow and firn cover in Antarctica by the thermal method]. Informatsionny Byulleten' Sovetkoj Antarkticheskoj ekspeditsii [Soviet Antarctic Expedition Information Bulletin] 21:16–18 (in Russian)

Ignatov VS (1960b) Opyt termicheskoy prokhodki ledyanikh skvazhin na stantsii Vostok [Experiment in the thermal drilling of holes in the ice at Vostok Station]. Informatsionny Byulleten' Sovetkoj Antarkticheskoj ekspeditsii [Soviet Antarctic Expedition Information Bulletin] 22:50–52 (in Russian)

Ignatov VS (1962) God na polyuse kholoda [A Year on the Pole of Cold]. Geografgiz, Moscow (in Russian)

Kamb B, Shreve RL (1966) Results of a new method for measuring internal deformation in glaciers. Trans Am Geophys Union 47:190

Kasser P (1960) Ein leichter thermischer Eisbohrer als Hilfsgerät zur Installation von Ablationsstangen auf Gletschern. Geofisica Pura e Applicllla 45(1):97–114

Kaufmann E, Kargl G, Kömle NI et al (2009) Melting and sublimation of planetary ices under low pressure conditions: laboratory experiments with a melting probe prototype. Earth Moon Planets 105:11–29

Kelty JR (1995) An in situ sampling thermal probe for studying global ice sheets. Dissertation presented to the Faculty of the Graduate College in the University of Nebraska in partial fulfillment of requirements for the degree of Doctor of Philosophy. Major: Interdepartmental area of engineering (Electrical engineering) under the supervision of Prof. D. P. Billesbach, Nebraska, Lincoln, USA, May 1995

Kharitonov VV (2005) Peculiarities of fractional composition of the Pechora Sea first-year ridges. In: Proceeding of the 18th international conferences on port and ocean engineering under Arctic Conditions (POAC), vol 2, 26–30 June, 2005, Potsdam, New York, pp. 907–916

Kharitonov VV (2008) Internal structure of ice ridges and stamukhas based on thermal drilling data. Cold Reg Sci Tech 52:302–325

Kharitonov VV, Morev VA (2005) Research of the internal structure of ridges in the Central Arctic by electrothermal drilling method. In: Proceeding of the 18th international conference on port and ocean engineering under Arctic conditions (POAC), vol 2, 26–30 June, 2005. Potsdam, New York, pp 917–926

Kharitonov VV, Morev VA (2009) Torosy v raione dreifuyushchei stantsii "Severnyi Polyus – 35" [Ice ridges in the region of drifting station "Severnyi Polyus – 35"]. Meteorologiya i gidrologiya [Meteorology and Hydrology] 6:68–73 (in Russian)

Kharitonov VV, Morev VA (2011) Metod issledovaniya vnutrennego stroeniya torosov i stamukh s pomoshchyu tekhnologii termobureniya [Research of inner structure of ice ridges and stamikhas with thermal drilling technology]. Meteorologiya i gidrologiya [Meteorology and Hydrology] 7:49–58 (in Russian)

Koechlin R (1945) Procédé pour sonder les glaciers et installation pour sa mise en oeuvre. FR Patent 910,034

Konstantinidis K, Martinez CLF, Dachwald B et al (2014) A lander mission to probe subglacial water on Saturn's moon Enceladus for life. Acta Astronaut 106:63–89

Korotkevich YeS (ed.) (1965) Pyataya kontinental'naya ekspeditsiya 1959–1961 gg. Obshchee opisanye [Fifth Continental Expedition of 1959–1961. General description]. Trudy Sovetskoy Antarkticheskoy Ekspeditsii (Transactions of Soviet Antarctic Expedition) 36. (in Russian)

Korotkevich YeS, Kudryashov BB (1976) Ice sheet drilling by Soviet Antarctic expeditions. In: Splettstoesser JF (ed.) Ice-core drilling: proceeding of the symposium, 28–30 August, 1974. University of Nebraska, Lincoln, USA. University of Nebraska Press, Lincoln, pp 63–70

Kotlyakov VM (ed) (1985) Glyatsiologiya Shpitsbergena [Glaciology of Spitsbergen]. Nauka, Moscow (in Russian.)

Kowalski J and 33 others (2016) Navigation technology for exploration of glacier ice with maneuverable melting probes. Cold Reg Sci Tech 123:53–70

Kömle NI, Kargl G, Steller M (2002) Melting probes as a means to explore planetary glaciers and ice caps. In: Proceeding of the first European workshop on exo-astrobiology, 16–19 Sept 2002, Graz, Austria, ESA SP-518, pp 305–308

Kömle NI, Treffer M, Kargl G, et al. (2004) Development of melting probes for exploring ice sheets and permafrost layers. J Glaciol Geocryol (Issue z1): 310–318

Kömle NI, Tiefenbacher P, Weiss P et al (2018a) Melting probes revisited—ice penetration experiments under Mars surface pressure conditions. Icarus 308:117–127

Kömle NI, Tiefenbacher P, Kahr A (2018b) Melting probe experiments under Mars surface conditions—the influence of dust layers, CO_2 ice and porosity. Icarus 315:7–19

LaChapelle E (1963) A simple thermal ice drill. J Glaciol 4(35):637–642

Lane AL, Carsey FD, French GD et al (2001) Development of extreme environment systems for seeking out extremophiles. In: Proceeding of OCEANS, 2001. MTS/IEEE Conference and Exhibition, 5–8 November, 2001, vol 4. Honolulu, USA, pp 2036–2042

Limeburner R, Harwood D, Webb P (2006) ANDRILL SMS Mooring Deployment Report. Available on-line at http://www.whoi.edu/science/PO/ANDRILL_Mooring/pdfs/deployment.pdf

Lorenz RD (2012) Thermal drilling in planetary ices: an analytic solution with application to planetary protection problems of radioisotope power sources. Astrobiology 12:799–802

Mae S, Wushiki H, Ageta Y et al (1975) Thermal drilling and temperature measurements in Khumbu Glacier, Nepal Measurements Himalayas. J Jpn Soc Snow Ice 37(4):161–169

Mathews WH (1957) Glaciological research in Western Canada in 1956. Can Alpine J 40:94–96

Mathews WH (1958) Glaciological research in Western Canada in 1957. Can Alpine J 41:90–91

Mathews WH (1959) Vertical distribution of velocity in Salmon Glacier, British Columbia. J Glaciol 3(26):448–454

Meier MF, Rigsbyt GP, Sharp RP (1954) Preliminary data from Saskatchewan glacier, Alberta, Canada. R.P. Arctic 7(1):3–26

Mellor M (1986) Equipment for making access holes through arctic sea ice. USA CRREL Special Report 86–32

Miller MM (1953) The application of electro-thermic boring methods to englacial research with special reference to the Juneau Icefield investigations in 1952–53. Arctic Institute of North America, Report No. 4, Project ONR-86. Dec 1952 (supplemented Dec 1953)

Morev VA (1966) Opity po bureniyu l'da elektroteplovym sposobom v Mirnom [Experimental electrothermal ice drilling at Mirny]. Informatsionny Byulleten' Sovetkoj Antarkticheskoj ekspeditsii [Soviet Antarctic Expedition Information Bulletin] 56:52–56 (in Russian)

Morev VA (1976) Elektrotermobury dlia bureniia skvazhin v lednikovom pokrove [Electric thermal drills for glacier core drilling]. Akademiya nauk SSSR. Institut geografii. Materialy gliatsiologicheskikh issledovanii [Academy of Sciences of the USSR. Institute of Geography. Data of Glaciological Studies] 28, pp 118–120. (in Russian)

Morev VA, Pukhov VA (1981) Eksperimental'nye raboty po bureniiu kholodnykh pokrovnykh lednikov termoburovymi snariadami AANII [Using AARI thermodrills in experimental drilling of cold ice sheets]. Trudy Arkticheskogo i Antarkticheskogo nauchno-issledovatel'skogo instituta [Transactions of Arctic and Antarctic Research Institute] 367:64–68 (in Russian)

Morev V, Kharitonov V (2001). Definition of the internal structure of large ice features by thermal drilling methods. In: Proceeding of the 16th international conference on port and ocean engineering under arctic condition POAC'01, vol 3, 12–17 August 2001. Ottawa, Ontario, Canada, pp 1465–1472

Morev VA, Pukhov VA, Yakovlev VM et al (1984) Equipment and technology for drilling in temperate glaciers. In: Holdsworth G, Kuivinen KC, Rand JH (eds.) Proceeding of the second international workshop/symposium on ice drilling technology, 30–31 August 1982. Calgary, Alberta, Canada. USA CRREL Special Report 84–34, pp 125–127

Morev VA, Klement'ev OL, Manevskii LN, et al. (1988) Glyatsio-burobye raboty na lednike Vavilova v 1979–1985 gg. [Ice drilling on Vavilov Glacier in 1979–1985]. Geograficheskie i glyatsiologicheskie issledovaniya v polarnikh stranakh [Geographical and glaciological investigations in polar regions]. Leningrad, Gidrometeoizdat, pp 25–32. (in Russian)

Morton BR, Lightfoot RM (1975) A prototype meltsonde probe-design and experience. Australian Antarctic Division, Department of Science, Tech Note No 14

Negre B (1950) Sondages Thermiques. Rapport Preliminaire de la Campagne au Groenland, 1949, Serie Scientifique, No. 10, Expéditions Françaises, pp 28–30

Nizery A (1951) Electrothermic rig for the boring of glaciers. Trans AGU 32(1):66–72

Nizery A, Terrier M (1952) Sonde thermique pour glaciers. FR Patent 1,011,327

Northwood TD (1947) Drill for determining thickness of ice. Can J Res 25:196–197

Orvig S, Mason RW (1963) Ice temperatures and heat flux, McCall Glacier, Alaska. Union Geodesique el Geophysique Internationale. Association Internationale d'Hydrologie Scientifique. Assemblee generale de Berkele. 19–31 March 1963. Commission des Neiges et des Glaces, pp 181–188

Paige DA, Wood SE, Vasavada AR (1993) Drill/borescope system for the mars polar pathfinder. Workshop on advanced technologies for planetary instruments, 28–30 April 1993. Fairfax, Virginia, USA. LPI Technical Report 93-02, Part I, pp 18–19

Philberth B (1956) Beseitigung radioaktiver Abfallsubstanzen. Atomkern-Energie 11(12):396–400

Philberth K (1962) Une méthode pour mesurer les témperatures à l'intérieur d'un Inlandsis. Comptes Rendus Hebdomadaires des Séances de l'Académie des Sciences 254(1962):3881–3883

Philberth K (1964) Über zwei Elktro-Schmelzsonden mit Vertikal-Stabilisierung. Polarforschung 34(1–2):278–280

Philberth K (1966) Sur la stabilisation de la course d'une sonde thermique. Comptes Rendus Hebdomadaires des Séances de l'Académie des Sciences 262:456–459

Philberth K (1976) The thermal probe deep-drilling method by EGIG in 1968 at Station Jarl-Joset, Central Greenland. In: Splettstoesser JF (ed) Ice-core drilling: proceeding of the symposium, 28–30 August 1974. University of Nebraska, Lincoln, USA. University of Nebraska Press, Lincoln, pp 117–132

Philberth K (1984) Die thermische Tiefbohrung in Station Jarl-Joset und ihre theoretische Auswertung. Polarforschung 54(1):43–49

Project SCIN (n.d.) Motivation-sea ice challenges. Retrieved 1 Feb 2017 from http://scini.mlml.calstate.edu/SCINI_2007/seaice.html

Rejcek P (2015) Lifeblood of a glacier. Probe melts its way into river of ice to retrieve brine that feeds red-stained feature. The Antarctic Sun. United States Antarctic Program. Posted 4 Mar 2015. Retrieved 26 Jan 2017 from https://antarcticsun.usap.gov/science/contentHandler.cfm?id=4122

Remenieras MG, Terrier MM (1951) La sonde électrothermique E.D.F. pour le forage des glaciers. Union Géodésiques Géophys. Intern. Assoc. Hydrologie Sci., Assemblée Générale de Bruxelles [Proc. of the Int. Assoc. of Scientific Hydrology, Union of Geodesy and Geophysics, General Assembly of Brussels,] 1, pp 254–260

Rogers JC, LaChapelle ER (1974) The measurement of vertical strain in glacier bore holes. J Glaciol 13(68):315–319

Ryumin AK, Nozdryukhin VK, Emel'yanov YuN et al (1974) Stroenie lednika Abramova po dannym radiolokatsionnogo zondirovaniya

[Structure of Abramov Glacier according with radar sounding]. Trudy SARNIGMI [Transactions of Middle-East Regional Research Hydro-Meteorological Institute] 14(95), 27–35. (in Russian)

Savage JC, Paterson WSB (1963) Borehole measurements in the Athabasca Glacier. J Geophys Res 68(15):4521–4536

Schuler CG, Winebrenner DP, Elam WT et al (2018). In situ contamination of melt probes: Implications for future subglacial microbiological sampling and icy worlds life detection missions. In: 67th Annual Southeastern GSA section meeting, 12–13 April, 2018. Knoxville, Tennessee, USA. Geological Society of America, Abstracts with Programs 50(3), Paper No. 23–10

Schüller K, Kowalski J (2019) Melting probe technology for subsurface exploration of extraterrestrial ice—critical refreezing length and the role of gravity. Icarus 317:1–9

Sharp RP (1951) Thermal regimen of firn on upper Seward glacier, Yukon territory, Canada. J Glaciol 1(9):476–487, 491

Sharp RP (1953) Deformation of a vertical bore hole in a piedmont glacier. J Glaciol 2(13):182–184

Shreve RL (1961) The borehole experiment on Blue Glacier, Washington. Union Geodesique et Ge-ophysique Internationale. Association Internationale d' Hydrologie Scientifique. Assemblie generale de Helsinki, 25 July–6 August, 1960. Commission des Neiges et Glaces, Publ. No 54, pp 530–531

Shreve RL (1962a) Theory of performance of isothermal solid-nose hotpoints boring in temperate ice. J Glaciol 4(32):151–160

Shreve RLP (1962b). Review on P. Kasser. Ein leichter thermischer Eisbohrer als Hilfsgerät zur Installation von Ablationsstangen auf Gletschern. Geofisica Pura e Applicata, vol 45, No 1, 1960, 97–114. J Glaciol 4(32):234–235

Shreve RL, Sharp RP (1970) Internal deformation and thermal anomalies in Lower Blue Glacier, Mount Olympus, Washington, U.S.A. J Glaciol 9(55):65–86

Siegel V, Hogan B, Stone WC et al (2016) Development and field testing VALKYRIE—a prototype cryobot for clean subglacial access and sampling. In: SCAR biennial meetings & open science Conference, 20–30 August 2016, Kuala Lumpur, Malaysia, Abstracts, p 52

Stacey JS (1960) A prototype hotpoint for thermal boring on the Athabaska Glacier. J Glaciol 3(28):783–786

Stone WC, Hogan B, Siegel V et al (2014) Progress towards an optically powered cryobot. Ann Glaciol 55(65):1–13

Stone W, Hogan B, Siegel VL, et al (2016) SPINDLE: a 2-stage nuclear-powered cryobot for ocean world exploration. In: American geophysical union, fall general assembly, 12–16 December, 2016. San-Francisco, California, USA, Abstract no C51E-07

Stone W, Hogan B, Siegel V et al (2018) Project VALKYRIE: laser-powered cryobots and other methods for penetrating deep ice on ocean worlds. In: Badescu V, Zacny K (eds) Outer solar system. Springer, Cham, pp 47–165

Sukhanov LA (1975) Sposob opredeleniya plotnosti snezhno-ledyanogo pokrova (Method of estimation of snow-ice layers density). USSR Patent 468,133

Sukhanov LA, Morev VA, Zotikov IA (1974) Potativnye ledovye elektrobury [Portable thermo-electric ice drills]. Akademiya nauk SSSR. Institut geografii. Materialy gliatsiologicheskikh issledovanii [Academy of Sciences of the USSR. Institute of Geography. Data of Glaciological Studies] 23, pp 234–238. (in Russian)

Suto Y, Saito S, Osada K et al (2008) Laboratory experiments and thermal calculations for the development of a next-generation glacier-ice exploration system: development of an electro-thermal drilling device. Polar Sci 2:15–26

Talalay PG (2016) Mechanical ice drilling technology. Geological Publishing House, Beijing and Springer Science+Business Media Singapore

Talalay PG, Markov AN, Sysoev MA (2013) New frontiers of Antarctic subglacial lakes exploration. Geogr Environ Sust 6(1):14–28

Talalay PG, Zagorodnov VS, Markov AN et al (2014) Recoverable autonomous sonde (RECAS) for environmental exploration of Antarctic subglacial lakes: general concept. Ann Glaciol 55(65):23–30

Talalay P, Yang C, Cao P et al (2015) Ice-core drilling problems and solutions. Cold Reg Sci Tech 120:1–20

Talalay PG, Li Y, Sysoev MA et al (2019) Thermal tips for ice hot-point drilling: experiments and preliminary thermal modeling. Cold Reg Sci Tech 160:97–109

Taylor PL (1976) Solid-nose and coring thermal drills for temperate ice. In: Splettstoesser JF (ed) Ice-core drilling: proceeding of the symposium 28–30 August, 1974. University of Nebraska, Lincoln, USA. University of Nebraska Press, Lincoln, pp 167–177

These priests' invention could help us drill into icy alien worlds someday (2015). WIRED. Retrieved 10 Dec 2016 from https://www.wired.com/2015/01/philberth-priests-probe-icy-moons-nuclear-waste/

Tibcken M, Dimmler W (1997) Einsatz einer Durchschmelzsonde (SUSI) zum Transporteiner kommerziellen CTD-Sonde unter das Schelfeis. Die Expedition ANTARKTIS-XII mit FS "Polarstern" 1995. In: Jokat W, Oerter H (eds) Bericht vom Fahrtabschnitt ANT-XII/3, Berichte zur Polarforschung, Ber. Polarforsch. 219, pp 106–112

Treffer M, Kömle NI, Kargl G et al (2006) Preliminary studies concerning subsurface probes for the exploration of icy planetary bodies. Planet Space Sci 54:621–634

Tyler SW, Holland DM, Zagorodnov V et al (2013) Using distributed temperature sensors to monitor an Antarctic ice shelf and sub-ice shelf cavity. J Glaciol 59(215):583–591

Ulamec S, Biele J, Drescher J, et al. (2005) A melting probe with applications on Mars, Europa and Antarctica. In: 56th international astronautical congress, IAC-A1.7.08, 17–21 October, 2005, Fukuoka, Japan, pp 1–8

Ulamec S, Biele J, Funke O et al (2007) Access to glacial and subglacial environments in the solar system by melting probe technology. Rev Environ Sci Biotechnol 6:71–94

VALKYRIE 2015 Matanuksa Glacier, Alaska (2015) Posted 16 June 2015. Retrieved 26 Jan 2017 from http://valkyrie2015.weebly.com/expedition-log/matanuska-or-bust

Vallelonga P, Svensson A (2014) Ice core archives of mineral dust. Mineral dust: a key player in the earth system. In: Knippertz P, Stuut J-BW (eds) Springer Science+Business Media Dordrecht, pp 463–485

Vasilenko EV, Gromyko AN, Dmitriev DN et al (1988) Stroenie lednika Davydova po dannym radiozondirovaniya i termobureniya [Structure of the Davydov Glacier according with radio sounding and thermal drilling data]. Akademiya nauk SSSR. Institut geografii. Materialy gliatsiologicheskikh issledovanii [Academy of Sciences of the USSR. Institute of Geography. Data of Glaciological Studies] 62, pp 208–215. (in Russian)

Wade FA (1945) The physical aspects of the Ross Shelf Ice. Proc Am Philoso Soc 89(1):160–173

Ward WH (1952) The glaciological studies of the Baffin Island Expedition, 1950. Part Ill: equipment and techniques. J Glaciol 2(12):115–121

Ward WH (1961) Experiences with electro-thermal ice drills on Austerdalsbre, 1956–59. Union Geodesique et Geophysique Internationale. Association Internationale d' Hydrologie Scien-tifique. Assemblie generale de Helsinki, 25 July–6 August, 1960. Commission des Neiges et Glaces, Publ No 54, pp 532–542

Weinberg B (1912) Der elektrische Eisbohrer. Zeitschrift für Gletscherkunde 6:214–217

Weiss P, Yung KL, Ng TC et al (2008) Study of a thermal drill head for the exploration of subsurface planetary ice layers. Planet Space Sci 56:1280–1292

Weiss P, Yung KL, Kömle N et al (2011) Thermal drill sampling system onboard high-velocity impactors for exploring the subsurface of Europa. Adv Space Res 48:743–754

Winebrenner DP, Elam WT, Miller V et al (2013) A thermal ice-melt probe for exploration of Earth-analogs to Mars, Europa and Enceladus. In: 44th lunar and planetary science conference, 18–22 March 2013. The Woodlands, Texas, USA, Abstract no 2986

Winebrenner DP, Elam WT, Kintner PMS et al (2016) Clean, logistically light access to explore the closest places on Earth to Europa and Enceladus. In: AGU Fall Meeting, 12–16 December, 2016. San-Francisco, California, USA, Abstract no C51E-08

Wirtz M, Hildebrandt M (2016) IceShuttle Teredo: an ice-penetrating robotic system to transport an exploration AUV into the ocean of Jupiter's Moon Europa. In: Proceeding of the 67th international astronautical congress (IAC), 26–30 Sept 2016. Guadalajara, Mexico, no. IAC-16-A3.5.2

Zacny K, Paulsen G, Bar-Cohen Y et al (2016) Drilling and breaking ice. In: Bar-Cohen Y (ed) Low temperature materials and mechanisms. CRC Press, pp 271–347

Zagorodnov VS (1981) Issledovanie stroeniia i temperaturnogo rezhima shpitsbergenskikh lednikov s pomoshch'iu termobureniia [Using thermal drills in studying temperature regime of Spitsbergen glaciers]. Akademiya nauk SSSR. Institut geografii. Materialy gliatsiologicheskikh issledovanii [Academy of Sciences of the USSR. Institute of Geography. Data of Glaciological Studies] 41, pp 196–199. (in Russian)

Zagorodnov VS (1988) Recent Soviet activities on ice core drilling and core investigations in Arctic region. Bull Glacier Res 6:81–84

Zagorodnov VS, Zotikov IA (1981) Kernovoe burenie na Shpitsbergene [Core drilling at Spitsbergen]. Akademiia nauk SSSR. Institut geografii. Materialy gliatsiologicheskikh issledo-vanii [Academy of Sciences of the USSR. Institute of Geography. Data of Glaciological Studies] 40, pp 157–163. (in Russian)

Zagorodnov VS, Zinger EM (1982) Gliatsiologicheskie raboty na Severo-Vostochnoi Zemle [Glaciological investigations on North East Land]. Akademiya nauk SSSR. Institut geografii. Materialy gliatsiologicheskikh issledovanii [Academy of Sciences of the USSR. Institute of Geography. Data of Glaciological Studies] 43, p 30. (Text in Russian)

Zagorodnov VS, Zotikov IA, Barbash VR et al (1976) O termoburenii na lednike Obrucheva [Thermal drilling on the Obruchev glacier]. Akademiya nauk SSSR. Institut geografii. Materialy gliatsiologicheskikh issledovanii [Academy of Sciences of the USSR. Institute of Geography. Data of Glaciological Studies] 28, pp 112–118. (in Russian)

Zagorodnov VS, Samoilov OYu, Raikovsii YuV et al (1984) Glubinnoe stroenie lednikovogo plato Lomonosova na o. Zap. Spitsbergen [Deep structure of the glacial Lomonosov Plateau on Western Spitsbergen]. Akademiya nauk SSSR. Institut geografii. Materialy gliatsiologicheskikh issledovanii [Academy of Sciences of the USSR. Institute of Geography. Data of Glaciological Studies] 50, pp 119–126. (in Russian)

Zagorodnov VS, Arkhipov SM Bazhev AB et al (1991) Stroenie, sostav i gidrotermicheskii rezhim lednika Garabashi na El'bruse [Structure, composition and hydrothermal regime of the Garabashi Glacier on Elbrus] Akademiya nauk SSSR. Institut geografii. Materialy gliatsiologicheskikh issledovanii [Academy of Sciences of the USSR. Institute of Geography. Data of Glaciological Studies] 73, pp 109–117. (in Russian)

Zagorodnov V, Mosley-Thompson E, Mikhalenko V (2013) Snow and firn density variability in West Central Greenland. IN: 7th international workshop on ice drilling technology: abstracts. Pyle Center, University of Wisconsin, Madison, USA, 9–13 September 2013, p 64

Zagorodnov V, Tyler S, Holland D et al (2014) New technique for access-borehole drilling in shelf glaciers using lightweight drills. J Glaciol 60(223):935–944

Zeibig M, Delisle G (1994) Drilling into Antarctic ice—the new BGR ice drill. Polarforschung 62(2/3):147–150

Zimmerman W, Bonitz R, Feldman J (2001) Cryobot: an ice penetrating robotic vehicle for Mars and Europa. In: IEEE aerospace conference, vol 1, 10–17 March 2001. Big Sky, Montana, USA, pp. 311–323

Zimmerman W, Anderson FS, Carsey F et al (2002) The Mars'07 North polar cap deep penetration cryo-scout mission. In: IEEE aerospace conference, 9–16 March 2002, vol 1. Big Sky, Montana, USA, pp 1-305–1-315

Zotikov IA (2006) The Antarctic Subglacial Lake Vostok. Glaciology, biology and planetology. Springer-Verlag, Berlin, Heidelberg, New York

Electric Thermal Coring Drills

2

Abstract

Ice-coring electric thermal drills are designed for continuous or discrete sampling in glaciers, ice sheets, and sea and lake ice covers. Coring is performed using a heated annular coring head that melts the ice around the desired core. Meltwater may be left down in shallow holes in temperate glaciers, can be removed from the borehole into a chamber above the core barrel or via intermittent bailing, replaced with non-freezing drilling fluid or hydrophilic antifreeze must be added and mixed with the meltwater. Electric thermal coring drills are particularly effective in temperate and polythermal glaciers.

Keywords

Annular thermal head • Drilling load • Pendulum steering • Ice-core quality • Hydrophobic and hydrophilic drilling fluids

Ice-coring electric thermal drills were designed for continuous or discrete sampling in glaciers, ice sheets, and sea and lake ice covers. Coring is performed using a heated annular coring head that melts the ice around the desired core. The first prototype of a thermal corer was developed in the Meteorological and Geophysical Institute of the University of Vienna, Austria (Schwarzacher and Untersteiner 1953) to obtain oriented ice samples from a glacial surface. This was a portable rim-heated device that enabled the extraction of cores 80 mm in diameter and up to 20 cm in length. In the summer of 1952, it was used in Pasterze Glacier, Eastern Alps with quite satisfactory results.

In the 1960–1970s, thermal ice-coring drills were highly developed; however, they had become practically obsolete since the 1980s. The main limitations of these type of drills are relatively high power requirements, poor core quality, and low penetration rate. Ice-coring cable-suspended electromechanical drills (Talalay 2016) replaced thermal coring drills in most applications. However, there are a few thermodynamic conditions in certain glaciers (e.g., temperate and polythermal glaciers) in which the use of thermal drills is still more effective than the use of electromechanical drills.

Temperate glaciers are at their melting point throughout the year from their surface to their base. Pressure-induced melting during mechanical drilling can cause the refreezing of meltwater on the drill, which then easily gets stuck in the borehole. Additionally, ice cuttings freeze together and ice forms on the cutters. In temperate ice, liquid water may fill the chip chamber, preventing the transport and removal of drilling chips. Polythermal glaciers are thermally complex, containing both warm ice at the pressure-melting point and cold ice. In the relatively warm intervals, the same complications as in temperate glaciers happen. All these issues are the motivation behind the use of thermal coring drills within this framework.

Depending on the ice temperature and borehole depth, several methods can be implemented regarding meltwater: (1) meltwater may be left down in shallow holes in temperate glaciers, where the refreezing time is enough for safe tripping; (2) to avoid refreezing, meltwater can be removed from the borehole into a chamber above the core barrel or via intermittent bailing (the borehole remains open); (3) to prevent borehole closure at greater depths, meltwater must be either replaced with non-freezing drilling fluid or hydrophilic antifreeze must be added and mixed with the meltwater.

During the first stages of development, steel ropes and independent electric lines to supply power to the downhole unit were used. Later on, systems used an armored cable with a winch to provide power to the thermal drill and to retrieve it from the hole. In some instances, more lightweight reinforced tough-rubber or plastic-sheathed cable was used for shallow drilling. The principal components of cable-suspended ice-core drilling systems are (Fig. 2.1): (1) a thermal drill, (2) electromechanical cable, (3) a hoisting winch, (4) a mast, (5) a control system (not shown), and (6) a power supply (typically an electrical generator) (Talalay 2016).

© Geological Publishing House and Springer Nature Singapore Pte Ltd. 2020
P. G. Talalay, *Thermal Ice Drilling Technology*, Springer Geophysics,
https://doi.org/10.1007/978-981-13-8848-4_2

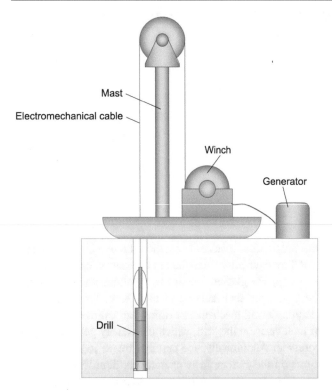

Fig. 2.1 Principal components of cable-suspended ice-coring drilling systems (Talalay 2016)

The drilling efficiency and rate of penetration depend on the shape and design of the annular thermal head, input power, drilling load, and ice temperature. Pendulum steering is essential for obtaining a plumb hole. So, this requires a controlled drilling load.

2.1 Thermal Coring Drills Without Meltwater Removal Systems

In minimal configurations, these drills contain an annular thermal head, a core barrel with core catchers at the lower end, and a cable termination. Thus, they are mechanically very simple with no moving parts and compact in size and weight. These drills are quite effective in glaciers with temperatures close to the melting point and can provide ice cores of very good quality with no thermal cracking. The produced hole is a few millimeters larger than the drill head, which provides little leeway for the refreezing of meltwater in the hole. In some cases, meltwater fills the whole hole, which is somewhat useful for preventing hole closure due to ice flow in response to ice-overburden stresses. In other cases, meltwater drains through the glaciers' channels and thus partly fills the hole. In colder ice (<-10 °C), where the refreezing rate of meltwater is expected to be higher than the rate of closure of open holes, these drills can be used with

intermittent drilling and bailing of the meltwater or a simple bailer can be installed at the top of the drill. From those listed in Table 2.1, the three last thermal drills are still in use.

2.1.1 Caltech Thermal Drill

In 1962, a portable and relatively simple electrically heated thermal core drill was constructed in the University of California and the California Institute of Technology, USA (Shreve and Kamb 1964). The thermal element was a 6.6-mm-diameter, 300-W, 150-V tubular heater with a stainless-steel jacket bent to form an annulus with an OD of 50 mm (Fig. 2.2). The ends of the heater were sealed with ceramic plugs cemented with a hard, heat-resistant epoxy resin and coated with synthetic rubber. The thermal element was mounted on the lower end of a 1.2-m-long core barrel made from stainless-steel tubing with an ID and an OD of 41 and 50 mm, respectively. To prevent the melting of the core, a Teflon liner was inserted inside the core barrel. Two core catchers were installed at the lower end of the core barrel.

To preserve the core orientation relative to the drill, a sliding cylindrical weight rested on top of the core inside the core barrel and held it down on the teeth. However, this system presented practical difficulties during operation, especially if the core was broken. A limit switch was mounted on the upper end of the core barrel and added a resistor to the power circuit when the core barrel was full. The resultant increase in current served as a signal that drilling was completed. Located just above the limit switch assembly was a single-shot inclinometer that recorded the positions of a compass needle, a pendulum, and an azimuthal mark indicating the orientation of the core barrel.

The 20 kg core drill was attached to a 3.2-mm-diameter 300-m-long galvanized steel rope weighing 14 kg, which was raised and lowered using a hand-cranked winch mounted on a wooden tripod. Power for the inclinometer was carried by a two-conductor rubber-covered cable. Power for the core drill was supplied by a commercial portable 2.5-kW 230/115-V 60-cycle generator driven by an integrally mounted two-cycle gasoline engine and was transmitted to the drill by two single-conductor rubber-covered cables.

In late August 1962, this core drill was set up at Blue Glacier, Mount Olympus, Washington, USA and used to drill a test hole 5 m deep, from which four cores that were 25 mm in diameter by 1.2 m in length were obtained. At a depth of 5 m, the drill penetrated the crevasse, allowing water to drain from the hole. The resulting overheating did not affect the thermal element, which was operating at its rated power, but it did cause the warping and partial destruction of the Teflon liner, which had to be replaced.

Table 2.1 Thermal coring drills without meltwater removal systems

Institution or Drill name	Year	Glacier type	Power/kW	Thermal head ID/OD/mm	Penetration rate in ice/ (m h^{-1})	Max depth reached/m	References
University of California and California Institute of Technology, USA	1962	Temperate glacier	0.3	36.8/50	1.2	137	Shreve and Kamb (1964)
University of Iceland	1968–1972	Temperate glacier	0.25–0.35	49*/61*	~1.0	108	Theodorsson (1976)
University of Washington, USA	1971	Temperate glacier	1.6–1.7	158*/172*	1.7	90	Taylor (1976)
ETB-1	1970s	Temperate glacier	1–3	84/108	2–6	NA	Morev et al. (1984)
LKTBM-1	1972	Temperate glacier	4	88/116	3.5–7	113	Sukhanov et al. (1974)
MTBS-76	1980s	Temperate and polar glaciers	1.0	66/76	1.5	93	Vasiliev et al. (1993), Ueda and Talalay (2007)
PICO/IDDO electrothermal drill	Since 1983	Temperate glacier	NA	76/90	NA	152	Koci (2002), Klein et al. (2016)
Icedrill.ch thermal drill	Since 2008	Temperate glacier	2.0	84/100	2.8	101	Schwikowski et al. (2014)
Hot Jay drill	Since 2015	Polar glacier	2	NA	NA	55	Goetz, personal communication

Note NA—data are not available; *—author's estimation

Fig. 2.2 Thermal core drill's head (Shreve and Kamb 1964)

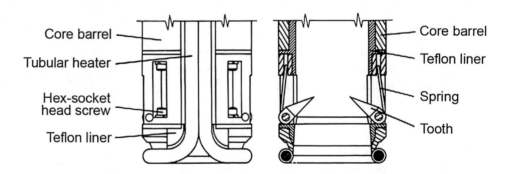

Core barrel — Tubular heater — Hex-socket head screw — Teflon liner

Core barrel — Teflon liner — Spring — Tooth

In early September, the core drill was moved to another site and a hole 137 m deep was drilled, reaching the base of the glacier (Fig. 2.3). With a power input of 300 W, the core drill penetrated the ice at 1.2 m/h; it was used alternately with a non-coring hot point whose speed was ~8 m/h (see Sect. 1.2.8). The hole drilled was ~55 mm in diameter and remained within 1° of the vertical at all depths. A total of 16 cores were obtained, 14 of which were oriented. Most of the cores were elliptical in cross-section and were broken into 2–4 pieces.

2.1.2 Thermal Core Drill of the University of Iceland

A simple thermal drill was constructed in the Science Institute, University of Iceland (UI) in 1968 (Fig. 2.4)

(Theodorsson 1976). The heating element was made from a 0.6-mm-diameter heating wire that was wound tightly into a helix and then pressed into a 6-mm-wide circular groove in a heater housing with thin mica insulation between the housing and the wire. The free space in the groove was then filled with alundum cement. One end of the wire was soldered to the housing and the other end was fed through a Teflon insulator in the upper annular plate of the heater housing. The plate was soft-soldered to the lower part of the heater housing. An annular thermal insulator made of Teflon was inserted between the heater housing and the rest of the thermal corer. The total length of the drill was 1.2 m.

In the first drill, a polyethylene liner was inserted inside the core tube for thermal insulation. However, in two subsequent thermal drills, this tube was omitted because it made no apparent difference. A power of approximately 250–350 W was usually applied to the heater, this being the

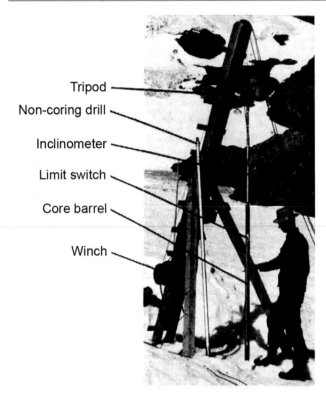

Tripod

Non-coring drill

Inclinometer

Limit switch

Core barrel

Winch

Fig. 2.3 General arrangement of a core drill and surface equipment at Blue Glacier, September 1962 (*Photo* R.L. Shreve) (Shreve and Kamb 1964)

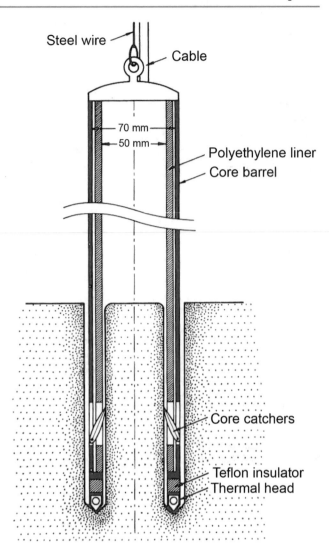

Steel wire

Cable

70 mm

50 mm

Polyethylene liner

Core barrel

Core catchers

Teflon insulator

Thermal head

Fig. 2.4 Icelandic thermal drill (Theodorsson 1976)

power that the generator could produce. The heater could withstand more power but, with 600 W, the useful lifetime of the heating element became too short.

In 1968, test drilling operations through firn were quite stable; however, meltwater likely disturbed the isotopic ratio of the core samples. The penetration rate was ∼2 m/h and the core diameter was ∼4 mm less than that of the thermal drill head. The next summer, more serious attempt were made to drill deeper holes into the Bardarbunga glacier, Iceland. The drill worked well in firn but, when solid ice was reached at a depth of ∼30 m, the penetration rate fell abruptly and the core diameter decreased to 35 from 45 mm in firn. Sometimes no core was recovered at all, and core recovery averaged less than 50%. Surprisingly, during a few runs in solid ice, the drilling speed was quite high compared with the average run and a good core was recovered, with a diameter of 35–40 mm along its full length of 1 m. The reason behind this unstable drilling performance was likely volcano ash layers that slowed down the melting process.

When a depth of 102 m was reached, the penetration rate drastically decreased and, at a depth of 108 m, penetration ultimately stopped. No core was recovered below a depth of 102 m, probably because of excessive melting caused by a thick ash layer (drilling operations three years later using an electromechanical drill revealed a thick ash layer at this depth).

This drill was used once in Sweden to drill through firn, but the penetration rate was much lower. When the drill reached solid ice at a depth of 30–40 m, the rate fell abruptly to less than 1 m/h and varied considerably from one run to the next.

In order to increase the drilling-melting efficiency, the thermal head was changed to a single circular turn of bare wire with a diameter of 3 mm. The diameter of the circle was 115 mm. Two short wires were soldered vertically to the circular heater, which served both for fastening the heater and as electrical leads for the low-voltage supply. A 3-kW transformer was fastened to the upper end of the core barrel. This drill was tried in solid ice at the Bardarbunga glacier in 1972, but the penetration rate was very slow and this attempt was soon abandoned, along with any prospect of using a thermal drill in the glaciers of Iceland.

2.1.3   Thermal Core Drill of the University of Washington

A lightweight, portable thermal coring drill for temperate ice and firn was developed in the University of Washington (Taylor 1976). This ~2.6-m-long, 170-mm-diameter fiberglass-epoxy core barrel with a small wall thickness was lightweight, strong, and electrically insulated, and its low thermal conductivity was useful for protecting the core.

The drill was handled by suspending it from a foldable 3.7-m-high tripod made of aluminum tubing (Fig. 2.5). A winch drum carrying 150 m of armored electrical cable was mounted between two legs. The winch was hand-cranked for breaking off the core and for approaching the end point slowly. A variable-speed drill motor was used through a speed reducer to power the winch when over tripping. A meter wheel indicated the depth of the drill. The complete drilling system weighed ~180 kg, including a 3-kW gasoline-powered electrical generator.

The replaceable thermal drilling head was made from a ring of heavy nichrome wire operating at a low voltage and a high current (6 V, 360 A) and directly exposed to the water. Special attention was given to the mechanical support and to the electrical connections to the heater ring to achieve a uniform temperature distribution around the ring, eliminating all cold spots. Power was fed to the heater ring through two sheet-copper conductors mounted on the outside of a fiberglass-epoxy core barrel. These sheet conductors were connected to the secondary winding of a step-down transformer rated at 2.2 kW and mounted in the upper section of the core barrel. The transformer's primary winding received 220-V, 60-Hz single-phase power at ~10 A through the armored cable. The transformer and all the connections to it were designed to be water-tight.

The thermal heads for wet, impermeable ice and for dry drilling in firn could be readily exchanged and were electrically identical, but differed in their mechanical construction because the firn head operated at a higher temperature and caused negligible side melting owing to the lack of water. When drilling in firn, the power input was reduced to ~70% of its full capacity to prevent damage to the heater ring in air. As soon as the water table was reached, the more efficient ice head was installed and full power was applied.

A short loop of heavy resistance wire located on the inside wall of the core barrel near the drill head melted a small longitudinal groove in the core as the drill advanced. This groove defined the orientation of the core in the barrel, even though the core may fracture or move during retrieval. As the core filled the space available in the barrel, a spring-loaded push rod activated a switch, which signaled the operator through a lamp circuit in the control box.

The core catchers were fixed near the drill head. The top end of the core barrel consisted of a tapered aluminum cone containing a heater, which could be activated from the surface in case the drill became stuck in the borehole during retrieval. This back-out heater remained untested as no difficulties ever occurred. To determine the in situ orientation of the core barrel, a remote-reading two-axis inclinometer and a magnetic compass using separate conductors in the cable were located in the upper section of the core barrel.

In the summer of 1971, this thermal coring drill was used to obtain two continuous cores of firn and ice down to the depths of 40 and 90 m at Blue Glacier. The penetration rate at a generator output of 210–220 V and 8.5 A (1.8–1.9 kW)

Fig. 2.5 Core removal from a thermal core drill, Blue Glacier, 1971 (Taylor 1976)

averaged ~ 1.7 m/h, within the range of 1.5–2.0 m/h. The voltage drop along the cable was estimated to be ~ 23 V, indicating a power at the drill transformer of 1.6–1.7 kW. At the reduced voltage input required at the generator (180 V) for snow and firn, the penetration rate varied from 1.5 to 2.5 m/h, depending on firn density and the presence of ice layers. The core diameter was ~ 152 mm and the diameter of the hole was ~ 178 mm. Core length was usually 1.65–1.7 m. In August 1972, drilling operations were continued to depths of 60 m (Raymond and Harrison 1975).

2.1.4 ETB-1 and LKTBM-1 Thermal Drills

Both of these drills were designed in AARI, USSR for core drilling in temperate ice in the beginning of 1970s. ETB-1 (Russian abbreviation for electro-thermal drill) had a very simple design (Fig. 2.6) and included a core thermal head, a 1 to 3-m-long core barrel, and a cable termination (Morev et al. 1984). Depending on the core barrel length, the drill weighed between 15 and 40 kg. The drill was suspended on 8.6-mm-diameter single-core armored cable. Unfortunately, the results of the field tests were not published.

Another thermal coring drill LKTBM-1, which was similar to ETB-1, was designed for high-mountain application (Sukhanov et al. 1974; Ryumin et al. 1974). The length

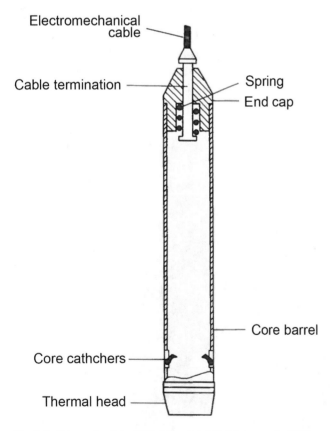

Fig. 2.6 Schematic of the ETB-1 thermal drill (Morev et al. 1984)

of the drill was 1.5 m (maximum core length of 1.4 m) and it weighted 25 kg. The total weight of the drill set (without the generator and fuel) was 150 kg. In July of 1972, the drill was used for coring two holes at Abramov Glacier, Alai Ridge (Pamirs-Alai Range, South Kyrgyzstan). The deepest hole reached bedrock at a depth of 113 m.

2.1.5 MTBS Thermal Drills

To drill shallow holes (< 100 m) in temperate and polar glaciers, a series of lightweight simple thermal drills of the MTBS-type (Russian abbreviation for portable thermal drill) were developed in the LMI. Depending on the OD of their thermal head, these drills were called MTBS-66, MTBS-76, or MTBS-100 (Vasiliev et al. 1993; Ueda and Talalay 2007).

All MTBS drills had quick-release cable terminations for connecting them either to the drill or to the bailer. After each or several runs, the drill was disconnected from the cable. Then, a bailer with a bottom valve was connected to a cable and lowered into the hole for picking up the meltwater. The drilling set for remote and high mountain glaciers included a lightweight surface arrangement mounted on aluminum sledges (Fig. 2.7). The winch, which had 100 m of KG1-40-180 cable, was driven by a 0.6 kW electric motor. The total weight of the drilling equipment was ~ 130 kg.

The first MTBS-66 drill was tested at Vavilov Glacier, Russian Arctic in 1979, where seven holes were drilled with a total depth of 35 m (maximum depth of 13.4 m). In the 1980s, the MTBS-76 and MTBS-100 drills were used by the Glacier Party of the Basic Research Laboratory of Snow Slides and Debris Flows (Moscow State University) and the North-Caucasus Geographic Expedition of Institute of Geography, USSR Academy of Sciences (IGAS) on the glaciers of the Caucasus, Tien-Shan, and Kodar Range. A total of nine holes were drilled on mountain glaciers with a total cumulative depth of 419 m. The deepest hole, which was 93 m deep, was drilled on the Dzhantuganskoye Firn Plateau on the northern slope of the Main Caucasus Ridge in August 1983. This hole was drilled by two men over 127 h (the total penetration time for drilling this hole was 68 h). The average length of the runs was ~ 2 m, and the rate of penetration was 1.5 m/h at a drill head power of ~ 1 kW.

2.1.6 Electrothermal Drill of the University of Minnesota/PICO/IDDO

Originally, this electrothermal drill was developed in the beginning of the 1980s at the University of Minnesota for a particular high-altitude drilling project at Quelccaya Ice Cap, located at an altitude of 5700 m asl in the Cordillera Oriental section of the Andes mountains of Peru (Koci 2002). This

Fig. 2.7 Lightweight thermal drilling system of the LMI (modified from Vasiliev et al. 1993)

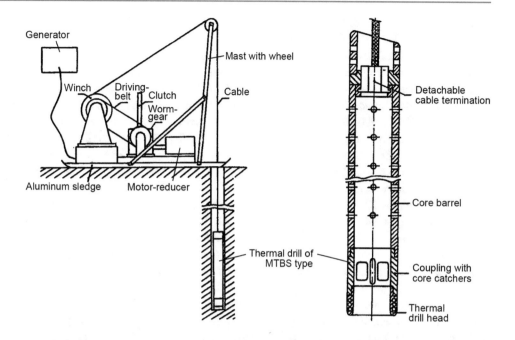

Fig. 2.8 Thermal head of the IDDO-3″ electrothermal drill (*Credit* T. Wendricks from Drilling ice cores, n.d.)

drill had very simple design and could collect a 3″ core. For that project, the drill used Kevlar-reinforced electromechanical cable, reducing cable weight by a factor of three, and power was supplied by a 2-kW array of solar voltaic panels (Koci 1985). In 1983, two holes were drilled at Quelccaya Ice Cap down to bedrock to the depths of 163 and 154 m at a rate of 14 m/day. An electromechanical drill was used to make two starting holes through the firn to a depth of 35 m. They were the first ice cores ever obtained from the tropics.

Afterwards, the drill was transferred to the PICO at Nebraska-Lincoln University and later to the PICO of the University of Alaska-Fairbanks. Then, the drill was moved to the storage section of the Ice Drilling Design and Operations (IDDO) group, University of Wisconsin-Madison, and was thereafter referred as the IDDO-3″ electrothermal drill (Fig. 2.8). Over many years, the drill was quite successfully used, particularly in ice close to the pressure melting point.

In August 1994, this drill was used for coring at South Cascade Glacier, USA (Giles 1994). The first hole was terminated at a depth of 20.6 m because, for unknown reasons, penetration became very slow; the obtained core was only 50 mm in diameter. The drill set was moved away for ~3 m and drilling was restarted. The last core was recovered from a depth of 158.4 m, where penetration rate slowed down further to almost zero. Thus, it was declared that the hole had reached bedrock. Core breakings below 104 m were very hard. On two occasions, the drill became stuck and was recovered with some difficulties.

In April 2008, the IDDO-3″ electrothermal drill was used for core sampling at McCall Glacier, eastern Brooks Range of Alaska (Klein et al. 2016). The ice core was collected using an electromechanical drill until water was reached ~70 m below the surface, at which point a thermal drill was used because it provides a cleaner core with less water-related mechanical damage. Thermal drilling was stopped at 152 m, where the drill hit a rock at what the team believed was the bottom of the glacier based on radar measurements of ice depth.

In July 2010, the IDDO electrothermal drill was used to drill a hole from 55 m to a final depth of 141 m at Combatant Col, 3000 m asl, in Mount Waddington, Coast Mountains, British Columbia, Canada (Fig. 2.9) (Neff et al. 2012). The hole was predrilled using a PICO-4″ electromechanical drill; however, the presence of water in the borehole prevented the evacuation of drill chips and it was possible to continue drilling using the thermal drill. The ice temperature at the site was between −3 and 0 °C at depths below 20 m, with consistent temperatures of 0 ± 1 °C below

40 m. The thickness of the ice at the site was estimated to be ~250 m. A second borehole was terminated at a depth of 90 m when difficulties resulting from glacial meltwater filling the borehole halted drilling (Neff 2010).

A scaled-down version of the IDDO-3″ electrothermal drill with a small tripod setup for lowering by hand using a short cable was tested on McCall Glacier, Alaska, in the spring of 2012 (Fig. 2.10) (Nolan 2012). The general idea was to develop a lightweight and compact ice drilling system that could be used on glaciers to reach depths of ~40 m. Unfortunately, the two available thermal heads burned out shortly upon actuation. In both cases, when thermal head burned out, it was going through snow rather than ice. Perhaps the heat conduction was not enough to transfer heat away from the thermal head and it overheated and burned out.

The drill was most recently deployed and used successfully to drill through firn aquifer layers in southeastern Greenland during the spring of 2013 and the spring of 2015. The depth to the water table in firn aquifers depends on elevation, surface slope, and glacier topography and ranges between 0 and 40 m (Miège et al. 2016). In April 2013, two cores were drilled into firn and ice (30 and 65 m deep) in southeastern Greenland with the IDDO-3″ electrothermal drill set along with the PICO electromechanical drill winch and tower (Fig. 2.11). The upper portion of the firn above the water table was drilled using PICO/IDDO hand augers and a sidewinder driven by an electric drill (Kyne 2013).

In April 2015, another hole from 19.7 to 56 m was drilled in the same area with the IDDO-3″ electrothermal drill (the hole was also predrilled using a hand auger) (Goetz 2015).

Fig. 2.9 Removal of a 2-m-long ice core from the IDDO-3″ electrothermal drill, Combatant Col, July 2010 (*Photo* D. Clark from Neff 2010)

Run lengths were ~1.05 m deep; however, extracting cores from the drill was an awkward procedure. Two cores could not be extracted from the barrel on the surface and had to be melted out with ethanol. From depths of 52 m onwards, it became necessary to ream a few meters at the bottom of the borehole in every drill run.

In 2015 and 2016, the Greenland Aquifer Expedition team employed the more lightweight Hot Jay Drill (see Sect. 2.1.8). The IDDO plans to redesign the thermal head in the near future because the exact model of the heating elements is now out of production and to upgrade the hoisting equipment to allow for coring down to 300 m (Long Range Drilling Technology Plan 2016).

2.1.7 Icedrill.ch Thermal Drill

For high alpine ice-core drilling, Icedrill.ch AG, Switzerland constructed and manufactured a thermal drilling system that can be used as a standalone system or in combination with the FELICS electromechanical drill (Schwikowski et al. 2014). The entire system consisted of a tower, a winch, cable, and an energy supply, and was alternatively equipped with lightweight thermal or electromechanical drills. The total weight of the standalone thermal drill was 200 kg, and the combined system weighed 289 kg. This did not include the drilling tent and fuel for the power supply.

Fig. 2.10 IDDO-3″ electrothermal drill and driller in action, McCall Glacier, Alaska, May 2012 (Nolan 2012)

Fig. 2.11 Collecting some of the water dripping from an IDDO-3″ electrothermal drill, southeastern Greenland, April 2013 (Kyne 2013)

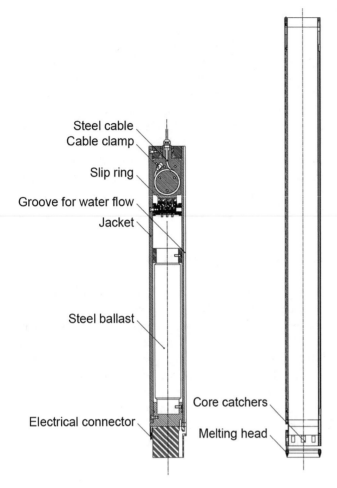

Fig. 2.12 Schematic of the Icedrill.ch thermal drill (Schwikowski et al. 2014)

cable windings. The slip-ring was generally not required for thermal drilling but provided flexibility to the entire system. External grooves in the jacket allowed for the passing of meltwater, facilitating the lowering and lifting of the drill in meltwater-filled boreholes. Steel ballast was added to increase the drag on the cable, with the potential of increasing the penetration rate by supplying pressure to the thermal head.

The core barrel consisted of a 4-m-thick aluminum tube coated with fiberglass-reinforced epoxy resin. The barrel included the electrical connections for the thermal head. Three designs of the thermal head with the same size (ID/OD: 84/100 mm) and the same heating element (Hot-spring coil heater with a diameter of 3.3 mm and a heated length of 610 mm) were tested (Fig. 2.13): (1) an open heater with electrical connectors at both ends of the coil, resulting in a crossing-over of the coil; (2) a simple aluminum heating annular head, in which the connectors of the coil were single-ended, allowing for a spiral shape; and (3) an aluminum crown with 90 heat-spreader fins, also with single-ended connectors. The last thermal head achieved penetration rates 1.5 times faster than the other heads.

Although the heater was rated 650 W at 230 V, it was supplied with 400 V DC at an actual power of 2 kW. Operation under such high power requires constant contact with the ice for cooling; otherwise, the heater would burn out. The drill system was equipped with two gasoline Honda EU2000i generators with a nominal output of 2 kW each, which together produced 1.5 kW at altitudes of ~3000 m asl. The 80–265 V AC output of the generators was transformed using ten AC/DC converters in series, producing 50 V each, i.e., a maximum direct voltage of 500 V.

This thermal drill was used for ice-core drilling in two temperate glaciers in the Swiss Alps. In April 2011, a 101-m-long surface-to-bedrock ice core was collected from the Silvretta glacier (Fig. 2.14). The borehole temperature was ~0 °C, and the borehole was filled with meltwater. The upper 12 m were drilled using an electromechanical drill before switching to thermal drilling. The penetration rate

The thermal drill produced a borehole with a diameter of 103 mm and ice cores with a length of 70 cm and a diameter of 80 mm (minimum of 75 mm). The drill consisted of an upper module and a core barrel (Fig. 2.12). The upper module included an aluminum jacket, steel ballast, a slip-ring, and a cable termination. The 4.8 mm armored cable was fixed to the module using the friction of three

Fig. 2.13 Drill heads of the Icedrill.ch thermal drill (Schwikowski et al. 2014)

Fig. 2.14 Ice core recovered from the Icedrill.ch thermal drill at the temperate Silvretta glacier, Swiss Alps (*Credit* M. Schwikowski)

Fig. 2.15 Hot Jay drill and a recovered ice core, southeastern Greenland, August 2016 (Miller 2016)

was ∼1.8 m/h. Alkylate fuel was used to tank up the generators. This fuel has a higher specific energy than ordinary gasoline and is more environmental- and user-friendly because it does not contain dangerous substances, such as benzene and aromatic hydrocarbons. Alkylate fuel consumption was 0.8 L/m of drilled hole.

In June 2012, two parallel ice cores with lengths of 56 and 20 m were drilled in Glacier de la Plaine Morte. This glacier was snow- and firn-free and the thermal drill was used from the surface. The penetration rate was higher than before at ∼2.8 m/h because the new and more efficient thermal head with heat-spreader fins was used. Alkylate fuel consumption was 0.5 L/m. In both glaciers, the thermal drill produced non-fractured ice cores of excellent quality with an average length of 0.68 m.

2.1.8 Hot Jay Drill

In 2015, a new thermal drill was especially developed by J. Kyne for the Greenland Aquifer Expedition (see Sect. 2.1.6). The main design idea behind this drill was that it was supposed to be a smaller and lightweight version of the IDDO-3″ electrothermal drill (J. Goetz, personal communication, 2016). The whole drill set weighted ∼23 kg (Miège 2015). The core diameter was ∼2.5″, the length of the core barrel was 1 m, and the drill head power was ∼2 kW.

The first tests were done in April 2015. The drill was deployed without a tripod, which is why it was setup with the sidewinder (Goetz 2015). Drilling began in a hole from a depth 17 m, which had been initiated using a Felix electromechanical drill. Progress was smooth until a depth of ∼27 m with a run length of ∼0.75 m. At this point, the drill became stuck in the borehole on its trip up at ∼20 m. Several liters of ethanol were dumped downhole with no

effect. It was therefore determined that the drill was most likely wedged into the borehole wall because the borehole was crooked. A metal pipe on a rope was dropped down the borehole to break the drill free. After doing this multiple times and pulling on the drill cable in a variety of opposing directions, the drill was eventually extracted from the borehole. This procedure damaged the top bracket of the Hot Jay drill to the point that it was deemed unsafe to continue drilling with it.

The lightweight tripod was added for subsequent trials in September 2015, and a 55 m firn/ice core was successfully extracted (Miège 2015). In July-August 2016, shallow thermal drilling was performed at three sites to measure firn density and hydraulic conductivity and to take chemical samples (Fig. 2.15) (Miller 2016; Miller et al. 2017).

2.1.9 Inductive Electro-Thermal Under-Ice Corer

This corer was designed in the National Institute of Ocean Technology, Chennai, India (Vedachalam et al. 2017) and could be fixed on the top of a remotely operated vehicle (ROV) and take cores below polar ice shelves. As the corer melted ice, the vehicle's buoyancy helped the corer to penetrate into the ice. The inductive thermal head was made of ferromagnetic material and it housed a multi-turn coil made of 4 mm^2 low self-inductance Litz wire (Fig. 2.16). The coil transferred power via mutual inductance and heated the head when energized with a high-frequency power supply.

The inductive thermal head had an ID of 63 mm and was threaded to an outer metal sleeve. The inner sleeve was used for fixing the core catcher, and the annular gap between the inner and the outer sleeve served as conduit for the power cable to pass from the enclosure of the power-electronics inverter to the head section. The corer had an adjustable

Fig. 2.16 Inductive
electro-thermal corer, its power
converter, and its power supply
(Vedachalam et al. 2017)

electrical input of 0.5–1.25 kW and operated in the frequency range of 10–100 kHz. Lab tests showed that the unit was able to obtain a 250–mm-long core with an average diameter of 51 mm with an input power of 1 kW at 20 kHz. At that power, the penetration rate was 0.84 m/h.

2.1.10 Summary

Thermal coring drills without meltwater removal systems are simple and lightweight and require relatively low power, but can only operate mainly in temperate ice. They are also suitable for coring cold firn at any temperature; however, firn cores are saturated with meltwater and are thus unsuitable for many kinds of analyses. Drills of this type have an open-top core barrel and therefore undergo low hydraulic drag, and have demonstrated high travelling speeds when moving through boreholes. The low logistic burden associated with this type of drills makes it possible to core in high-altitude glaciers (Koci 2002; Schwikowski et al. 2014). To make them smaller and lightweight, these drills can be equipped with a simple tripod without a winch; feeding, tripping, and core braking can be accomplished by hand.

The crucial technical challenge associated with these drills is the design of the thermal head that is mounted onto the core barrel. Most of these drills use winding tubular heating elements, either open or casted in an aluminum annular head. An interesting concept was realized in Icedrill. ch, whose coring head had heat-spreader fins. The drill head showed high penetration rates owing to effective heat transfer. Special attention must be paid to the use of thermal heads in firn, where heat removal is much lower than in ice, resulting in the possible overheating of the head.

Some thermal coring drills can work either in temperate glaciers without meltwater removal or in cold glaciers with meltwater sucking via air circulation (for example, the LGGE thermal drill). For the latter, an air vacuum pump is installed in the water tank of the drill. This type of ice-coring drills will be reviewed in the next section.

2.2 Electrothermal Coring Drills for Open-Hole Drilling

These drills use different methods to remove meltwater from the kerf produced by their heating thermal annular head. There are three types of open-hole thermal ice core drills (Fig. 2.17): (1) simple thermal drills with intermittent drilling and bailing of the meltwater; (2) thermal drills with a bailer installed on top of the drill (in fact, this type of thermal core drill was never designed or used for drilling in glaciers and ice sheets; however, they can be considered for open-hole ice core drilling as well); (3) thermal drills that can remove meltwater from the kerf and store it in a container above the core barrel (Talalay et al. 2018). All these types have advantages and disadvantages. In all cases, meltwater is recovered from the hole and can be used for analysis together with the ice core. Theoretically, meltwater can be removed from the borehole using an electric deep-well water-submersible pump, but such instrument are

Fig. 2.17 Options for open-hole thermal core drilling in ice: **a** simple thermal core drill and bailer for the periodic removal of meltwater; **b** thermal core drill combined with the bailer; **c** thermal core drill with reverse air circulation (Talalay et al. 2018)

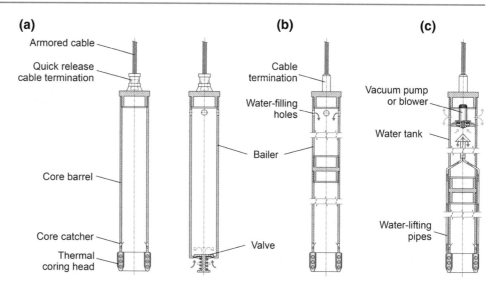

(a) Armored cable — Quick release cable termination — Core barrel — Core catcher — Thermal coring head

(b) Cable termination — Water-filling holes — Bailer — Valve

(c) Vacuum pump or blower — Water tank — Water-lifting pipes

bulky and power-consuming and require additional power lines and a water lifting hose or pipe string. Therefore, this was not considered as a practical option for open-hole thermal core drilling in ice.

The first two drill types do not have a forced water circulation system and are simpler than the third type. Meltwater is removed from the lower portion of the hole in batch actions. These drills have a temperature limit because, in cold ice, the meltwater can refreeze in the water-filling holes of the bailer and the water remaining at the bottom of the hole can also refreeze, causing problems with continuous runs. In most cases, meltwater is sucked using a vacuum pump in film mode and delivered into a water-collection tank above the core barrel.

Both these options suffer from serious shortcomings because the hole remains open and undergoes significant closure depending on ice temperature, overburden pressure (hole depth), and drilling time. The thermal head produces a hole larger in diameter than itself depending on the drill head design, input power, and ice temperature. The diameter of the borehole is typically 1.05–1.08 times larger than the diameter of the drill head. The hole diameter, which shrinks due to closure, should be always larger than the diameter of the drill. Otherwise, the drill gets jammed during tripping and finally becomes stuck.

Almost all the drills listed in Table 2.2 are not used now because they are quite slow, require considerable power input, and their operation requires substantial time for servicing on the surface. The latest drilling operation with this kind of tools was carried out using a large-diameter Australian National Antarctic Research Expedition (ANARE) thermal drill near Law Dome to a depth of 258 m in the season of 2005–2006. The deepest open hole drilled was

952.5 m deep, drilled using the TELGA-14 thermal drill at Vostok Station, East Antarctica in 1972 (Korotkevich and Kudryashov 1976). For drilling at greater depths, it is necessary to prevent borehole closure by filling the borehole with a fluid.

2.2.1 CRREL Deep Thermal Drill

At the end of the 1950s, the CRREL started to develop a new thermal drill for the polar deep-drilling program: a thermal core drill with a downhole vacuum pump that drew the meltwater up through heated tubes into a tank above the core barrel (Ueda and Garfield 1968). The thermal drill head was made into a copper annular shape with an ID and an OD of 124 and 162 mm, respectively, with 18 cartridge heaters rated at 9 kW. Spring-loaded core catchers were located just above the head. A 3.05-m-long steel core barrel, a laminated plastic meltwater collection tank, switch housing, and a transformer comprised the remainder of the drill. A vacuum gear pump located inside the meltwater collection tank pumped the meltwater through heated nichrome tubes up into the collection tank. The drill was 9.15 m long and weighed 410 kg. A hydraulic winch with a length of 3650 m and 25.4 mm in diameter, twelve-core armored cable, and a 9-m-high tower were used to raise and lower the drill. The overall weight of the winch with the cable was 19 tons.

The first stage of the CRREL deep core drilling project was carried out at Camp Tuto, at the edge of the Greenland ice sheet on the trail to Camp Century in the summers of 1959 and 1960 (Langway 2008). Here, the preliminary testing of the new concept of a cable-suspended deep thermal drill was conducted. The operation then moved to Camp

Table 2.2 Electrothermal coring drills for open-hole drilling

Institution or Drill name	Years	Thermal head (ID/OD)/mm	Drill head power/kW	Drill site	Minimum temperature/°C	Penetration rate in ice/ (m h^{-1})	Maximum depth reached/m	References
CRREL 'deep' thermal drill	1961–1965	124/162	9.0	Camp Century, Greenland	−24.6	NA	264	Ueda and Garfield (1968)
CRREL portable thermal drill	1965–1973	124/162	NA	Canadian ice caps	−23	NA	299	Paterson (1976)
			2.9	Byrd station, Antarctica	−29*	1.9	335	Ueda and Garfield (1969)
ANARE thermal drill	1968–1982	124/162	3.9	Amery Ice Shelf, Law Dome, Antarctica	−21.5	2.4	474	Bird (1976), Morgan et al. (1984)
CRREL large-diameter drill	1966	NA/203	NA	Byrd station, Antarctica	−29*	NA	75	Ueda and Hansen (1967)
CRREL large-diameter drill	1979	181/219	11.2	Dye-3, Greenland	−20.5	4.0 in firn; 2.3 in ice	75.5	Rand (1980)
ANARE large-diameter drill	1987–1989; 2005–2006	200*/255*	3.5	Law Dome, Antarctica	−20.4	1.0	234; 258	Etheridge and Wookey (1988)
JARE thermal drill	1969-1984	132/168	3.0	Mizuho Station, Antarctica	−35.5	1.55	700.6	Narita et al. (1994)
ETBLK-1 M thermal coring drill	1963–1966	100/146	5.0-5.5	Mirny Station, Antarctica	−11*	2.9–3.6	77	Sekurov (1967)
LGGE thermal drill	1968-1983	108*/132*	4.1	Dome C, Antarctica	−53.5	4.0–5.5	905	Lorious and Donnou (1978a)
TELGA-2 thermal drill	1967–1968	90/150	3.04	Mirny Station, Antarctica	−24	1.29–1.65	212.5	Morev and Shamont'ev (1970), Morev (1972)
TELGA-14 thermal drill	1968–1990	130/178	3.5	Mirny, Vostok, traverse Mirny – Vostok, Severnaya Zemlya	−57	1.2–1.8	952.5	Ueda and Talalay (2007)

Note *—Author's estimations; *NA*—data are not available

Century in the fall of 1960 in preparation for drilling the next summer, in which the drill rig was installed in an undersnow trench (Fig. 2.18).

Three ice cores were recovered at Camp Century using this thermal coring system. The first hole, recovered in 1961, reached a depth of 186 m, where the drill got stuck and the hole was abandoned. A new hole was started in 1962 and reached a depth of 238 m, where the heating head was lost and the hole was also abandoned. A third hole was started in 1963, which reached a depth 264 m. In 1964, this last hole was extended to 535 m using the thermal drill in a fluid-filled hole (see Sect. 2.3.1). The minimum temperature was −24.6 °C at 154 m (Hansen and Langway Jr 1966). In mid-1965, because of the unsatisfactory slow rate of core recovery and complications during operation, the thermal drill was replaced with a newly acquired and modified electromechanical drill (Ueda and Garfield 1968).

2.2.2   CRREL Portable Thermal Drill

General design

After more than five years of testing, CRREL specialists concluded that for relatively shallow holes (<450 m) in clean ice and when fluid is not, the thermal drilling concept could be feasible. Thus, in the mid-1960s, a 4.6-m-long 80 kg portable thermal drill with the ability to remove meltwater via an air vacuum pump was developed (Fig. 2.19) (Ueda and Garfield 1969). The drill head was made from aluminum and was less power-consuming (3.5–4 kW) than the heads of deep drills of the same size (Fig. 2.20).

The tank could hold ∼15 L of meltwater produced from a 1.6 m coring run. The vacuum pump and a motor were housed above the tank. Two limit switches on a spring suspension at the top of the drill triggered indicator lights on

Fig. 2.18 Thermal coring rig in Trench #12 at Camp Century, 1964 (Langway 2008)

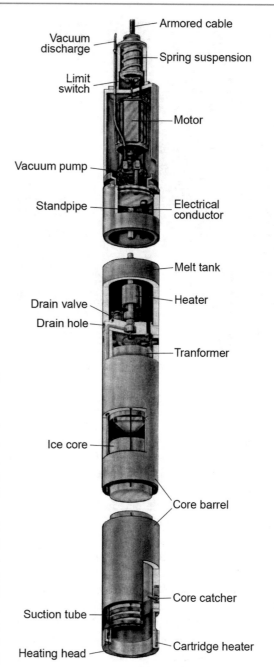

Fig. 2.19 CRREL portable thermal drill (Ueda and Garfield 1969)

the control panel and allowed the operator to manually operate the winch drum and always keep the drill in suspension, which is essential for maintaining a plumb hole. According to the authors' estimations, to drill a plumb hole at the optimum rate, ~ 160 N (20% of the drill weight) should be applied on the ice face.

The other pieces of equipment consisted of 450 m of 12-mm-diameter six-core cable, a generator, a hoist, and a tower (Fig. 2.21). An orthocyclic winding principle provided even cable spooling without the use of a level winding device. A 5 kW gasoline generator supplied electrical and mechanical power (the fuel consumption rate was ~ 3 L/h).

A flexible shaft transmitted power mechanically from the generator motor to the hoist. In the winch, power was transmitted through a clutch and a V-belt to a speed reducer and finally through a chain drive to the cable drum. The 6.7 m tower consisted of three sections of aluminum tubing, two of which were split and served as shipping containers for the drill. The entire drive unit, cable drum, and tower assembly were mounted on a single aluminum frame. Depth was read on a counter that was mechanically operated through a flexible shaft extending the length of the tower and driven by the sheave at the top of the tower. The tripping rate averaged at 0.3 m/s. The total weight of all drilling equipment was ~ 1100 kg.

The first version of this drill was built in 1963 for the Canadian Department of Mines and Technical Surveys. Two other drills were built in 1966 for the ANARE and for the US Antarctic Research Program.

Drilling in Canadian ice caps

The first CRREL portable thermal coring drill was used to make four holes in the Canadian Arctic: a 121 m-deep borehole through the Meighen Ice Cap (June 1965) and three

Fig. 2.20 CRREL thermal drill head. (*Credit* H. Ueda)

other holes with depths of 230, 299, and 299 m (in May 1971, May 1972, and May 1973, respectively) in the ice cap on Devon Island (Paterson 1976). Three of the four holes reached bedrock (in the 230 m-deep hole, the drill became ultimately stuck). The top 60 m of each borehole were drilled through firn. The 3–4 m uppermost portion of each borehole was cased with a plastic pipe to keep out surface meltwater. To decrease the pulling force required for core breaking, penetration was stopped for 1–2 min to melt a depression in the core 20–25 cm away from the end of the drill run.

For the first drilling operation, there was only a little shelter around the rig; as a result, bad weather caused frequent interruptions. Subsequently, all the equipment was mounted inside a Parcoll building, with part of one roof section removed to make space for the tower. The main improvements in a later drill model were a modified mechanism for suspending the drill from the cable, a better pump motor, the inclusion of a water-level indicator in the water tank, and the elimination of valves in the tubes that carry the meltwater from the drill head to the tank. The pump was expected to be a possible weakness in the drill and needed to be flushed with alcohol after each run.

Core recovery ($\sim$1.6 m/run) from the boreholes was achieved, but many of the cores were badly fractured. The filtered meltwater was used for chemical analyses. The

measured closure rates near the base of the ice were less than 0.3 mm/day. No dirt bands were encountered in the ice on Devon Island. On the Meighen Ice Cap, pockets of dirt reduced penetration rates by up to 25%. The boreholes showed little tendency to deviate from the vertical.

In the 230 m hole on Devon Island, the drill became frozen-in and was lost. This probably resulted from a leakage of water from the water tank while the drill was downhole. The first sign of trouble was a reduction in cable tension. Attempts to bring the drill to the surface failed because the motor on the hoist stalled repeatedly. When this was prevented by boosting the output of the generator, the main sprocket on the hoist broke. This put the hoist out of action and greatly reduced the chances of saving the drill. Approximately 70 L of an ethylene-glycol solution were pumped downhole by means of a rubber hose to ensure that the solution reached the top of the drill, and an electric heater was kept running in the solution. However, these attempts to free the drill were unsuccessful and were abandoned after three days. The cable was then cut a short distance above the drill to leave the hole free for measurements.

On the Meighen Ice Cap, the temperature profile was measured ten days after drilling was completed in July 1965, and again in June 1966 and May 1967 (Paterson 1967). Below 20 m, the temperatures recorded in 1965 were consistently higher than the others; the differences ranged from 0.03 to 0.12 °C. These differences almost certainly resulted from the heating of the ice by the drill. In 1967, the temperature at a depth of 21.34 m was −17.12 °C, and it was −15.96 °C at a depth of 121.16 m (bottom of the hole).

Drilling at Plateau Station and Byrd Station, Antarctica

In the season of 1966–1967, the CRREL portable thermal drill was used for coring at Plateau Station, Antarctica (Koerner and Kane 1967). To protect the people and the equipment from harsh weather conditions, it was erected in a Jamesway tent. Most of the core obtained from the upper 26 m of firn was not usable because of the production of considerable meltwater during the drilling process. Almost all of the core drilled below that depth was recovered in good condition. The maximum depth reached was 71 m. At this point the drill froze in the ice, preventing further penetration. Fortunately, two weeks later it was finally freed and returned to the surface.

On the next season, in 1967–1968, the drill was used at Byrd station to core five holes from 57 to 335 m in depth (Ueda and Garfield 1969). The following averaged data

Fig. 2.21 D. Garfield with the thermal drill in the front of the CRREL laboratory (*Credit* H. Ueda)

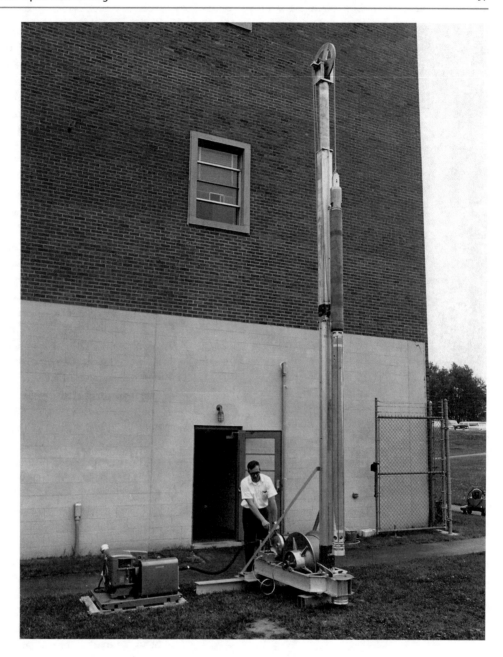

were obtained in the deepest hole (335 m): a penetration rate in firn of 2.4 m/h, a penetration rate in ice of 1.9 m/h, a total consumed power of 4 kW, an actual power consumed by the thermal drill head of 2.9 kW, a drill load of 180–300 N (drilling pressure was 21–35 kN/m^2), and average core diameter of $\sim$122 mm, and a total time to complete the hole of 260 h. Core quality was good to a depth of 100 m. From 100 to 130 m, the cores were shattered; from 130 to 335 m, the cores remained whole but were extremely brittle. Numerous fractures in planes approximately perpendicular to the core axis and spaced approximately 1 cm apart occurred over the length of the core. Near the bottom of the hole, the cores gained in strength but were still fractured.

2.2.3 ANARE Thermal Drill Developments and Experience

Amery Ice Shelf, Antarctica

Experience on thermal drilling began to be acquired by the ANARE in 1968, following the acquisition of the CRREL portable thermal drill (Bird 1976). That year, they planned to drill two boreholes at Amery Ice Shelf, Antarctica at points G1 and G3, which are respectively 67 and 240 km inland from the ice front. However, it took five months overall to drill a 310-m-deep hole at G1 (Fig. 2.22), because of delays caused by inclement weather and drill malfunctions. This loss of time and the consumption of spares prevented drilling

Fig. 2.22 The CRREL thermal drill used for coring at Amery Ice Shelf at site G1, 1968 (*Photo* M. Corry)

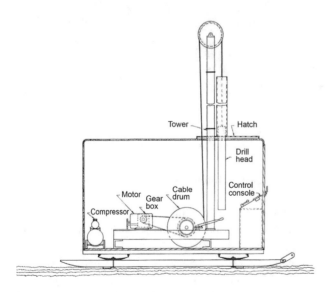

Fig. 2.23 Schematic of the ANARE drilling cabin (Bird 1976)

at G3. A malfunction in the meltwater pump caused the drill head to be frozen-in at a depth of 60 m; it was finally recovered after several days of applying a cable tension of 4.5 kN and jerking and after the application of several liters of alcohol and the intermittent operation of the heaters. Because future Australian operations were intended in similar harsh conditions, a modified drill set was designed (Bird and Ballantyne 1971).

Drill set modifications

The surface drilling equipment were installed within a sledge mounting cabin 3.7 m × 2.15 m × 2.15 m (Fig. 2.23). A hatch in the cabin roof allowed the tower to be lowered and then sectioned and housed inside for transportation. The winch was equipped with an automatic servo-controlled feed system that switched the winch motor current with a relay activated by the lower limit switch in the cable termination, indicating the lower limit of the drill load. The drill head was gently lowered in ~5-mm increments, thus automatically maintaining the correct drilling load for pendulum steering.

The core barrel of the drill was enlarged to 2 m. The cartridge thermal drill head was changed to a triple-coiled tubular heating element (230 V, 3.9 kW) casted integrally with the aluminum or copper annulus (Fig. 2.24). The control system was enhanced with protective devices: a solid-state vacuum transducer to monitor the vacuum level in the water tank; a meltwater level indicator for the tank; a thyristor control circuit for the drill-head heater power and temperature; and an inclinometer to control the borehole's azimuth and inclination. A drill with this modified CRREL design was constructed in 1968 and, during the following years, the Australian Antarctic Division drilled dozens of intermediate-depth boreholes using it.

To determine the azimuth and inclination of the borehole, a Pajari borehole surveying instrument was mounted at the top of the drill. A stylus fixed at the inner surface of the drill head grooved a reference line along each core; this line could be easily associated to the azimuth measured by the Pajari instrument and formed a permanent record of core orientation in the drill. The stylus was retractable and did not hinder the removal of the core.

Law Dome

All further Australian drilling activities have been undertaken in the Casey region on Law Dome (Fig. 2.25). Law Dome is an ice cap ~200 km in diameter situated between 110 and 115°E on the periphery of the main Antarctic ice sheet. This ice cap is approximately circular, with a

Fig. 2.24 ANARE thermal drill head: **a** general schematic (Bird 1976); **b** in the course of operations (core catchers were removed) (*Photo* J. Wilson)

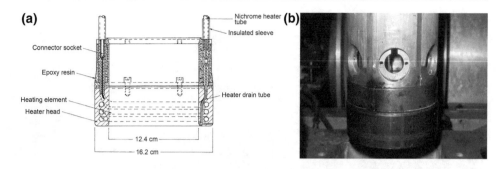

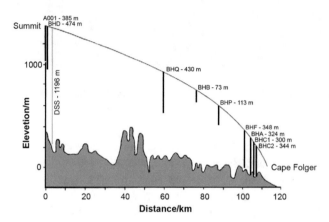

Fig. 2.25 Profile of the flow line from Law Dome Summit to Cape Folger, showing the location and depths of ten ice core-drilling sites; DSS (Dome Summit South) hole was drilled using the ISTUK electromechanical drill in 1988–1993 away from the flow line (4.6 km south-southwest of the summit) and is shown for comparison

maximum elevation of 1389 m asl at point A001 (Hamley et al. 1986). Cape Folger was the first drilling site BHA (Fig. 2.26). In May–August 1969, drilling proceeded normally to a depth of 324 m (within 40 m of bedrock), where borehole closure became a problem and hot reaming was necessary to maintain drill clearance. The drill eventually cut a new hole at a depth of 232 m and drilling was terminated.

In October–November 1969, a second 385-m-deep hole at A001 was drilled at the summit of Law Dome (approximately one third of the total ice thickness). It took 21 days to complete the hole on the basis of a two-shift 24-hour day, which represented a total of 500 working hours including maintenance periods. At the final depth, the drill became frozen-in because of an overflow of the water tank (the operator neglected to empty the previously collected meltwater). All attempts to release the drill by pulling with a

Fig. 2.26 Drilling at Cape Folger, Law Dome, 1969: **a** winch system of the ANARE thermal drill (*Photo* S. Little); **b** ice core being removed from the drill (*Photo* R. Wiggins)

(a)

(b)

Fig. 2.27 ANARE overwintering drilling campaign of 1977: **a** setting up the drill van at BHQ with ice core containers on the tow bar, Law Dome, May 1977 (*Photo* D. Robinson); **b** drill van and instrument van mounted on an Otago sled, October 1977 (*Photo* J. Wilson)

maximum cable load of 40 kN, adding alcohol (a total of 220 L), jerking, and switching the drill head and the drill's inner heaters were unsuccessful. Because there was no indication of movement from the drill during 18 days, the decision to cut the cable was made.

Explosive charges, comprising quarter-sticks of gelignite, were attached to the cable and dispatched to the base of the borehole. Six such charges appeared to not ignite; the seventh, however, got jammed 15 m down the hole and severed the cable at that point. Thus, both the drill head and 370 m of cable were lost, and the presence of the cable in the borehole prevented further measurements.

In 1972, the drilling equipment had been refurbished and was returned to Law Dome. Four holes with depths in the range of 48–113 m were drilled: BHS1 (48 m), BHP (113 m), BHB (73 m), and BHJ (112 m). The main reason for the termination of drilling operations was a technical problem with the winch driving system: the worm gear reducer that coupled the motor to the winch had no lubrication, and thus transmission efficiency was extremely low. A block of timber fell into the second borehole when its depth was 73 m and it could not be dislodged. Thus, this borehole was also abandoned.

In 1974, two other holes were drilled at Cape Folger. Drilling of the first hole (BHF) continued normally down to a depth of 348 m (the ice was approximately 385 m thick), at which borehole closure was becoming an increasing problem and thermal reaming was commenced. The rate of penetration averaged 2.4 m/h with a heater dissipation of 3.9 kW in ice with a temperature of −20 °C. Drilling was terminated because the head got jammed at a depth of 300 m during tripping and the heated annulus broke. Attempts to dislodge the thermal head proved unsuccessful. The second borehole was drilled 100 m away from the first to a depth of 345 m.

In 1977, the fourth wintering-over ice core drilling program was undertaken, and two intermediate-depth ice cores were obtained. In May and June of 1977, the first hole (BHQ) was drilled to a depth of 430 m at a midway site between Cape Folger and the summit of Law Dome (the ice thickness was estimated to be 790 m) (Etheridge 1989). A drill van and an instrument van were mounted on separate sleds in the configuration used (Fig. 2.27a). They were later mounted together on a large Otago sled, which proved more practical (Fig. 2.27b). The mast could be hinged—it was very quick to setup to commence drilling. The second hole (BHD) was drilled at the summit of Law Dome to a depth of 474 m, which, at the time, was the limit of the thermal drill's winch cable and was near the dry-hole drilling limit in this region of Antarctica.

The 1981–1982 austral summer season was particularly successful. For the first time, a summer-only ice core drilling operation was carried out at two sites at Cape Folger (Morgan et al. 1984). The drilling cabin was changed to single larger van and the drill tower was completely enclosed within it (Fig. 2.28). A hydraulic motor was used to raise and lower the drill and a hydraulic ram, which lowers the cable sheave, was used to feed the drill while coring. Better quality cores and a somewhat faster penetration rate were obtained using this feed system.

Drilling was stopped in the first hole (BHC1) at a depth of 300.2 m, where bedrock material and borehole closure prevented further penetration. Ice-radar soundings showed that bedrock was not more than 10 m below (Etheridge 1989). The second hole (BHC2) was drilled to a depth of 344.3 m, where an inclined band of bedrock material was encountered and closure rates of 2 mm/h prevented further penetration. The ice thickness was not more than 358 m. After drilling was completed, all these holes were filled with

Fig. 2.28 ANARE thermal drill
van at BHC1, Law Dome, 1982
(*Photo* A. Blake)

low-temperature fluid up to the firn–ice transition depth in
order to keep them open for future measurements.

2.2.4   CRREL Large-Diameter Thermal Drill

To drill large-diameter boreholes, the CRREL developed an
oversized thermal drill with a thermal head OD of 203 mm
(Ueda and Hansen 1967). The structure of the drill was
basically the same as that of a deep thermal drill. In
December 1966, this drill melted two holes 30 cm apart to
depths of 75 and 35 m at Byrd Station in Antarctica, starting
from the tunnel floor 12 m below the surface level
(Fig. 2.29). The deeper hole was cased in preparation for
subsequent drilling to greater depths, and the other was
reserved for the temporary storage of the assembled drill.
After the hole was cased and the drilling equipment was
installed, an electromechanical drill was used to penetrate
the ice sheet.

A modified version of the large-diameter CRREL thermal
drill was used to drill a pilot hole within the Greenland Ice
Sheet Program (GISP) project at Dye-3, Greenland in the
1979 summer season (Rand 1980). The drill was 4.34 m long
and weighed ∼70 kg. In the same way as all CRREL drills,
an aluminum thermal head with an ID and an OD of 181 and
219 mm, respectively, was heated using 18 cartridge heaters.
These individual heaters were 9.5 mm in diameter, 100 mm
long, and rated at 625 W with an internal resistance of 70 Ω.
They were electrically connected in parallel series for an
equivalent resistance of 18 Ω. The electrical power dissipated
by the drill head was adjusted by varying the voltage output of
the generator. Six spring-loaded core catchers were mounted

Fig. 2.29 Thermal drilling of a pilot hole at Byrd Station, Antarctica,
December 1966 (*Credit* H. Ueda)

at the bottom of the core barrel. The aluminum core barrel had
space for a 1.7-m-long core. The water tank was produced
with a laminated plastic tube and had a capacity of 22.7 L.

Fig. 2.30 Core removal from the large-diameter CRREL thermal drill, Dye-3, Greenland, May 1979 (Rand 1980)

Two diaphragm vacuum pumps connected in parallel were housed above the water tank. The pumps were rated at 115 V AC and 1.5 A. Before each run, a suction test was performed by immersing the thermal head in a bucket of alcohol and actuating the vacuum pumps. The drill was lowered until the vacuum pumps were accessible and the pumps were lubricated with ethyl alcohol to ensure that both the pumps and lines were free from ice. Then, the drill was lowered to the bottom and the drilling run started.

The drill set also included 11.9-mm-diameter seven-core armored cable, a winch holding 550 m of cable, a 5.5-m-high tower, a 35-kW generator available at the site, and a control desk to monitor the voltage and current sent to the drill, to activate the pump motors and heaters, to monitor the weight-indicator lights, to control drill penetration rates, and to switch the hoist motor for lifting or lowering operations. The winch was driven by a 2.2-kW, three-phase, 208 V AC motor. This motor was coupled to a clutch assembly that provided hoisting control. From the clutch assembly, power was transmitted to a gear reducer and finally through a chain-and-sprocket assembly to the drum. The hoisting speed was ~0.5 m/s.

Drilling of the pilot hole was accomplished over four days (the actual drilling time was 45.2 h) in May 1979 (Fig. 2.30). As drilling progressed, the heater voltage was gradually increased from the initial 240–280 V at the final depth of 75.5 m. The average core length was 1.3 m; penetration rate decreased from 4.0 m/h in firn to 2.3 m/h in solid ice.

2.2.5 ANARE Large-Diameter Thermal Drill

Another large-diameter thermal drill was developed in the ANARE (Etheridge and Wookey 1988). This drill was called

colloquially as "Errol Drill" after the Tasmanian-born film star, Errol Flynn, known for his romantic swashbuckler roles in Hollywood films (I. Allison, personal communication, 2016). The drill produced 260-mm-diameter holes and recovered 195-mm-diameter cores, which had many advantages for core analyses. The thermal head included one twisted heating element, vacuum-brazed to a copper annulus. The heating element could dissipate power up to 3.5 kW. Six core catchers were housed in the head, which was threaded onto a fiberglass core barrel capable of taking a maximum core length of 1.95 m. Meltwater was sucked up heated tubes into an evacuated water tank above the core barrel. The vacuum pump and electronics for drill telemetry (for the head's current and temperature and the water tank vacuum) and drill feed were contained in a module on top of the water tank. A borehole inclinometer was also housed in the module. The 12-mm-diameter armored cable used had seven conductors. Feeding of the drill was done according to the signal coming from a switch in the drill cable termination by continuously triggering a valve, which lowered the hydraulic ram during drilling.

To assist in breaking off cores with such large cross-sections, a prototype device was incorporated on top of the core barrel. It consisted of a powerful solenoid with a massive iron core. When tension was applied to the cable to withdraw the drill and core, the solenoid was energized, creating an impact force that was transferred through the core dogs to sever the ice. The core barrel was suspended on rubber mounts to amplify this effect. Although useful in some applications, the deeper and more brittle ice cores were often damaged by severing them in this way.

At the surface, the winch, hydraulics, and drill controls were housed inside a fully enclosed cabin, which was set on a sled. The drill tower laid flat on the cabin roof until the drill site was reached, where it was raised hydraulically. Drilling operations were carried out in January–February 1988 at site DE08, 16 km east of the summit of Law Dome, Antarctica (Fig. 2.31). This site has an extremely high snow accumulation rate of ~1.1 m water equivalent (perhaps one of the highest rates on Earth), which minimizes the smoothing of gas records. The first hole was drilled to a depth of 52 m, when a bolt and washer came adrift from the drill and got lodged in the bottom of the hole. Efforts to drill past these parts failed (the bolt was retrieved but the washer prevented further drilling), and a new hole was begun 2 m aside of the blocked one.

The second hole was drilled during 17 days, and drilling ceased at a depth of 234 m. The temperature at the bottom was measured to be −20.4 °C two days after drilling termination. Penetration rate decreased from 3.3 m/h in the upper firn to 1.0 m/h in solid ice. The core and borehole inclination was typically within 3° of the vertical, but it occasionally increased to 8°. The suspected cause was the

Fig. 2.31 DE08 drilling camp, east side of Law Dome, Antarctica, already starting to be drifted by the huge snow accumulation, January 1988 (*Photo* D. Etheridge)

flexible middle joint of the drill. Decreasing the feed rate brought the hole back to the vertical. The top 20 m were heavily infiltrated by meltwater. Core quality below 20 m was excellent. Typically, each 2-m core was retrieved with a full, continuous cross-section and two or three horizontal breaks. Cores from depths of 160 m or more required careful handling because they were quite brittle.

In the subsequent field season of 1988–1989, this large-diameter thermal drill was used to make a pilot hole down to a depth of 84 m for casing installation prior to deep drilling with the ISTUK electromechanical drill at a site 4.6 km south-southwest of the Law Dome summit (Morgan et al. 1994). When the fiberglass casing was set up, thermal drilling was continued with a 140-mm drill to a depth of 96 m to make a suitable pilot-hole for the electromechanical drill.

In November 2005, a 258-m-deep ice core was drilled out using the Errol thermal drill 0.27 km south of Dome Summit South (DSS) on Law Dome (Burn-Nunes et al. 2011). Drilling was conducted in two-shift mode with break of almost 7 h (Fig. 2.32).

2.2.6 JARE Thermal Drills

The Japanese Antarctic Research Expedition (JARE) thermal drill was also based on the concept of the CRREL drill but was totally redesigned in order to be more lightweight and portable (Suzuki 1976). This project was started in May 1969. The first version of the JARE-140 drill (for ~140-mm-diameter holes) was 2.5 m in length, weighed only 30 kg, and was capable of extracting 1-m-long cores

(Suzuki and Takizawa 1978). A subsequent enhanced version JARE-160 (for ~170-mm-diameter holes; at first, the hole diameter was planned to be ~160 mm, which led to the final misleading name) designed for extracting 1.5-m-long cores was 3.4 m in length and weighed 50 kg.

Armored cable was fixed to the termination of the inner cylinder with a screw clamp instead of via epoxy cementing, as done in the CRREL drill (Fig. 2.33). It is possible that this was the reason that cable got disconnected easily from the drill during field testing. Three values of acceptable cable tension in the range of 140–180 N (small, normal, and large) were indicated on the control desk by signals coming from a microswitch in the cable termination.

A diaphragm-type 20-W vacuum pump (IWAKI AP 220) capable of producing a vacuum of −60 kPa was used. However, the diaphragm was often stiff, making it difficult to start the pump; in later versions, a heater was added to warm it. The water tank was made of a 2.1-mm-thick stainless-steel pipe with an OD of 114.3 mm in JARE-140 and with an OD of 139.8 mm in JARE-160. A simple buoy water gauge was mounted in the tank. The tank was covered with foam plastic and was inserted in a 4.5-mm-thick fiberglass reinforced plastic (FRP) pipe with an OD of either 139 mm (JARE-140) or 159 mm (JARE-160).

The core barrel of the JARE-140 was an extension of the FRP pipe used as water tank casing. A core-breaking ring was glued to its lower end. The turning of a special catcher-releaser could be done to disengage all catchers so that the core could be easily taken out. The core barrel of the JARE-160 drill was a 2.1-mm-thick stainless-steel pipe with an OD of 139.8 mm, core catchers welded at the lower end, and a flange at the upper end. The barrel had no

Fig. 2.32 Core removal from an Errol thermal drill at a depth of nearly 250 m, Low Dome, November 2005; the hole closure was likely caused by the problems with meltwater removal during the previous coring run (*Credit* A. M. Smith)

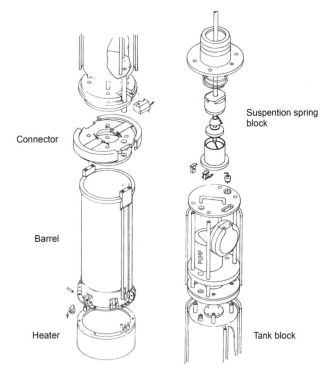

Fig. 2.33 Schematic of the JARE-160 thermal drill (Suzuki 1976)

catcher-releaser. The cores were taken out by detaching the barrel from the water tank. The thermal head had two twisted tubular heaters with diameters of 8 or 9 mm with a stainless-steel sheath molded in an aluminum-alloy annulus (Fig. 2.34). Heating elements provided clearance for vertical water holes to be machined after molding.

An improved thermal head, made in late 1974, had a spiral heating element with one end of resistive wire earthed to the sheath so that it had only one apparent electric connector. Two stainless-steel tubes with an ID of 1.5 mm were molded in the new head for making water holes instead of machining them afterwards. Several horizontal holes were drilled near the bottom for air circulation and a few round grooves were machined on the bottom face to be used as water channels. The ID and OD of the JARE-140 thermal head were 105 and 142 mm with a power of 2.4 kW (100 V), whereas they were 134 and 168 mm in the JARE-160 thermal head with a power of 3 kW (200 V).

The capability of the winch was 400 m for a 12.8-mm-diameter cable having seven signal lines and one power line (winch capacity was later increased to 800 m for 11.8-mm-diameter cable). The power drive used, which had a 1.5-kW three-phase motor, was similar for both winch systems (Fig. 2.35).

All holes were drilled at Mizuho Station, Enderby Land, East Antarctica, located 270 km southeast of Syowa Station (ice thickness was estimated to be 2095 m). During the overwintering of the JARE-12 expedition, the drilling set was installed on the floor of 4-m-deep snow trench. The drilling of the first hole started in November 1971 and proceeded rather smoothly with a penetration rate of up to 1.8 m/h. However, during hard core breaking or drill sticking at a depth of 71 m, the cable slipped from the

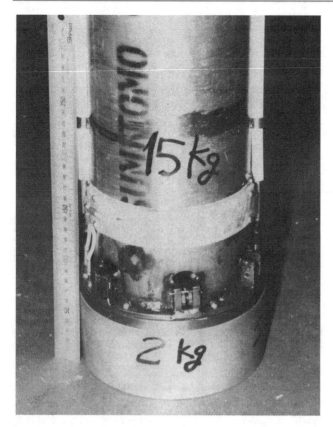

Fig. 2.34 Bottom side of the JARE-160 thermal drill (Suzuki 1976)

termination and the drill was lost. The retrieved cores had good quality; however, cores from down to 26 m were considerably infiltrated with meltwater.

A drill, called JARE-140 Mk II, which was a modified version of JARE-140 for easy disassembling, was used in the subsequent JARE-13 expedition. The winch was moved several meters away and drilling was started in July 1972. Despite many problems, a depth of 104.5 m was reached in one and a half months, time at which the drill got stuck. A hand-made container with a valve at the bottom was used

to pour antifreeze to the bottom of the hole. In total, 50 L of hot antifreeze had been poured when the drill was finally recovered, with severe damage done to the pump. Drilling was restarted in November. However, after reaching a depth of 147.5 m, the drill got stuck again, probably because meltwater had leaked from the tank. The cable was pulled with a lever block, resulting in the breakage of the drill between the pump housing and the water tank. Drilling was then terminated. The average penetration rate achieved with the JARE-I40 Mk II drill was 1.3 m/h in firn and 1.1 m/h in solid ice.

In the few years afterwards, a new thermal drill based on a modification of the JARE-160 drill and a large-diameter Type-300 drill with a head with an ID and an OD of 252 and 285 mm, respectively, were designed. Both drills were tested at ice island T-3 in the Arctic in November 1973, and both reached the island's bottom at depths of 31 and 30 m, respectively.

The new 800-m winch was installed in a 5.2-m-deep trench ~50 m north from the first trench in August 1974 (JARE-15) and, at the beginning of December, a third hole was started. To prevent severe imbalance of the generator output, a rectifier unit to convert 200-V three-phase AC into 0–230 V DC power was installed to supply the drill. The hole advanced very slowly (~3.3 m/day, working 6–9 h/day). This slow advance could partly be the result of the inadequate design of the drill, which required the detaching of the core barrel to take out each core. It was troublesome and time-consuming to reattach the barrel to the swinging drill.

When the hole reached 142.6 m, a sudden stop of the winch during drill lowering caused the cable to slip off the termination and the drill was lost (it is most likely that the screw clamp in the cable termination was loosened). It was decided to deviate the hole, and a new drill was attached to the cable. The drill was lowered to the narrowing interval at 106.7 m and slowly advanced with a penetration rate of 1.6 m/h with the thermal head operating at half of its rated

Fig. 2.35 Schematic of the JARE-15 winch system with an 800-m-long cable (Suzuki 1976)

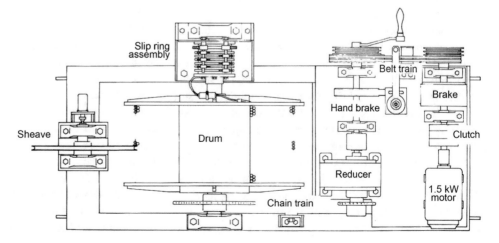

power (~ 1.5 kW). No core was recovered in this run. In the next run, the first crescent-cross-section core was recovered. In subsequent runs, the cross-section of the recovered cores became larger and larger and finally, at a depth of 124.7 m, a completely cylindrical core that was 0.8 m long was recovered. However, between 135.8 and 137.1 m, the cores lacked up to 10% of the area of a circle, showing that the two holes intersected.

At a depth of 145.7 m, the rectifier burned out and the power supply was changed to AC. In the first AC-powered run, the drill advanced very slowly at 0.75 m/h and a 1.18-m-long narrow core was recovered. The penetration rate in the next run was even slower at 0.5 m/h and, after 0.5 m of penetration, the operators decided to pull up. The drill was stopped for 3 min to cause the recession of the core for easy breakage, but then it was found that the drill was stuck. A total of 60 L of hot antifreeze were poured into the hole through a 140-m-long rubber hose. Several pulling efforts were made without success. At the same time, a fire broke out in the generator room, ruining the generator. With little time left in the summer season, the drilling operations were terminated.

In the beginning of 1980s, thermal drilling at Mizuho Station was restarted within the Japanese Glaciological Research Program in East Queen Maud Land (Narita et al. 1994). The thermal drill system was considerably modified by Geo-Tecs Co. Ltd. in order to provide better drilling monitoring and control and to increase the mechanical strength of the drill's parts and connectors. Two types of drills were designed: a 2.8-m-long ID-140 drill with a thermal head with an ID and an OD of 128 and 145 mm (2.5 kW), respectively, weighing 40 kg, and a 3.5-m-long ID-160 drill with a thermal head with an ID and an OD of 132 and 168 mm (3 kW), respectively, weighing 60 kg. A linear-type A-1017 vacuum pump capable of providing an air pressure of −40 kPa was used. The pump was able to remove meltwater with a flow of at least 1.2 L/min.

The winch driving system was equipped with an auxiliary motor for slow feeding at ~ 3.6 m/h during drilling. This motor and the powered brake were automatically and alternately activated depending on the signal coming from the drill load sensor. The operator could monitor the temperature of the thermal head and the water tank, the drill load, the input voltages of the thermal head and vacuum pump, the fullness of the core barrel (limit switch), and the fullness of the water tank (float switch). A downhole computer continuously sent these data to the surface. The optimum drill load was found to be in the range of 135–235 N.

These new thermal drills were first tested on ice blocks in Japan and on the ice cap at Ellesmer Island, Canadian Arctic Archipelago and were then transported to Antarctica. A drilling hut measuring 5 m × 5 m × 5 m was constructed at Mizuho Station by personnel of the JARE-23

Fig. 2.36 Top view of the drill set at Mizuho Station, 1983 (Deep Ice-Coring Project at Dome Fuji, East Antarctica 1992)

expedition (1982). The hut was warmed up with heat from the generator cooling system, which was located in the neighboring room, keeping the temperature in the hut near −10 °C. The staff of the JARE-24 expedition began drilling in April 1983 (Fig. 2.36). At a depth of 358 m, drill jamming started and, at a depth of 411.7 m, deep drilling was terminated exactly three months after the beginning of the drilling operations because of the risk of the drill getting stuck and be lost in the narrowing intervals of the hole (the estimated closure rate was ~ 0.33 mm/day at a depth of 400 m). The average rate of penetration was ~ 1.25 m/h and the average tripping speed was 0.7 m/s. Actual drilling took 54% of the total time spent at the site, tripping took 22%, and surface servicing, including equipment repairing, took 24%.

Caliper measurements made in March 1984, eight months after drilling termination, showed that borehole diameter decreased by 1.0, 10.8, and 27.6 mm at depths of 200, 300, and 400 m, respectively. Thermal and electromechanical reaming was attempted without much effect, and a new deviated hole was started at a depth of 133.5 m. In order to speed up drilling from a depth of 200 m onwards, operations were switched to two-shift mode and 16 h/day. The distance of the meltwater inlet holes in the thermal head was raised from 4 mm to 14 mm from the end of the head in the hope that increasing the water level at the bottom would increase the melting hole diameter.

Because of frequent occurrences of jamming, the TD-160 drill was swapped with the TD-140 drill at a depth of 625 m, and finally on August 1, 1984 the hole had reached 700.63 m when the entire length (~ 730 m) of the armored cable had been used. Drilling took a total of 1014 h (approximately two months); the average penetration rate was 1.55 m/h. Ice core recovery was almost 100%. The core quality obtained from the surface down to a depth of 96 m was perfect; however, cores recovered from greater depths contained many cracks. In the interval from depths of 280 to 540 m, an almost uniform temperature ranging from −35.2 to −35.5 °C was measured (Naruse et al. 1988).

2.2.7 TELGA Thermal Drills

First prototypes

The first concept of a cable-suspended electric-powered thermal drill with an annular heating head capable of producing a core and having a pumping system to remove meltwater and store it in a chamber within the drill was proposed by N.I. Barkov of the AARI (Fig. 2.37) (Barkov 1960). During August–November 1961, a drill prototype with a thermal head consisting of two turns of a single 3-kW tubular electric heater was tested at Mirny Station, Antarctica (Barkov 1963). The drill produced a 150-mm-diameter hole and a 75-mm-diameter core at a penetration rate of 7.5 m/h in a dry hole (Morev 1966).

Tests at Mirny Station were continued in December 1963 with an end plate heater at the lower end instead of an annular thermal element. Meltwater was removed using a centrifugal BTsN-1 pump and stored in a 22-L water collection tank. The power consumed by the drill was 4.5 kW.

The diameter of the hole was 150 mm, the rate of penetration was 2.2 m/h, and the length of a run was 1.3 m. Drilling was discontinued because of a breakdown of the hydraulic winch.

In the summer season of 1966, a new ETBLK-1M thermal coring drill, designed and built by the Moscow Mining Institute and the Moscow Institute of Radioelectronics and Mining Electromechanics, was tested again at Mirny Station (Sekurov 1967). The drilling complex included a wooden shelter (3.45 m × 5.75 m) with a mast mounted on metal sledges. The winch, which had a 26-mm-diameter electromechanical rubber-covered cable (with 6 power lines and 14 signal lines), was driven through a reduction gear and a DC motor (8 kW) that was regulated by a converter controller.

The drill was 5.25 m long and weighed ~140 kg. It consisted of three sections: a 25-L heated water tank with a float-type switch for the thermal head in case the tank was full; a pump with a maximal suction pressure of −25 kPa; and a 1.2-m-long hollow transformer with an ID of 100 mm that served as a core barrel. A few sucking tubes were installed in the transformer, which was terminated by the thermal head with core catchers.

The thermal head comprised eight parabolic low-resistance segments inserted into annular isolator that was attached to two concentric current-conducting cylinders (Fig. 2.38) (Sekurov et al. 1965). In order to increase resistance and to ensure uniform heating, the segments had a labyrinthine slot system. The segments were fixed using screws to the current-conducting cylinders. The resistance of each segment was 60 mΩ. The thermal head was able to dissipate power up to 8 kW at a feed voltage of 7.6 V.

Three dry holes were drilled, with the deepest one reaching bedrock at 77 m. Hole diameters were ~150 mm and core diameters were ~90 mm. A total of 101 m were drilled. The penetration rate was in the range of 2.9–3.6 m/h at a thermal head power of 5.0–5.5 kW. Surface servicing took 5–10 min, and tripping speed did not exceed 0.25 m/s.

TELGA-2 thermal drill

A new cooperative venture between the AARI and the LMI was established in 1967, resulting in the development of the

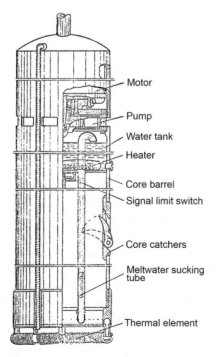

Fig. 2.37 Barkov's thermal drill (1960)

Motor

Pump

Water tank

Heater

Core barrel

Signal limit switch

Core catchers

Meltwater sucking tube

Thermal element

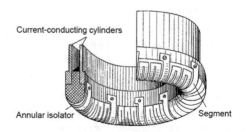

Current-conducting cylinders

Annular isolator

Segment

Fig. 2.38 Thermal head with low-resistivity segments (Sekurov et al. 1965)

new type of thermal drill TELGA-2 (Russian abbreviation for "Electrothermal drill constructed by the LMI and the AARI"; the number 2 is the modernization number) (Kudryashov et al. 1973a; Morev and Shamont'ev 1970; Morev 1972). Water filled the bottom part of the hole to a height of 8–14 cm above the bottom, from where it was pumped into a water tank. The drill had a heated head (3.04 kW) with an ID and an OD of 90 and 150 mm, respectively.

Testing of the TELGA-2 thermal drill was carried out during the wintering of the 13th SAE (1967–1968) at Mirny Station and along the Mirny–Vostok route. A total of six holes with a cumulative length of 386.5 m were drilled. The major part of the holes (306.5 m) was drilled using one thermal head. The deepest hole was 212.5 m deep and was drilled at the 26-km mark on the Mirny–Vostok route in August–September 1968. The drilling of this hole was terminated because of steady drill sticking in the hole and the possibility of losing it. The rate of penetration was 1.37 m/h at a depth of 117 m and 1.65 m/h at a depth of 190 m. Core length varied in the range of 1.5–2.0 m. The temperature at a depth of 210 m was measured at −24 °C.

TELGA-3 (TELGA-14) thermal drill

To decrease temperature losses and inefficient borehole enlarging due to melted water, the drill was modified (TELGA-3) in such a manner that meltwater was removed almost immediately from the bottom of the hole, and the distance between the thermal head body and the hole walls was enlarged using a ring on the lower end of the head (Fig. 2.39). This thermal drill head, which had an ID and an OD of 130 mm and 178 mm, respectively, consisted of a copper body which contained a 1.1-mm-diameter nichrome wire element insulated by ceramic beads (Kudryashov et al. 1973a). The wire resistance of the thermal head was 15 Ω.

The melted water was removed through six pipes fixed on the outer surface of the core barrel. The upper ends of the pipes were fixed on the connector, which allowed for air flow into the inner pipe of the water tank. The connector had a port to drain water after the drill was raised. One or a pair of two-stage vacuum pumps connected in series were located in the upper part of the tank. The drill's length and weight were 8 m and 200 kg.

The first tests of the TELGA-3 drill were conducted during the 14th SAE at Mirny station and on the Mirny–Vostok route. The wooden drilling shelter measured 8 m 4 m × 2.5 m and had a 9-m-high mast installed on steel sledges. The drill was suspended from an 18-mm-diameter cable reeled on a conventional SKP-1500 logging winch.

Two test holes were drilled: one at the Mirny meteorological grounds (34.5 m deep) and another one near the

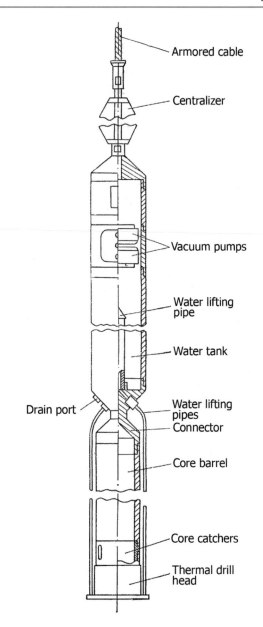

Fig. 2.39 TELGA-3(14) thermal drill (Kudryashov 1989)

runway (34.7 m deep). Then, the drilling shelter was transported to the 50-km mark of the Mirny–Vostok route where, in October–November of 1969, a dry test hole that was 250 m deep was drilled over 36 days. Drilling was organized in one shift and roughly five runs for retrieving cores almost 2 m long were conducted each day. The power of the drill head was adjusted in the range of 2–5 kW, with the optimal range being between 3 and 3.5 kW. It was found that one vacuum pump was enough to remove the melted water. The temperature of the drilled ice was between −16 and −21 °C. The water tank was not heated and the temperature of the melted water drained at the surface was 6–14 °C. The penetration rate in ice was 1.6–1.8 m/h with

10–11 L of meltwater recovered per meter and the cores had a diameter of 124–129 mm.

Afterwards, the TELGA-3 thermal drill was transported from Mirny to Vostok Station and was renamed to TELGA-14 (the number 14 means that the drill was modified by members of the 14th SAE). Then, after enlarging the core barrel to 2.5 m and some other modifications made in the 17th SAE, the letter "M" was added to the name, making it TELGA-14M.

Vostok Station: Hole No. 1
15th SAE (1970)

In the summer season of the 15th SAE, a drilling shelter was constructed at Vostok on two steel sleds (Fig. 2.40), providing a work area measuring 15 m × 2.9 m × 2.5 m (Barkov 1970; Barkov et al. 1973). An 11-m-high mast, measured from the shelter floor and adapted from a conventional URB-150A drill rig with an additional section, was constructed in the center of the work area and was surrounded by a round framework with a height of 9.5 m above the shelter roof. It was wrapped with rubberized cloth for protection (Kudryashov et al. 1973a, b, 1983; Korotkevich and Kudryashov 1976; Kudryashov, 1989). All the necessary equipment, namely an SKP-1500 winch with 700 m of 18-mm-diameter KGTN-10 cable, a 7.5-kW driven motor, a control box, and auxiliary equipment, were installed inside the drilling shelter (Fig. 2.41).

Drilling operations with the TELGA-14 thermal drill began in April 1970 and a dry Hole No. 1 reached a depth of 506.9 m by September 1970. In total, 293 drilling runs were completed. Drilling was conducted on a one-shift basis with two to four runs in one day depending on hole depth. At the surface, the voltage applied was 370–380 V, resulting in 240–260 V applied to the drill with an operating current of 14–17 A. The penetration rate in ice was 1.34 m/h with 13 L of meltwater recovered per meter. The average core length was 1.73 m. The measured diameter of the hole varied in the range of 180–183 mm. The temperature in the hole decreased from −57 °C at a depth of 20 m to −53.5 °C at 500 m.

The stator and rotor windings of the motor of the vacuum pump were additionally saturated with K-55 type silicone lacquer. The thermal head was sealed hermetically with KLT-30 heat-resistant paste, resulting in its lifetime increasing to 183.5 m on average. The number of sucking pipes was increased from six to seven.

To avoid meltwater infiltration into the snow/firn core, the sucking pipes were connected to the lower ring on the drill head and air was pulled from under the borehole bottom from snow/firn. Below 70 m, when air circulation became difficult, two 4-mm-diameter holes were drilled for each sucking pipe at a distance of 15 mm from the bottom. To prevent water from freezing in the sucking pipes, 200-W PEN heating wires were fixed along the pipes. From a depth of 270 m onwards, the power of the heating wires was increased to 500 W. During hoisting operations, the heating wires worked at full capacity, whereas the thermal head operated at reduced capacity (5 A). The water tank was also equipped with a heater so that water temperature was maintained in the range of 5–10 °C.

The most serious difficulty encountered during drilling was the contamination of the hole by impregnated graphite escaping from beneath the cable armor. The graphite settled

Fig. 2.40 The first drilling shelter at Vostok station under construction in the summer season of the 15th SAE, 1970 (Vasiliev et al. 2007)

Fig. 2.41 Schematic diagram of the first drilling shelter at Vostok station: (1) electrical winch board; (2) control panel; (3) window aperture; (4) oil furnace; (5) auxiliary hoist; (6) rig; (7) tower rings; (8) drill positioned over the hole; (9) supporting framework; (10) bench and instruments; (11) fire-prevention apparatus; (12) steel sled; (13) hoist motor; (14) two-speed gear reducer; (15) main hoist; (16) cable; (17) generator–motor system; (18) lathe; (19) collapsible balance block; (20) drill in the hole; (21) conductor (Kudryashov et al. 1973b)

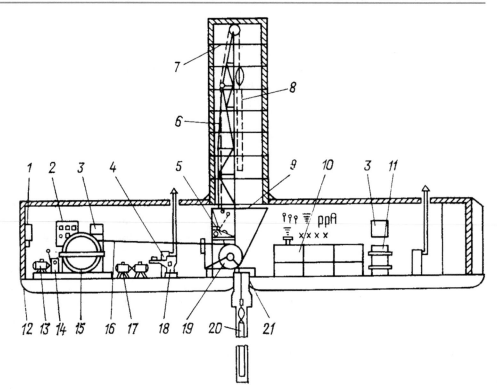

to the bottom of the hole and impaired the heat transfer from the heating head to the ice, reducing the penetration rate by 20–30%. A special conically shaped 3-kW heating head was used to melt a depression at the bottom of the hole, into which the graphite settled (Fig. 2.42). After the meltwater had frozen, the ice containing the contaminant was removed as a part of a core. The contaminant also coated the walls of the water suction tubes and impaired the operation of the pump; both the suction tubes and the pump had to be periodically cleaned.

During drilling two serious incidents happened (Fisenko et al. 1974). The first occurred at a depth of 15.7 m owing to a switch malfunction. A section of the drill broke off and fell down the hole. It was retrieved using an adapter on the end of a column of rods, which was threaded into the broken section. Another incident happened at a depth of 497.4 m when a 20-cm steel rod was accidentally dropped to the bottom of the hole. First, a large heater was used to melt out the rod from the bottom and then a small heater was used to melt a small central hole, which repositioned the rod from an inclined to a more vertical position (Fig. 2.43). In this position, the rod would fit within the annular heating head of the TELGA drill. It was then recovered atop the next core. This operation took 14 days.

16th SAE (1971)

The drilling of Hole No. 1 was continued. At the beginning of winter, because of carelessness, the TELGA-14 drill, unattached to a cable, was dropped to the bottom of the hole.

Fig. 2.42 Assembly with conical thermal head (Fisenko et al. 1974)

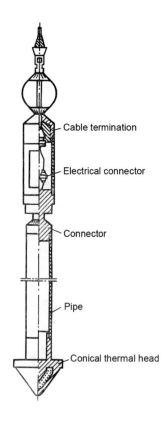

It was recovered using special grips (Fisenko et al. 1974). Later, at a depth of 560 m, as the drill was being hoisted, it became stuck near the bottom of the hole, probably because of the inadequate removal of meltwater. A whipstock that

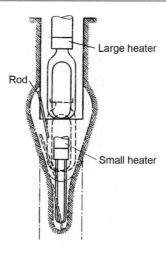

Fig. 2.43 Schematic of steel rod recovery operations (Fisenko et al. 1974) *16th SAE (1971)*

was 4.5 m long was set at a depth of 505 m, and directional drilling began using a short assembly (total length of 2 m) with a conical drill head.

A new offset hole was started from a depth of 518 m (Fig. 2.44). To ensure the free passage of the longer thermal drill, the deviated interval of the hole was enlarged, but unfortunately the old hole was reopened in the process. From that time on, the drill sometimes entered the new hole and sometimes the old one. At a depth of 534.4 m, the drilling of the offset hole was stopped. The bottom part of the hole was filled with ice up to a depth of 471 m and a new directional drilling effort was started without using a whipstock. The deviation was accomplished using a thermal drill with a special asymmetrical thermal head. By the end of winter, the depth of Hole No. 1 had reached 625.2 m.

17th SAE (1972)

The drilling of Hole No. 1 was continued using the modified TELGA-14M thermal drill (Fig. 2.45) (Kudryashov et al.

1973b, 1977a; Korotkevich and Kudryashov 1976). From a depth of 920 m onwards, drilling was conducted using a spliced cable because the necessary length of cable was not available at the station. Lowering and hoisting was accompanied by repetitive jerking and sticking in the narrow parts of the hole (occurring as much as 30 times during one run). Nevertheless, the hole reached 952.5 m in May 1972, making it the deepest dry hole in ice. In the final run, while the drill was being lowered to the bottom of the hole, the winch-brake mechanism failed (Fisenko et al. 1974; Ueda and Talalay 2007). The drill dropped to the bottom of the hole followed by the cable, which eventually became detached from the clamp holding it to the winch drum and fell into the hole. The drill and cable were eventually left in the hole. The average rate of penetration was 1.2 m/h at a power of 2.9 kW transmitted to the drill head; the average core length was 2.3 m (Kudryashov et al. 1991).

A new offset hole was then started. A cylindrical piece of ice was dropped from the surface into the hole, but unfortunately it got stuck at a depth of 30 m and efforts to dislodge it failed (Fisenko et al. 1974). From this depth, a new offset hole, called Hole No. 1-bis, was started and, by the end of the year, it reached a depth of 774 m.

18th SAE (1973)

Drilling in Hole No. 1-bis was continued through the depth interval of 774–780.2 m, at which point the hole became inaccessible owing to an incident during hoisting. The

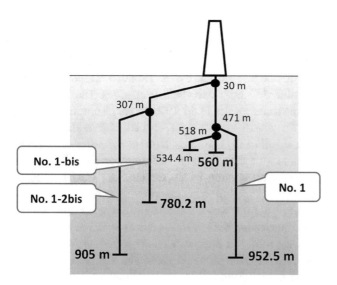

Fig. 2.44 Schematic of the deepest dry hole made in ice, Hole No.1, with several deviations at Vostok station (1970–1973)

Fig. 2.45 Winch operations at Hole No. 1, Vostok station, 17th SAE, 1972 (*Credit* V. Chistyakov)

thermal head got stuck near the bottom and was separated from the drill at a depth of 762 m. The drill itself then became stuck at a depth of 308 m (Vartykian et al. 1977). The main reason for the incident was the inexperience of the new team of the 18th SAE, who did not recognize problems with the pumping of meltwater at the bottom of the hole during one of the first runs.

A new offset hole, called Hole No. 1–2bis, was started at a depth of 307 m in Hole No. 1-bis and reached a depth of 905 m (Kudryashov et al. 1977a, 1991). Further drilling in this hole was stopped owing to the high risk of the drill getting stuck. Hole No. 1–2bis was available for several years for conducting observations of temperature, deviation from the vertical, and wall deformation (Dmitriev et al. 1978; Dmitriev and Vostretsov 1979).

TELGA-17 thermal drill

TELGA-17 thermal drill was a modified version of the TELGA-14M drill, made in order to determine the in situ orientation of core (Kudryashov et al. 1975, 1977b). An orientation sensor was fixed to the water tank and it was made out of nonmagnetic material. In the upper part of the drill body, a horizontal bolt was screwed and it protruded from the inner side. During drilling, this bolt melted a groove on the core surface. During the wintering of the 17th SAE (1972), three oriented cores were recovered at depth intervals of 734–735 m, 736–738 m, and 770–771 m from Hole No. 1-bis.

Vostok Station: Holes No. 2, 3G, 4G, and 5G

The TELGA-14M thermal drill had a very simple and reliable design and became the workhorse of the shallow drilling operations at Vostok station and in Antarctic scientific traverses. To drill through permeable snow/firn at Vostok, this drill was used for the pilot-hole drilling of Hole No. 2 (16th SAE, 1971, 108 m) and for deep holes 3G (25th SAE, 1980, 112 m), 4G (28th SAE, 1983, 120 m), and 5G (35th SAE, 1990, 119.7 m, Fig. 2.46) (Kudryashov et al. 1991, 1998a). Once these were done, the boreholes were filled with liquid and a TBZS-152M thermal drill was used to advance (see Sect. 2.3.2).

In addition, the TBZS drill lost in Hole 4G was bypassed with an offset hole, called Hole 4G-1, started at 120 m using a TELGA-14M thermal drill. It was continued to a depth of 167 m, and then drilling was again carried out using a fluid-filled TBZS-152M thermal drill (Kudryashov et al. 1991). The average drilling parameters achieved using the TELGA-14M drill were: a rate of penetration of 1.3–1.8 m/h, a core length of 1.9–2.0 m, and a core diameter of 122 ± 2 mm.

Fig. 2.46 Author (on the left) and N.I. Vasiliev with one of the first cores recovered with the TELGA-14M thermal drill in Hole 5G, Vostok Station, 35th SAE, February 1990

Antarctic scientific traverses

The TELGA-14M drill was used multiple times in Antarctic scientific traverses. For this purpose, a series of movable drilling shelters called PBUs (Russian abbreviation for "movable drill rig") were built (Fig. 2.47). The first shelter was designed and built in 1972 (Kudryashov et al. 1991) and, for almost two decades, a total of five shelters with slight modifications were constructed for different projects. The names of these shelters had numbers according to the order of their coming into operation, i.e., PBU-1, PBU-2, etc.

Each PBU hut was installed on two steel sledges and its inner dimensions were 9.0 m × 3.3 m × 2.5 m (Fig. 2.48). The wall, roof, and floor were composed of sandwich panels with outer Bakelite plywood plates with a thickness of 5 mm and an inner 100-mm-thick thermal insulator (foamed plastic). The basic configuration of the PBU hut consisted in

Fig. 2.47 Movable drilling shelter PBU-2 at temporary base Gornaya, at the 73-km mark of the Mirny–Vostok route, 24th SAE, 1979 (*Credit* N. E. Bobin)

Fig. 2.48 Schematic of a PBU movable drilling shelter: (1) fuel tank; (2) diesel generator; (3) sleeping berths; (4) heater; (5) tables; (6) winch; (7) electric box; (8) control box; (9) legs of the mast; (10) thermal drill for movement; (11) borehole mouth; (12) gas cylinders; (13) workbench; (14) drill–press; (15) welding set (Kudryashov et al. 1991)

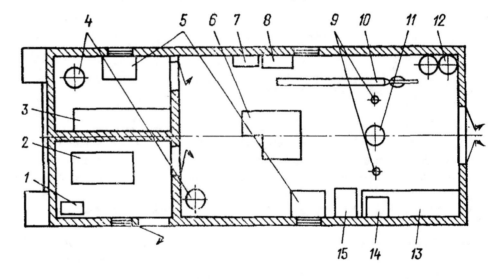

three rooms: the drill rig itself, a living room for two persons, and a generator room. Above the mouth of the borehole, a pyramidal tower with a height of 7.0 m was mounted. PBU-1 was equipped with a SKP-900 winch with a double-speed gear and a 5-kW driven motor, an E8R diesel generator capable of generating 8 kW of electric power at sea level (later replaced by a 16-kW generator), a control box, and other equipment.

The TELGA-14M thermal drill was shortened by using a 1.5-m-long core barrel (instead of a 2–2.5-m long one) and a shorter water tank capable of holding 30 L of melted water (Bobin and Fisenko 1974; Kovalenko et al. 1981; Bobin

et al. 1986). The first traverse on the Mirny–Vostok route was conducted during the summer of the 17th SAE (1972). The unit drilled down to 50.5 m at the 57-km mark and 56.6 m at the 153-km mark. Drilling rates varied from 2.0–3.75 m/h depending on snow/firn density, and core diameters were 123–125 mm.

Over the years, 19 boreholes were drilled in the Mirny–Vostok and Mirny–Pionerskaya Dome C traverses using movable PBU shelters (Table 2.3). The cumulative length of the cores retrieved was almost 4500 m.

The longest traverse took place from January 19 to March 19, 1981. At the 1045-km mark on the Mirny–Pionerskaya–

Table 2.3 LMI drilling on the Mirny–Vostok and Mirny–Dome C traverses (Kudryashov et al. 1991; Ueda and Talalay 2007)

SAE #	Summer season	Site	Depth/m	Type of drill	Drilling time/days
17	1971–1972	57 km	50.5	TELGA-14M	3
		153 km	56.5	TELGA-14M	4
18	1972–1973	353 km	76	TELGA-14M	7
20	1974–1975	Vostok-1 (647 km)	105.3	TELGA-14M	7
21	1975–1976	Vostok-1 (647 km)	0–181	TELGA-14M	10
22	1976–1977	Vostok-1 (647 km)	181–304	TELGA-14M	12
23	1977-1978	Vostok-1 (647 km)	304–430	TELGA-14M	15
24	1979*	Base Gornaya (73 km)	0–430.3	TELGA-14M	36
			430.3–721	TBZS-152M	43
			500–750	TBZS-152M	
24	1978–1979	Pionerskaya (375 km)	137.5	TELGA-14M	14
25	1979–1980	Pionerskaya (375 km)	375	TELGA-14M	21
26	1980–1981	1045 km on the Mirny–Dome C route	306	TELGA-14M	14
33	1987–1988	105 km	0–527	TELGA-14M	35
			527–740	TBZS-152M	28
34	1988–1989	60 km	150.5	TELGA-14M	10
		140 km	150	TELGA-14M	8
		200 km	150.8	TELGA-14M	11
		Vostok Station	180	TELGA-14M	21
35	1989–1990	Vostok Station	138	TELGA-14M	11
		400 km	150	TELGA-14M	7
		325 km	150	TELGA-14M	7
		260 km	150	TELGA-14M	7

Note *—Drilling was conducted during the winter

Dome C route, a 306-m hole was drilled using a TELGA-14 M drill. Drilling took 14 days, with an average ice core production rate of 21.9 m/day. On the way back on March 16, drilling shelter PBU-4 was destroyed by fire, and, tragically, one crew member died (Kudryashov et al. 1991; Savatyugin and Preobrazhenskaya 2000).

Microbiological drilling

One of the PBU drilling shelters was modified for microbiological studies. Besides the drilling facilities, it contained a laboratory section for the microbiological sampling of ice cores (Kudryashov et al. 1978; Bobin et al. 1994, 1998). In November 1974, this microbiological shelter was transported from Mirny to Vostok and set at a site 400 m to the west of Vostok Station (Fig. 2.49). Over four seasons, a special hole MB-1 for microbiological investigations was drilled using a TELGA-14M thermal drill in the following depth intervals: 0–105.1 m (February 1975), 105.1–206.7 m (January–February 1976), 206.7–312 m (February 1977), and 312–320.8 m (December 1979–February 1980). The drill was sterilized in an ethanol bath at the beginning of every season

and then using a flame burner before each run. The mean rate of penetration was 1.57 m/h, the mean core length was 1.3 m, and the core diameter was 125 mm. A total of 169 core segments were analyzed (Abyzov et al. 1988, 2007).

Severnaya Zemlya

The TELGA-14M thermal drill was modified so that the outer diameter of the drill head was reduced to 152 mm; this drill was referred to as TELGA-152. It was used for drilling several holes at Kupol Vavilova station in the northern part of the Vavilov ice cap, Severnaya Zemlya. In September 1981, a 52-m-deep hole was drilled using TELGA-152 to sample the carbon dioxide of the air bubbles trapped in the ice (Zemtsov and Men'shikov 1988; Chistyakov et al. 1988). During April–June of 1983, two other dry holes were drilled down to depths of 200 and 100 m for testing the carbon analysis sampler. In the course of drilling, the following average parameters were obtained: a penetration rate of 2.7 m/h, a core length of 2.1 m, and a consumed power of 3.4 kW. In 1984, 1986, and 1988, TELGA-152 was used to drill through permeable and crumbling ice to the depths of

Fig. 2.49 General view of the microbiological drilling shelter at Vostok Station (*Credit* S. Abyzov)

30–37 m for further testing of the KEMS-112 electromechanical drill (Kudryashov et al. 1994).

2.2.8   LGGE Thermal Drill

The thermal drill designed in the LGGE had an OD of 130 mm and could work in cold glaciers as well as in temperate glaciers, where the holes are full of water (Gillet et al. 1976). In the latter cases, the suction assembly for meltwater was removed. The drill was then 5.9 m long and weighed 135 kg. For cold ice, on the other hand, its total length was 8.2 m and its total weight was 170 kg. The armored cable was attached using Araldite epoxy resin to a piston that was spring-linked to a linear potentiometer for drill load monitoring (Fig. 2.50). A 600-W heating resistance at the top of the termination could be used in case of jamming when the drill was being pulled up. A switch at the top enabled the drill to change over to this resistance from the thermal head. A three-phase oil-cooled transformer rated at 6.8 kVA rating converted 338-V primary power to 45-V secondary power.

A Marion membrane vacuum pump was used to create a vacuum up to −93 kPa and was driven by a single-phase 220-V 120-W electric motor. Meltwater was sucked through three stainless-steel tubes with an ID of 4 mm into a water tank. To prevent the water from freezing on the way up, the tubes were heated from the inside by a 1.25-mm-diameter wire with a heating power of 30 W/m. The heating resistances in the tubes and in the water tank were mounted in series, and thus the small variable transformer at the surface regulated the heating of the whole system.

Fig. 2.50 LGGE thermal drill (Gillet et al. 1976)

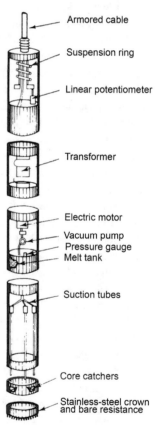

Armored cable

Suspension ring

Linear potentiometer

Transformer

Electric motor

Vacuum pump
Pressure gauge
Melt tank

Suction tubes

Core catchers

Stainless-steel crown and bare resistance

The water tank was made from a 1.75-long stainless-steel tube with an ID and an OD of 125 and 129 mm, respectively. The volume of the tank corresponded to the amount of water obtained after drilling a 2.8-m-long core with a diameter of 102–104 mm and a hole with a diameter of 140 mm. The proper functioning of the suction system could be checked with a vacuum gauge. A level indicator was used to control the fullness of the tank.

A 2.8-m long core barrel with a two-ply structure with two concentric pipes was used: the inner polyethylene tube had an ID and an OD of 110 and 114 mm, respectively, and those of the outer stainless-steel tube were 125 and 129 mm. Between them, the suction tubes for the meltwater and the electric lines for the thermal head were fixed. A stainless-steel crown (ID/OD: 110 mm/130 mm) was mounted on the lower part of the drill. It was tooled at the bottom to allow for a porcelain ring that housed the bare resistance to be attached. Three core catchers were used for breaking the core.

The thermal head was made in the form of an open bare resistance with a diameter of 1.3–1.6 mm and having enough mechanical strength to withstand the drill weight. The wire was wound into a helix and mounted on an annular resistance support. Because the resistance of the ~2.5-m-long wire was very low (2–3 Ω), it was necessary to work at low voltages with a transformer placed in the drill. Because considerable power had to be transmitted, on-line losses were reduced using three-phase current. A 500-m-long, 24-mm-diameter armored cable $(3 \times 6 \text{ mm}^2 + 9 \times 0.75 \text{ mm}^2)$ weighing 400 kg was used to suspend the drill.

During drilling and lifting, the winch was driven by a 2.2-kW motor connected via a Jaeger coupling to a gear reducer. The purpose of the coupling was to limit the torque in case of jamming and to allow for almost automatic feeding during drilling. Free-fall lowering was controlled by a foot-operated air-compressed disc brake. However, this device, after various incidents, was shown to be unsatisfactory and was replaced by a variable-speed motor. The mast was made from a duralumin pipe that was 5 mm thick and 200 mm in diameter. It was 8.8 m high and came apart in three pieces. At the top, the pulley was connected to a depth counter. The weight of the whole drilling set without packing materials, shelter, or spare parts was ~1.9 t.

The first drilling tests were carried out on the temperate Saint-Sorlin Glacier, Alps in 1968. Bedrock was reached at a depth of 67 m. In 1969, another coring operation was performed on the same glacier down to bedrock at 72 m.

In July 1971, a hole through Vallée Blanche in the Massif du Mont-Blanc was drilled to bedrock at 187 m. The penetration rate was in the range of 2.25–3.6 m/h down to 152 m. However, various problems, in particular the breaking of the drill-load potentiometer, reduced the rate at greater depths to 0.8–1.6 m/h. The power of the heating head was limited to 3.3 kW. This open-resistance head easily penetrated sand-laden ice several meters above bedrock. On several occasions, some spirals of the helix resistance were crushed and the resistance had to be changed. The quality of the retrieved ice cores was excellent.

In January 1972, the same equipment was used at Terre Adélie, Antarctica several kilometers from the coast. The drilling operations reached a depth of 44 m with a penetration rate of 4.5 m/h in ice. Various problems arose with the motor of the vacuum pump and with the generator.

In January 1974, another hole at site D10, Adélie Land, was drilled to a depth of 304 m and reached bedrock. The operation, on the whole, went smoothly and took only 16 days, including a five-day interruption at 86 m for a series of measurements and testing of a thermal reamer. The experiment with the thermal reamer showed that meltwater could not be sucked in by the vacuum pump. It run into the hole and refroze further down. Water that accumulated in this way as ice caused the drill to deviate from its initial path. Therefore, beyond 60 m and because of this deviation, a new hole was sidetracked. Although relatively high temperatures were observed below 200 m (> −11 °C), there were no problems with jamming, probably because of rapid drilling (the drilling of the interval of 200–304 m took only three days). The penetration rate at a maximum power of 4.1 kW was 4–5.5 m/h. On the whole, the quality of the retrieved cores was very good; however, at several intervals, a significant number of diagonal or horizontal fractures were found.

During the summer season of 1977–1978, one of the deepest open holes was drilled at Dome C. Drilling started on December 4, 1977; working in two shifts around the clock, it took 72 h to reach a depth of 141 m using an electromechanical shallow drill (Fig. 2.51) (Lorious and Donnou, 1978a, b). The LGGE thermal drill was then used. Drilling was stopped after 42 days at a depth of 905 m with an average core production of 21.5 m/day. This number

Fig. 2.51 Drilling shelter at Dome C in the season of 1977–1978 (Lorious and Donnou 1978a)

includes the time required to overcome technical problems, which occurred mainly at 620 m, depth at which almost no progress was made for 5 days. Although the cable was a little longer, it was impossible to drill deeper than 905 m, probably because of the closure of the dry hole. Fifteen persons spent 2000 working hours at the site. Core recovery was ~98%, but all cores had cracks. The ice temperature near the surface was −53.5 °C and, at a depth of 900 m, temperature increased to −47 °C.

In June 1980, the LGGE core thermal drill was used for drilling through Argentière Glacier in the French Alps down to 240 m, with operators working 20 h per day in two shifts for eleven running days (Hantz and Lliboutry 1983).

In 1981–1982, another dry hole was drilled down to a depth of 329 m again at Adélie Land (Donnou et al. 1984). Some modifications were made to the drill. The electrical insulation of the bare wire was applied with two types of support: a stainless-steel crown covered with a thin layer of chromium oxide and a machined ceramic piece fixed in a hollow stainless-steel crown. The OD of the drill was increased to 140 mm to make it compatible with the hole produced by the electromechanical drill used. After completion of the dry drilling operations, kerosene was poured into the hole and drilling was continued using a "wet" thermal drill down to a depth of 348 m (see Sect. 2.3.4).

2.2.9 ETB-140 (ETB-130) Thermal Drills

In the beginning of the 1980s, an electric thermal drill ETB-140 was designed in the AARI with an OD of the drill head of 140 mm for drilling dry holes in cold firn and ice. However, the drill parameters and schematics were not published. Later on, this drill was scaled down to a model called ETB-130. The working principle of both drills is unknown, and the only thing that was published is that the drills "pump meltwater from the borehole" (Morev et al. 1988a). ETB-140 was first tested near the runway at Mirny, Antarctica in 1980–1981 (Dyurgerov et al. 1987). Then, it was used at Komsomolskaya Station in the season of 1981–1982 for drilling in a snow–firn zone down to 120 m (Manevskyi et al. 1983). In this interval, 191 runs were carried out over 13 days with the following parameters: core length of 0.62–0.71 m; core diameter of 75–80 mm; and power consumption of 1.9–2 kW. Penetration rate decreased from 2.6 m/h in firn to 1.8–2 m/h in ice.

In January–March 1986, a 92-m-deep hole was drilled at Dome B with the ETB-140 and ETB-130 thermal drills (Savatyugin 2001). In the season of 1987–1988, the temporary base Dome B was reopened and a borehole from the surface down to 780 m was drilled using an ETB-130 thermal drill (Morev et al. 1988a). This hole was then filled with an ethanol antifreeze solution up to 120 m, which was

below the ice–firn transition depth. The minimum borehole temperature was −58 °C.

2.2.10 Summary

In 1970–1980s, open-hole thermal drilling was a quite effective method for shallow and intermediate drilling in cold glaciers. However, if meltwater is not removed immediately from the bottom, the core from the upper snow/firn layers can become considerably infiltrated with meltwater, resulting in large errors in chemical and isotope analyses. To avoid firn water saturation, water must be rapidly removed from the bottom, keeping the water level as low as possible. In addition, ice cores usually suffered from thermal shock during drilling and had visible cracks on their surface.

The drilling experiences in Antarctica clearly demonstrated the need for adequate housing of the thermal drill rigs, the desirability of automated drilling and tripping control, and the need for telemetry during the drilling process. The efficiency of thermal drills is determined primarily by the rate at which the heat generated within the thermal head is conducted into the annulus and the ice face. The cartridge elements used for heating the annulus had a number of undesirable features in practice, and a single tubular heating element casted in an aluminum or copper annulus proved more advantageous. The highest penetration rates (up to 6 m/h) were achieved by the thermal head with the bare resistance introduced in the LGGE drills.

The serious limitation of open-hole thermal drilling is borehole closure due to the lack of hydrostatic balance, and closure rates of up to 1.3–2.0 mm/h were encountered (this depends on ice temperature and hole depth). This introduced considerable difficulties during drilling and, towards the base of the holes, several reaming runs (in which the drill was operated at a slow feed rate, with both the heater and the vacuum pump on) were necessary before taking each core. Drill jamming and sticking in shrinking holes was "normal practice" during open-hole thermal drilling. Even though Russian drillers reached an amazing depth of 952.5 m in an open hole at Vostok station, borehole closure caused by ice creep limits the practical depth for thermal drilling in dry holes to approximately 450–500 m.

There is great risk in allowing any meltwater to remain in the borehole, where it may cause the freezing-in of drill. Unfortunately, drill incidents caused by the failure of the thermal drill head or the vacuum pump often led to meltwater refreezing and drill sticking. Therefore, open-hole thermal drilling needs reliable control methods and tools for incident prevention.

Another problem related to thermal drilling are penetration difficulties in dirty ice or in cases in which foreign particles drop into the hole, especially particles made of insulating material. Dust and small rock particles precipitate between the drill head and the bottom of the hole, inhibiting the melting-drilling process. All of these things have resulted in open-hole thermal drills not being currently in use. However, the experience gained is useful for showing us what is possible and what not to do in the future.

2.3 Electrothermal Coring Drills for Drilling with Borehole Fluids

To increase the maximum ice-coring depth, it is necessary to compensate for the overburden ice pressure (Talalay et al. 2014a; Zacny et al. 2016; Zagorodnov and Thompson 2014). This is achieved by filling the borehole with low-temperature drilling fluid having the appropriate density and viscosity properties, which allows for ice coring down to virtually any depth at any ice temperature (Talalay and Gundestrup 2002). In general, drills of this type are similar in design to open-hole type thermal drills, but the vacuum pump is changed to fluid pumps of different types (such as centrifugal, vibrational, piston, and gear). Meltwater is also pumped from the kerf, stored in the drill, and transported to the surface.

Depending on the type of drilling fluid used and the fluid's ice/water solubility, these drills can be divided into two groups: (1) drills working with hydrophobic drilling fluids and (2) drills working with hydrophilic fluids. Hydrophobic fluids are insoluble in water and are stable with ice, whereas hydrophilic fluids are able to mix with water in any concentration and dissolve ice at temperatures below zero. Boreholes are filled with hydrophobic fluids from the surface, whereas hydrophilic fluids are usually delivered to the bottom of the hole in the drill tank.

Hydrophobic fluids comprise low-temperature petroleum oil products for intermediate-depth drilling, where borehole closure is moderate, or are mixed with densifiers for deep ice-core drilling, where borehole closure should be prevented as much as possible. Only one type of hydrophilic liquid is used for thermal drilling, i.e., ethanol and its aqueous solutions, although theoretically other types of organic compounds (such as methanol and propylene glycol) could be utilized as well. Table 2.4 includes the main parameters of these drills and some characteristics of the associated drilling processes.

2.3.1 CRREL Fluid-Type Thermal Drill

In general, the design of the CRREL fluid-type thermal drill was the same as that of the drill for dry holes (see Sect. 2.2.1), but a gear pump was installed in the water tank instead of the vacuum pump (Fig. 2.52). In 1964, the third

open hole at Camp Century, Greenland was extended from 264 to 535 m using this thermal drill in a fluid-filled hole, but not without difficulties (Ueda and Garfield 1968). The hole was filled with a mixture of diesel fuel (DF-A) and trichlorethylene of the same density as ice (920 kg/m^3). Penetration rates in the wet hole averaged $\sim$1.5 m/h at power inputs of 5–6 kW and core recovery was 96%.

While drilling, numerous problems were encountered. The drilling fluid possessed strong solvent properties. It eventually removed a rust-inhibiting compound used on the cable, creating a residue which continually settled at the bottom of the hole, which reduced the melting rate, clogged the pumping circuit, and ultimately forced the discontinuation of drilling with thermal method. Increasing the power applied did not increase the penetration rate and only resulted in premature heater burnouts. Other problems included the breakage of electrical conductors within the cable, breakage of the cable's outer armor strands, and leaks in the winch hydraulic system. All these were the reasons for further trials of thermal drilling in fluid-filled holes not being pursued by CRREL.

2.3.2 TBZS-152 (132) Thermal Drills

The TBZS-152 (Russian abbreviation for "thermal drill for fluid-filled holes", where the number 152 signifies the outer diameter of the coring drill head) thermal drill was developed by the LMI in 1973 (Fig. 2.53). After modifications, the name of the TBZS-152 drill was followed by the letter "M". The drill head of the TBZS-152 drill contained two coiled electric tubular heaters with a total power of 2.5–3.2 kW, casted into a copper or aluminum ring (Kudryashov 1989; Kudryashov et al. 1991). The drill head had an ID and an OD of 114 and 152 mm, respectively. The thermal head was fixed with bolts to the sleeve, with three core catchers screwed to a 3-m-long steel core barrel (ID/OD: 118 mm/127 mm). Three water-lifting pipes (OD: 10 mm) and one planar pipe with electric lines were fixed on the outer surface of the core barrel. The core barrel was attached through a connector to a steel water tank (ID/OD: 136 mm/146 mm) with a volume of $\sim$35 L. The length of the central water lifting pipe was shorter than that of the water tank by 0.5 m. A centrifugal ETsN-91B pump with housing driven by a DC motor was fixed at the top of the water tank. The voltage of the electric-type DC pump motor could be changed from 15 to 27 V with a maximum current of 8 A. The components of the fluid in the tank were separated by gravity: water was collected at the bottom of the tank and the drilling fluid was pumped back to the borehole. The maximal flow rate of the pump was 30 L/min. The total length of the TBZS-152M drill was 7.5 m and its weight was 180 kg. The total power consumption (including the heaters of the water lifting pipes and the water tank) was 5–6 kW.

Table 2.4 Electrothermal coring drills for drilling with low-temperature fluids

Institution or Drill name	Type of drilling fluid	Years	Thermal head (ID/OD)/mm	Drill head power/kW	Drill site(s)	Min temperature/°C	Penetration rate in ice/(m h⁻¹)	Max depth reached/m	References
CRREL fluid-type thermal drill	DF-A + trichloroethylene	1965	124/162	5-6	Camp Century, Greenland	−24.6	1.5	535	Ueda and Garfield (1968)
TBZS-152 M (132) thermal drill	DF-A + CFC-11 (HCFC-141b)	1974–1993	114/152 (95/132)	3.5 (3.0)	Vostok station, Antarctica	−57.2	1.8–2.3 (1.8–2.0)	2502.7 (2755.3)	Vasiliev et al. (2007)
TBS-112VCh thermal drill	DF-A + CFC-11	1981–1986	92/112	4.5	Vostok station, Antarctica	−57.2	3.7–4.2	2201.7	Vasiliev et al. (2007)
LGGE thermal drill 'Forage 4000'	DF-A + CFC-11	1984–1989	125*/145	7.5	D47, Adélie Land, Antarctica	−25.8	NA	871	Marec et al. (1988)
ETB-3	Ethanol-water solution	1972–1993	84/108	1–4	Mountain glaciers, Antarctica, Arctic	−28	3.5–5.0	812	Morev and Raikovsky (1979)
ETB-5	Ethanol-water solution	1979–1982	82/100	1.5–3.0	Komsomolskaya station, Antarctica	−54	3.6–3.8	871.5	Morev et al. (1988a, 1990)
m-ATED	Ethanol-water solution	1997	105/124	3.3	Vetrennyi Ice Cap, Franz Josef Land	−11	2.7	315	Zagorodnov et al. (1998)
ETED	Ethanol-water solution	2002 up to now	102/127	2.25	Mountain glaciers, Antarctica	−24	up to 4.3	447.6	Zagorodnov et al. (2005)

Note *—Author's estimations; *NA*—data are not available

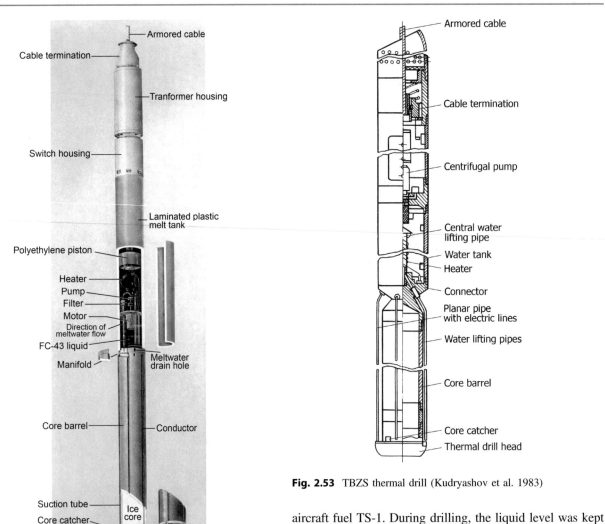

Fig. 2.52 CRREL fluid-type thermal drill (Ueda and Garfield 1968)

Fig. 2.53 TBZS thermal drill (Kudryashov et al. 1983)

First trials at Vostok station

During the first attempts carried out during the 19th SAE (1974) to drill in a fluid-filled hole at Vostok station using a TBZS-152 thermal drill and a water–ethanol solution as the hole fluid, many problems were encountered (Kudryashov et al. 1983, 1991; Kudryashov 1989). Owing to the dissolution of ice from the borehole walls, the concentration of the water–ethanol solution decreased and formed an icy slush. This may be attributable to an overly rich concentration dissolving the ice and the subsequent precipitation of the ice as the solution cooled to equilibrium temperatures. This hindered drill penetration. In total, the depth interval of 108–142 m of Hole No. 2 was drilled.

At the beginning of the 20th SAE (1975), the drilling liquid in Hole No. 2 was changed to a kerosene-based

aircraft fuel TS-1. During drilling, the liquid level was kept at a depth of 130–150 m. Drilling continued with the same TBZS-152 thermal drill down to 450.4 m, where the drill became stuck, most likely because of problems in the water removal system (Zagrivnyi et al. 1981; Kudryashov et al. 1983). The cable was ruptured during attempts to free the drill. In the depth interval of 142–450.4 m, 199 drilling runs were completed. The power of the copper drill head was 6 kW, and the power of the aluminum drill head was 5.3 kW (in later drill modifications, the power of the head was halved). The power of the heating elements in the water lifting pipes was 120 W/m; the power of the heating elements in the water tank was 1.5 kW per meter of water column (Zagrivnyi et al. 1980).

In the 21st SAE (1976), a new offset hole called No. 2-bis was started at a depth of 224 m in the original Hole No. 2, but the drill became stuck again. The final depth of Hole No. 2-bis was not reported. According to the drillers' personal recollections, only a few runs in the new offset hole were made.

Experiments with the TBZS-152 thermal drill continued during the 22nd SAE (1977). Two unnumbered holes with depths of 171 and 300 m were drilled. Drilling was difficult, and a few serious incidents occurred. Drilling at Vostok was suspended in order to improve and test the technology used for fluid-filled drilling. Such experimental drilling operations have been carried out on Vavilov Glacier in the Russian Arctic.

Vavilov glacier

After a series of bad results at Vostok station in Antarctica, a modified thermal coring drill TBZS-152M was tested on Vavilov Glacier, Severnaya Zemlya archipelago (Zagrivnyi et al. 1981; Chistyakov et al. 1988; Kudryashov et al. 1991). These tests mainly focused on the practical approval of new designs of the thermal head and the optimization of the flow delivered by the embedded pump. In 1977–1979, 10 boreholes with a cumulative depth of more than 1500 m were drilled in ice with a minimal temperature of −11 °C. Two holes with depths of 462.2 and 460 m reached the base of the glacier. The drilling of the deepest hole (462.2 m deep, drilled during September–October 1977) took 38 days, with an average rate of penetration of 2.03 m/h and a core length of 1.6 m per run. Owing to a shortage of drilling fluid, drilling was done with a minimal amount of fluid in the hole (∼15 m above the bottom of the hole). Experiments were continued at Gornaya base in Antarctica.

Gornaya base

During the wintering of the 24th SAE (1979), a temporary base Gornaya was established at the 73-km mark of the Mirny–Vostok route (Bobin et al. 1986; Kudryashov et al. 1991). All drilling equipment was installed inside the new mobile PBU-2 drilling shelter, which measured 3.5 m × 9

m with an A-shaped mast that was 8 m high (see Fig. 2.46). First, a dry hole was drilled down to 430.3 m using a TELGA-14 M thermal drill. Then, the hole was filled with TS-1 aircraft fuel and drilling was continued using a TBZS-152M thermal drill. When the hole reached a depth of 721 m, the drill head was changed. The outer diameter of the new head was 0.7 mm larger than the old one. During the enlarging or reaming process, a new offset hole was accidentally started at a depth of ∼500 m. After this deviation, the new hole was drilled to 750 m. The rate of penetration of the TBZS-152M drill was 1.75 m/h, the mean length of the cores was 1.9 m, and the power of the drill head was 2.5 kW.

Vostok station: Hole No. 3G

At the end of 1979, the mobile drilling shelter PBU-2 used for drilling at Gornaya base was towed to Vostok and used for drilling a new deep hole at Vostok station. The first 112 m of the new Hole No. 3G (G is the first letter of the Russian word "glubokaya", meaning "deep") was drilled as an uncased dry hole using a TELGA-14M thermal drill (Fig. 2.54). Below the firn–ice transition layer, the TBZS-152M thermal drill was used in the fluid-filled hole, with the fluid consisting of TS-1 aircraft fuel with densifier CFC-11 (Kudryashov et al. 1983, 1984a, b, 1991, 1989). During the 25th SAE (1980) and the 26th SAE (1981) the hole was extended to 1500 m with an average rate of penetration of 2.0 m/h and a core length of 2.0 m. The drill head power was 3.2 kW in average and the total power of the heaters in the water tank and water lifting pipes was 2.8 kW. Beyond this depth, the TBS-112VCh thermal coring drill was used (see the next section).

Fig. 2.54 Schematics of Hole No. 3G (1980–1985), Hole No. 4G (1983–1989), and Hole No. 5G (1990–present) with several deviations at Vostok station

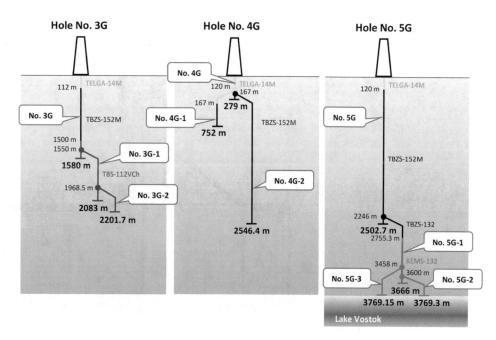

Vostok station: Hole No. 4G

The construction of a new stationary drilling complex was started during the 27th SAE (1982). It was built on two steel sleds with a 15-m mast (Fig. 2.55) and was equipped with a Kaspyi logging winch with 4000 m of KG-95-180 seven-conductor armored cable (Kudryashov 1989; Kudryashov et al. 1991). The new Hole No. 4G was started in July 1983. Drilling in Holes No. 3G and No. 4G overlapped for several years with the efforts of two different shifts. The pilot hole was drilled down to a depth of 120 m using a TELGA-14M thermal drill. After the hole was filled with liquid, a TBZS-152M thermal drill was used to advance it to 279 m; however problems with the winch (the bolts of the winch's drive-shaft coupling failed) resulted in the loss of the drill.

During the 30th SAE (1985), the lost drill was bypassed with an offset hole called Hole No. 4G-1, which was started at 120 m using a TELGA-14M thermal drill; drilling continued to a depth of 167 m and then to 752 m using a TBZS-152M thermal drill (Fig. 2.56). Unfortunately, the old hole (Hole No. 4G) was accidentally reopened because of the absence of an ice plug at the place of the deviation. This made it impossible to return to hole No. 4G-1, so a new offset hole, called Hole No. 4G-2, was started from Hole No. 4G at 159 m. Owing to inadequate centering of the drill, the inclination of Hole No. 4G-2 reached 18° in the depth interval of 500–600 m. In order to reduce hole inclination, the depth interval of 779–936 m was drilled with a shortened version of the TBZS-152M thermal drill.

During the 32nd SAE (1987) (Fig. 2.57), the 33rd SAE (1988), and the season of the 34th SAE (1989), this hole was advanced to 2428.5 m, where operations were switched to use an electromechanical KEMS-132 drill (Talalay 2016). The hole inclination at that depth did not exceed 4°. The average drilling parameters with the TBZS-152M drill were:

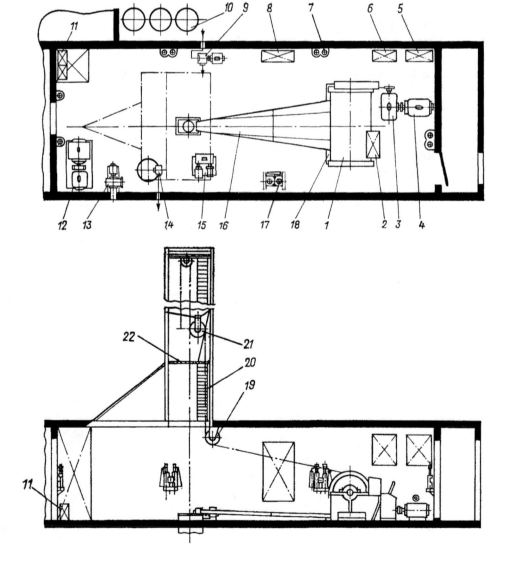

Fig. 2.55 Schematic diagram of the drilling building used for drilling Hole No. 4 and Hole No. 5: (1) winch drum; (2) control desk; (3) transmission; (4) winch motor; (5) winch control box; (6) drill control box; (7) lamps; (8) power control box; (9) drilling-fluid pump; (10) drums with drilling fluid; (11) workbench; (12) adjustable-potential system; (13) air exhauster; (14) water pump; (15) auxiliary winch; (16) drilling-fluid tray; (17) fire extinguishers; (18) drain pan; (19) guide pulley; (20) armored cable; (21) crown block; (22) stairway (Kudryashov et al. 1991)

Fig. 2.56 Drilling operator at Hole No. 4, Vostok station, 30th SAE, 1985 (*Credit* V. Chistyakov)

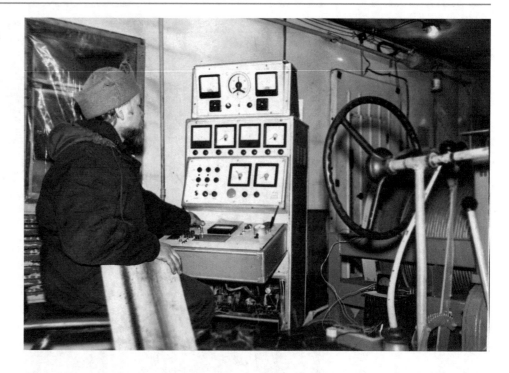

Fig. 2.57 Drilling operations at the 32nd SAE (1987): **a** mean daily core production: three pieces of core, each with a length of nearly 3 m (Vasiliev et al. 2007); **b** removal of an ice core from the TBZS-152M thermal drill (Giles 2004)

(a)

(b)

a penetration rate of 1.8–2.2 m/h, core lengths of 2.25–2.8 m, a drill head power of 3.6–3.9 kW, and a total power for the heating of the water lifting pipes and tank of 3.5 kW.

Vostok station: Hole No. 5G

In January 1990 (35th SAE), the drilling complex was moved 25 m to the west (Fig. 2.58). A new Hole No. 5G, was started in February 1990 using a TELGA-14M thermal drill (see Fig. 2.44). Beyond this depth, a TBZS-152 M drill

was used (Tchistiakov et al. 1994; Kudryashov et al. 1998a, b, 2002). In December 1991 (36th SAE), this hole reached a depth of 2502.7 m, where the drill became stuck at 2259 m during an uphole trip. This problem was caused by hole closure owing to insufficient fluid pressurization. All attempts to recover the drill failed and the cable was pulled out of the top of the drill (Dmitriev et al. 1995).

Densifier HCFC 141b was added to the hole, increasing the fluid density to 900 kg/m^3. Approximately 35 m of artificial core were dropped on top of the stuck drill, creating

Fig. 2.58 Drill building at Vostok station (summer season of the 35th SAE, 1990); in the corner: Russian post stamp with the Vostok drilling building (Ueda and Talalay 2007)

a new hole bottom at 2232 m. A shortened version of the TBZS-132 thermal drill, which was 6 m long, was used to drill a new offset hole called Hole No. 5G-1. From a depth of 2249.5 m, drilling was continued using a TBZS-132 drill with a normal length of 8 m down to a depth of 2755.3 m (September 1993). This depth set a new record for thermal drilling in ice. The main differences between the TBZS-132 and TBZS-152 thermal drills were the OD and ID of the drill head and the tubing used for the core barrel and water tank. The TBZS-132 drill was slightly shorter (7.2 m instead of 7.5 m) and lighter (120 kg instead of 180 kg). The average drilling parameters for the TBZS-132 drill were: a rate of penetration of 1.8–2.0 m/h, core lengths of 2.0–2.5 m, and a drill head power of 3 kW. Further drilling in this hole was done using a KEMS-132 electromechanical drill (Talalay 2016).

2.3.3 TBS-112VCh Thermal Drill

This drill used high-frequency current for power supply, which reduced the dimensions of the electric transformers located in the drill (Fig. 2.59). TBS-112VCh is the Russian abbreviation for "thermal drill with a high frequency", referring to its current power supply, and the number 112 is the outer diameter of the drill head (Kudryashov et al. 1991; Zagrivnyi et al. 1985). The OD/ID dimensions of the drill

head were 112 and 92 mm, respectively, whereas the drill had a length of 7.1 m and weighed 120 kg.

The drill was suspended using coaxial KG2-59-90 armored cable (the cross-area of the central line was 4 mm^2), which was 12 mm in diameter and had a breaking force of 58 kN. At the surface, the power supply had a frequency of 2500 Hz and a voltage of 800–1000 V. After passing through the cable, the voltage dropped to 700–900 V. At the bottom, this primary voltage was reduced using electric transformers to a voltage of 20–25 V, which was used in the drill. The water tank was not heated. Meltwater was collected in a tank and was allowed to freeze inside it. At the surface, the tank was removed and replaced with an empty tank. The core barrel was made from a thick-walled tube, with the water lifting pipes and the electrical lines hidden within the tube's wall.

Hole No. 3G (see Fig. 2.54) was continued from 1500 m using a TBS-112VCh drill during the overwintering of the 26th SAE (1981). The drilling rate was doubled to 3.5–4 m/h compared with that obtained with the TBZS-152 M drill with 4.5 kW of drill head power. The power of the heaters of the water-lifting pipes was 3 kW. Core length was in the range of 1.6–2.2 m, and core diameter was 87–90 mm (Fig. 2.60). After reaching a depth of 1580 m and owing to problems with the meltwater removal system, the drill became stuck near the bottom of the hole. A new offset hole, called Hole No. 3G-1, was started using a spare TBS-112VCh thermal

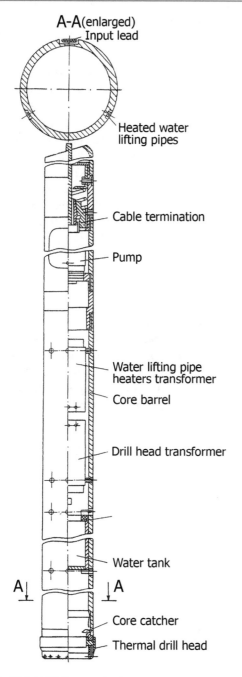

Fig. 2.59 TBS-112VCh thermal drill (Kudryashov et al. 1984b)

Fig. 2.60 Surface servicing of the TBS-112VCh thermal drill, 26th SAE, 1981 (*Credit* V. Chistyakov)

A mobile PBU-3 drilling complex, which was transported from Mirny during the 28th SAE (1983), was installed above the mouth of Hole No. 3G (Fig. 2.61), replacing the PBU-2 shelter. The drilling operations in Hole No. 3G-1 were resumed, but the lower part of the hole had closed during a long period of inactivity. During the reaming process, a new offset hole, called Hole No. 3G-2, was accidentally started at a depth of 1968.5 m. In this hole, the TBS-112VCh thermal drill used reached 2201.7 m in August 1985 and set a new record for ice-core drilling depth. Drilling was carried out with continual enlarging of the narrowing parts of the hole. The drill thermal heads used were not properly adapted to

Fig. 2.61 The mobile drilling shelter PBU-3 installed above the mouth of Hole No. 3G (summer season of 29th SAE, 1984) (Vasiliev et al. 2007)

drill with the same length of 7.1 m at a depth of 1500 m (Zagrivnyi and Moiseev 1988). In the depth interval of 1550–1560 m, a full-diameter core was recovered. The total run time at depths of 1600–1700 m was 2.2–2.5 h.

In early 1982, Hole No. 3G-1 was advanced to 2083 m; however, a fire in the power station of the generator room on April 12, 1982 tragically took the life of the chief mechanic and curtailed further drilling. On December 26, 1982, another fire destroyed the PBU-2 drilling shelter, which housed the drilling complex for Hole No. 3G, and drilling was suspended.

the high hydrostatic pressure of the drilling fluid, which led to the premature burning of heating elements. In total, nearly 75 thermal drill heads were used in the last depth interval of 2040–2201.7 m. Finally, the TBS-112VCh thermal drill got stuck in Hole No. 3G-2 during enlarging operations in the depth interval of 1935–1943 m (31st SAE, 1986) and further drilling activity in this hole was stopped.

2.3.4   LGGE Fluid-Type Thermal Drills

First prototype

The LGGE open-hole thermal drill (see Sect. 2.2.8) was taken as the basis for the first version of this drill that could work in a fluid. The vacuum pump was replaced by a small 30-W, 220-V vibration pump located at the top of the water tank so that no water flowed through it (Fig. 2.62). The flow could vary from 5 L/h to 300 L/h but was generally adjusted at 40 L/h. The heating of the suction tubes and of the water tank was increased and varied separately from 0 to 350 W for the tubes and from 0 to 500 W for the tank. Three different temperature measurements were made at the thermal head, at the top of the suction tubes, and in the water tank.

The signals were transmitted consecutively to the surface through two conductors using an electronic switch.

In the season of 1981–1982, the open-hole drilled to a depth of 329 m at Adélie Land, site D15, Antarctica was then filled with kerosene and was continued using a fluid-type thermal drill down to a depth of 348 m. With a 2.8-m-long core barrel and runs of 2–2.5 m, no problems with fluid circulation occurred. The penetration rate was the same as in the dry hole and varied from 5 to 7 m/h, corresponding to power on the head ranging from 3.5 to 5 kW. Core quality was very good, without any fractures. The diameter of the cores was the same as those obtained in the dry hole, namely ∼115–116 mm. The amount of meltwater recovered during each run was 14–18 L.

Forage 4000

In order to penetrate the Antarctic ice sheet to a depth of 3500–4000 m in a single summer season, the previous drill was then re-built to make it capable of taking cores with a length of 6–8 m. An ambitious project "Forage 4000" ("Drilling to 4000 m") had been started in 1984 (Augustin et al. 1988). This drill applied many innovations from that period. The total length of this thermal drill was ∼22.4 m,

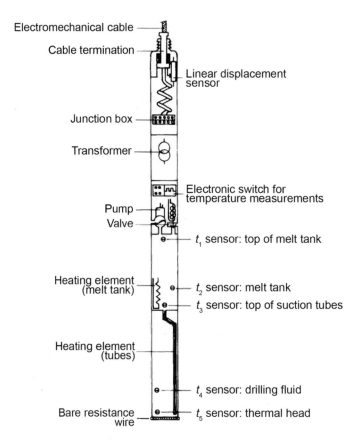

Fig. 2.62   Fluid-type LGGE thermal drill (Donnou et al. 1984)

Fig. 2.63 Bare-wire thermal head of Forage 4000 drill (Forage 4000 1986)

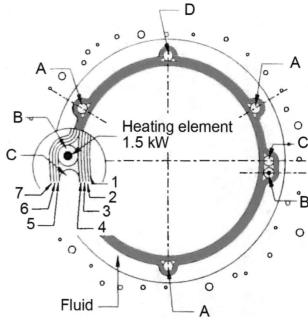

Fig. 2.64 Core barrel section of Forage 4000 thermal drill: **a** drill head power supply; **b** heating element; **c** water-sucking tubes; **d** tubes for temperature measurements; (1) fiberglass and Teflon; (2) Kevlar; (3) carbon; (4) Kevlar; (6) carbon; (7) Kevlar (Augustin et al. 1988)

and its weight was ∼410 kg. The design of the thermal head was still based on spiral bare-wire heating elements. A 1.3-mm-diameter Kanthal A-1 resistance wire (22% Cr, 5.5% Al, 75% Fe) was fixed with three pins on the glass ceramic insulator (Fig. 2.63). The total wire length was ∼3.8 m; the heating power was 7.5 kW at 50 V AC. The resistance was supplied with three-phase power at three points via copper pins soldered with silver. The insulator was made in six pieces from Macor machinable glass ceramic and inserted in a stainless-steel U-form piece, which protected it from mechanical damage.

The 8.3-m-long core barrel (ID/OD: 125 mm/145 mm) was made from composite materials, namely a combination of Kevlar and carbon fibers, and had seven fiber layers (Fig. 2.64). The first fiberglass layer was coated with Teflon from the inner side for easy releasing from the long vertical aluminum bar that was used during manufacturing (Fig. 2.65) and later for the smooth extraction of the ice cores from the barrel. The last Kevlar layer was coated with white polyurethane. Layers were fixed together with bi-component epoxy resin EPO 20-12 RESOLIN. To ensure a strong mechanical connection with the other parts of the drill, stainless-steel sleeves were molded into the core barrel on each end.

The 5.5-m-long water tank (ID/OD: 116 mm/130 mm) of the system was also made from composite materials and had eight layers: fiberglass with Teflon (inner layer)–Kevlar–carbon–Kevlar–epoxy resin with silica microbubbles (the thickness of this layer was ∼5 mm)–Kevlar–carbon–Kevlar. Along the water tank axis, a bare electrical resistance wire was spirally set up. It had variable power to set a higher temperature at the bottom of the tank. Its maximum power was 1.5 kW. Control of the water filling of the tank was

carried out using four temperature sensors located at different heights in the tank.

To reduce hydraulic resistance during drill moving in the fluid-filled hole and to increase the tripping speed, only the core barrel had a diameter of 145 mm; the other parts of the drill had a smaller diameter of 130 mm. Four holes at the top of the core barrel allowed fluid to pass inside the core barrel when the drill went down. Flap valves in the drill head allowed fluid to pass inside the core barrel when the drill went up with the ice core.

An electromagnetic pump was connected to the top of the tank and pumped drilling fluid, allowing meltwater to replace the fluid in the tank. The pump always pumped drilling fluid, never water. The suction head of the pump had a pressure of 40 kPa. The electric valve located below the pump had two functions. It allowed the air inlet at the top of the tank to empty it at the surface, and in turn this air inlet allowed the filling up of the tank when the drill went down. The pump flow was regulated via a manual valve.

Two transformers were used: a three-phase transformer that supplied power to the drill head (1236/60 V) and a single-phase transformer (1404/335 V) that supplied power to the electronics, the electric valve, the pump, the water tank heater, and the heater of the sucking tubes. These transformers were located in a 3-m-long oil-filled stainless-steel tube. A high-tension power unit allowed operators to manually adjust the supply voltage without disassembling the drill.

Fig. 2.65 Manufacturing of the core barrel of Forage 4000 thermal drill: fixing of the first fiberglass layer (Forage 4000 1986)

The electronic boards were fixed in 2.8-m-long stainless-steel water-tight chamber that could withstand outer pressures as high as 40 MPa. One linear displacement sensor in the cable termination was used to measure drill load, and a second sensor was used to measure the hole diameter through a link with a three-skate centralizer.

The 16-mm-diameter armored cable used included eight power lines (1.5 mm^2) for all the energy-consuming units (six lines were used to supply the thermal head and two lines were used to supply other electric devices and electronic components) and a twisted pair (2×0.34 mm^2) for the transmission of data signals (Marec et al. 1988). The breaking strength of the cable was 84.5 kN, and its weight was 0.976 kg/m.

Two different transmissions systems had been designed and built. The first one used frequency-division multiplexing (FDM) and the second one used time-division multiplexing (TDM). FDM transmission was more advantageous because all transmission channels were fully independent. That was one of the reasons why an FDM system was chosen for field

testing in Antarctica, where it proved to be reliable, accurate, and easy to operate.

Tripping operations were carried out with a winch having 4030 m of cable. During drilling, core breaking and core recovery at the surface was done using a 4-m-stroke hydraulic cylinder supporting the upper sheave (Perrin 1988). Both the winch and the hydraulic cylinder could work simultaneously, and the automatic drill positioning system provided an actual depth reference. The winch was the heaviest piece of equipment and weighed ∼6 t (Donnou et al. 1988). It was driven by a 75-kW hydraulic motor and allowed for tripping with a speed of 0.8–1.5 m/s. The winch was equipped with an emergency brake that operated in case of failure of the hydraulic circuit. Hydraulic power was supplied by four pumps driven by an air-cooled Deutz F8L413 engine rated at 135 kW.

The 26-m-high and 3-m-wide aluminum tower used had seven floors, each 3.7 m in height. The five upper floors were similar, whereas the two lower ones could be opened on one side to release the core. Each floor section weighed ∼350 kg. A hydraulic cylinder was mounted in the middle of the upper floor using four arms.

The control desk (Fig. 2.66) was mounted in a separate operator room that was kept warm and allowed workers to monitor all internal drill parameters (such as temperature at different points, electric power, drill load, and borehole diameter), to control the power of the sucking water tubes, water tank, drill head, electromagnetic circulation pump, and electric valve, to operate the winch and feed the hydraulic

Fig. 2.66 Control desk of Forage 4000 thermal drill (Perrin 1988)

cylinder, to display drilling depth, to detect faults with the hydraulic system and the power plant (with the possibility of stopping them in emergency cases), and to operate with auxiliary and safety devices (such as the winch chain, the tower door, etc.).

D47, Adélie Land, Antarctica

First, the drill components were checked separately. Then, the drill was tested at a laboratory in an 8-m-deep ice pit. In the season of 1987–1988, the drilling equipment was transported to site D47 (67°23′38″S, 138°43′31″E, 1560 m asl), Adélie Land, Antarctica, 107 km from Dumont d'Urville Station (Fig. 2.67). After the installation of the casing, the hole was filled with a mixture of diesel fuel and CFC-11 and the Forage 4000 drill and its control system were tested.

During the subsequent 1988–1989 season, 720 m of ice cores with lengths of up to 7 m were drilled over 20 productive days, and the hole reached a depth of 871 m. In average, retrieving 8-m cores required 2 h of drilling time with a power of 7 kW at the heating element of the drill head (Fig. 2.68). The overall operation was slowed down owing to very frequent and strong katabatic wind (the maximum observed speed was 34 m/s), with heavy snow drifts caused by low temperature, which varied between −25 and −10 °C (Fig. 2.69). One third of the time was used to maintain the camp operational owing to the snow drift burying process, which had a high impact on daily life. Heavy wind had a high impact on the drill itself (it got covered by snow while going up at the surface and banged against the edge of every floor of the tower) with the risk of damaging the thermal drill's fiber glass tubes.

Another problem was related with drill surface servicing. The core (approximately 100 kg for one 8-m piece) had to be retrieved vertically, which caused a lot of much trouble

Fig. 2.68 Drill head maintenance, site D47, Antarctica, in the season of 1988–1989 (*Photo* L. Augustin)

Fig. 2.69 Drilling pad during a storm, site D47, Antarctica, in the season of 1988–1989 (*Photo* L. Augustin)

during extraction, especially for brittle ice zones, with cores very often found in several wedge pieces inside the core barrel (Fig. 2.70). Unfortunately, owing to the overall drill and tower design, the set up did not allow for the retrieval of cores in any other position.

Fig. 2.67 General view of site D47, Antarctica, in the season of 1988–1989 (*Credit* J. Chappellaz; What Can Ice Reveal About Fire? 2010)

Fig. 2.70 Positioning of the core barrel, site D47, Antarctica, in the season of 1988–1989 (*Photo* L. Augustin)

Further on, system development was abandoned principally because of logistic issues. The French Antarctic Department, the French Polar Expedition (EPF) at that time, was not ready to take on the request for an inland deep-drilling operation to be logistically supplied from the French base at Dumont D'Urville, located near the coast (Jouzel et al. 2013).

2.3.5 Antifreeze Thermal Eclectic Drills

During the past few decades, many holes were drilled using antifreeze thermal eclectic drills that use an ethanol–water solution to fill the borehole to prevent it from closure through viscoplastic deformation. Ethanol–water solutions are hydrophilic liquids and can mix with water up to the equilibrium concentration at a given temperature. Meltwater goes into the solution and remains in the hole. Therefore, the amount of antifreeze delivered to a drilling site is less than the volume of the borehole. Depending on the temperature distribution in the glacier, the amount of pure ethanol required for borehole drilling is 5–75% of the borehole volume. The lower the ice temperature, the higher the ethanol concentration required.

ETB-3 thermal drill

The ETB-3 electro-thermal drill was designed by V. A. Morev from the AARI to be used with an ethanol–water solution and consisted of a double-core barrel made of stainless steel and an annular thermal head connected to the lower end of the outer barrel with bolts (Fig. 2.71a)

Fig. 2.71 Schematic diagram of the (**a**) ETB-3 and (**b**) ETB-5 electro-thermal core drills (Bogrodsky and Morev 1984; Bogordsky et al. 1984)

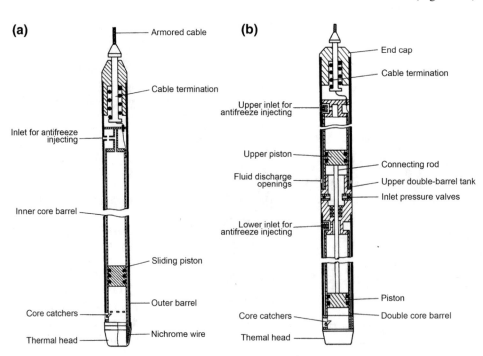

(Bogorodsky and Morev 1984). The thermal head consisted of two copper sleeves hermetically welded together at the upper and lower ends of the ring. A high-resistivity nichrome wire was located between these sleeves and was used for heating the ring. Electrical insulation was put between the wire and walls of the copper sleeves. A piston slid inside the inner core barrel. A spring-loaded cable termination was located at the upper end of the drill.

The drill was suspended in the hole by means of an 8.6-mm-diameter single-conductor armored electro-mechanical cable. Power was transmitted through the central conductor and the armor. The OD and ID dimensions of the drill head were 108 and 84 mm, respectively, and drill head power was 1.7 kW (Morev 1976; Kudryashov et al. 1983). The length of the core barrel was 1.5–2.7 m according to different modifications (usually 1.5–1.7 m); the overall drill length was only ~0.7 m longer than the core barrel.

At the surface, the piston was set in its lowest position and the core barrel was filled through an inlet nozzle with an ethanol–water solution of the required concentration before the drill was lowered into the hole. As drilling proceeded, the core passed up into the inner pipe and pushed the piston up. As a result, the solution was expelled from the inner barrel through holes in the upper end. It percolated down the space between the concentric barrels to the borehole, where it mixed with the meltwater.

The ETB-3 drill was first tested at drifting station Severnyi Polyus-19, located on an ice island in the Arctic Ocean, in 1972, where several shallow holes were drilled with a maximum depth of 34 m (Morev 1976). Then, starting from 1973, a few holes with a maximum depth of 137 m were drilled in Abramov Glacier (~4000 m asl), Alai Ridge, Pamirs. The mean rate of penetration was 2.5–3.5 m/h at a drill head power of 1.0–2.5 kW. In the summer of 1974, specialists of the IGAS and the AARI drilled a hole using an ETB-3 drill at Obruchev Glacier, Polar Urals to a depth of 86 m with a core recovery of 98% (Zagorodnov et al. 1976). Drilling operations took only four days. The rate of penetration was 3.5–4 m/h at a drill head power of 2 kW.

Drillings with ETB-3 in Antarctica

In subsequent years, ETB-3 thermal drills were used many times in Antarctica. In July–August 1975 (20th SAE), a 374-m-deep hole was drilled down to bedrock at a site approximately 50 km to the north of Schirmacher Oasis in Queen Maud Land, East Antarctica (Korotkevich et al. 1978; Morev et al. 1988a, 1990). The mean rate of penetration was nearly 5 m/h. Then, in October–November 1975, a borehole 25 km northwest of the first hole was drilled to a depth of 357 m, which penetrated through the Lazarev Ice Shelf and reached the ocean cavity. In June–August 1976 (21st SAE),

another hole was drilled at the Lazarev Ice Shelf and reached the ocean cavity at a depth of 447 m (Korotkevich et al. 1978; Morev and Raikovsky 1979).

In 1977 (22nd SAE), an ETB-3 drill was used to drill a hole with an ethanol–water solution down to a depth of 812 m, including 62 m of firn, at a site 40 km to the south of Novolazarezkaya Station (Morev and Raikovsky 1979; Kislov et al. 1983; Morev et al. 1988a). A special thermal head for dry drilling in firn was used down to 30 m. From this depth onwards, the drill was filled with a 20% ethanol–water solution. At a depth of 812 m, the thermal head burned out and, at the same time, the electric contact within the winch drive motor broke. The non-working drill was left at the bottom of the hole for 5–7 min, which was long enough for it to become frozen in and get stuck. Attempts to pull up on the cable resulted in the cable being separated from the drill, and further drilling ceased. The drilling operations consisted of three men working on one shift. They achieved an average ice core production rate of 14.3 m/day. The mean core length was close to 2 m, and the core diameters varied from 78 to 81 mm. The lowest temperature measured in the hole was approximately −18 °C.

Another borehole was drilled in February–April 1978 (23rd RAE) through the central part of the Shackleton Ice Shelf, 50 km to the south of its edge. This borehole reached the ocean cavity at a depth of 195.7 m (Savatyugin 1980). Drilling down to 43.5 m was carried out using a special firn thermal head. This interval was then cased with aluminum tubing (ID: 115 mm). The drill head power of the ETB-3 drill was reported as 1.2 kW. The maximum core length was 2.05 m. The rate of penetration varied from 3.5 to 4 m/h. The mean ice core production rate was 4.9 m/day.

In December 1978, an ETB-3 drill was used to drill 416 m through the Ross Ice Shelf at Camp J-9 (Fig. 2.72). The concentration of ethanol steadily decreased from 75% at the top to 40% at a depth of 400 m. The mean length of the runs was approximately 2.5 m. The penetration rate varied from 3.5 to 4.6 m/h. A total of approximately 2000 L of ethanol was needed to complete the drilling of this hole (Zotikov 1979; Zotikov et al. 1979, 1981). In 1988 (33rd SAE), a borehole with a depth of 252 m was drilled using an ETB-3 drill at the Druzhnaya-4 temporary summer station at the Amery Ice Shelf (Raikovskyi et al. 1990).

Drillings with ETB-3 in the Arctic

In the 1970s and 1980s, the ETB-3 thermal drill was often used in the Arctic at the Severnaya Zemlya archipelago and Spitsbergen. From 1976 to 1983, eleven boreholes were drilled in Vavilov Ice Cap, Severnaya Zemlya with a total cumulative length of approximately 2200 m (Morev and Pukhov 1981; Morev et al. 1981, 1988b). Five boreholes

Fig. 2.72 Drilling of an antifreeze core hole at J-9, Ross Ice Shelf, December 1978. **a** Inside the drill shelter: (1) ETB-3 thermal drill, (2) mast, (3) winch; **b** last core showing the ice formed by the freezing of sea water directly onto the bottom of the shelf (Zotikov 1979)

(a)

(b)

reached bedrock with a maximal depth of 556.6 m. During the spring–summer season of 1980, a new lightweight version of the ETB-3 thermal drill with a core barrel that was 0.7 m long and new drill heads was tested as well. The improved drill head provided the inflow of the ethanol solution closer to the bottom so that only the lower 10–12 mm (instead of 70–80 mm in the previous design) of the drill head was immersed in meltwater. In April 1983, during the drilling of a 467-m-deep hole down to the glacier's bedrock, glycerin in a quantity of 2% of fluid volume was added to the ethanol–water solution to decrease the amount of slush forming in the hole. However, drillers failed to get good results.

During June–October 1986, the drilling of a 561-m-deep hole using an ETB-3 drill was carried out in the central part of the Akademiya Nauk Ice Cap, which is the largest glacier on Severnaya Zemlya (Savatyugin and Zagorodnov 1987; Klement'ev et al. 1988; Zagorodnov 1989). Drilling proceeded on a one-shift basis over 42 days and was stopped because of a shortage of antifreeze. The average ice core production rate was 13.3 m/day. During drilling, the slush formed plugs, hindering round trips. That is why the drilling fluid was bailed out to a depth of 160 m. Because of imperfections in the control system, the inclination of the hole reached 25° at 360 m. During May–August 1987, the drilling operations with the ETB-3 drill in this hole were continued. The old 561-m-deep hole was reopened but could only be restored to a depth of 222 m. At this depth, a new hole was started, and it reached bedrock at a depth of 761 m. Due to borehole deviations from vertical, the true vertical depth of the hole was estimated to be 720 ± 10 m.

In the mid-1970s, the IGAS started comprehensive glaciological investigations of the Spitsbergen glaciers. Drilling operations were started in June 1975 at the ice-divide area between the East Grønfjord and Fridtjof glaciers, where the first hole was drilled to 201 m using an ETB-3 thermal drill (Zagorodnov and Zotikov 1981). The bottom of the glacier was reached with a hot-point drill at a depth of 211 m. The drilling site was located in the firn

percolation zone and the glacier sequence was temperate from the surface to the bottom. Therefore, the water table in the borehole at a depth of 27 m was naturally kept stable by the inflow of in-glacier meltwater. Apparently, the glacier had channels and internal cavities containing water. Sometimes the cores contained channels with cross sectional areas of 10–200 mm² and thin horizontal crevasses.

In July–August 1976, drilling operations with the ETB-3 drill were continued at Lomonosov Plateau. The hole was terminated at a depth of 201 m because of increasing slush formation at the bottom of the hole. The antifreeze solution was bailed out to a depth of 178 m and the hole began to close at a rate of a few millimeters per day. Drillers assumed that the reason for the slush formation was the unusual temperature distribution in the hole. From the surface down to 40 m, temperature increased from −2 to −0.3 °C. In the depth interval of 40–106 m, a reverse temperature gradient was found, as temperature decreased to −2.6 °C at 106 m. This led to gravity circulation within the fluid. The solution with a higher ethanol concentration (lower specific gravity) moved up, and the heavier weaker solution moved to the bottom and formed slush.

In 1979, three holes were drilled in Fridtjof glacier (48, 114, and 119 m) (Zagorodnov et al. 1985). The third hole was continued using a non-coring ETI-1 hot point from 119 to 220 m. Owing to significant hole inclination and limited cable length, bedrock was not reached. During May–June 1980, a deep hole was drilled in the central part of the Amundsen Plateau, at the southern part of Western Spitsbergen Island. The upper 386.5 m were drilled using an ETB-3 thermal drill (Zagorodnov 1981, 1988). Drilling was continued to bedrock at a depth of 586 m using the ETI-1 hot point. The inclination in the main part of the hole did not exceed 1°–2°. To bypass the disconnected hot point 30 m above the glacier bottom, a whipstock was placed above the hot point being used. The cavern melted above the whipstock, allowing the second hot point to deflect from the main hole at an angle of 30° and bypass the stuck hot point.

In June–July 1981, five holes (24, 24, 25, 24, and 208.2 m) were drilled using an ETB-3 drill in the Vestfonna ice cap (Kotlyakov 1985). The deepest hole reached bedrock at 208.2 m. The mean rate of penetration was 3.3 m/h and the core production rate was 40–45 m per 12–14 h of drilling. The maximum inclination of the holes was 7°.

In July 1982, two holes were drilled down to 31 and 135 m using an ETB-3 drill again at Lomonosov Plateau (Zagorodnov et al. 1984; Kotlyakov 1985). The second hole reached bedrock. Previous drilling operations at Lomonosov Plateau using ethanol–water solutions revealed that slush rapidly formed in the upper part of holes drilled in glaciers with an inverse temperature gradient. That is why, at the end of each drilling shift, the drilling liquid was bailed out of the hole. This method was used at depths less than 100 m, below which the hole was filled with antifreeze.

In June–July 1985, a drilling complex was built in the central part of Austfonna, Nordaustlandet (Fig. 2.73) and four antifreeze holes (32.4, 204.1, 53.7, and 60.9 m) were drilled (Anonymous author 1986; Arkhipov et al. 1987; Zagorodnov 1989). The drilling of the first hole was terminated at a depth of 32.4 m because of severe englacial water problems. The next hole reached a depth of 204.1 m; however, drilling was terminated again because englacial water

flowed into the hole and froze on the hole wall. The water inflow also caused slush to form in the upper part of the fluid column. Water continued to flow in spite of casing the hole to a depth of 7 m. Two other antifreeze holes were drilled to depths of 53.7 and 60.9 m. The drilling of both of these holes was stopped for the same reason—water inflow. The distances between the mouths of all of these holes were 0.5–2 m. Antifreeze drilling was carried out with an ethanol concentration in the range of 10–20%.

In June 1987, drilling operations at this glacier were continued (Zagorodnov 1989). In anticipation of the water inflow problem, the drilling site was moved to the ice-divide area, 2 km to the south of the camp built in 1985. A 566.7-m-deep hole was drilled down to bedrock using an ETB-3 drill. Drilling was conducted on one shift (two people) of 10–14 h. The core production rate was 30–33 m/day. To dissolve the slush at the top of the fluid column, an additional 5–10 L of ethanol were poured into the hole every 4–5 days. In total, approximately 1000 L of ethanol were used during drilling.

After the completion of the drilling operations, two additional cores were taken from the basal ice of the glacier. A 5-m-long whipstock was installed at the bottom of the borehole and it was fixed using a flat spring. The upper part

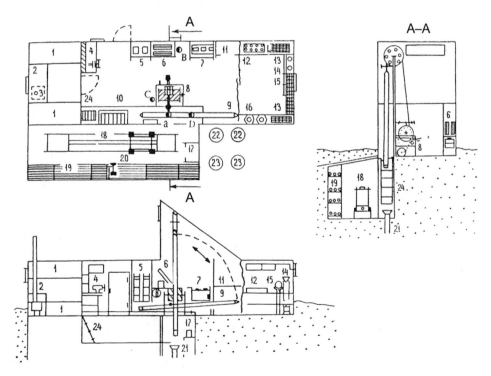

Fig. 2.73 Drilling complex on Austfonna, Spitsbergen: (1) bunks; (2) table; (3) solar heating; (4) metal-working bench; (5) apparatus for measuring gas pressure in the ice and for taking gas samples; (6) electro-thermal plane for preparing thin sections; (7) photographic table; (8) drilling winch; (9) drilling rig; (10) instruments for automatically recording parameters from the core; (11) table for cleaning the core; (12) water bath for melting geochemical samples; (13) isotope and geochemical samples; (14) filtering apparatus; (15) mechanical sampler; (16) hydrothermal sampler; (17) balance; (18) apparatus for automatically examining the core; (19) shelves with cores; (20) circular saw for cutting cores transversely; (21) mouth of the borehole; (22) containers for preparing the drilling fluid; (23) containers for antifreeze; (24) stairs (from Anonymous Author 1986; Arkhipov et al. 1987)

of the whipstock had a wedge that allowed the drill to deviate from the main borehole and start a directional hole (Fig. 2.74). After the first attempt, a 1.65-m core was extracted. During the second deviation, the drill was stopped, probably by a wedge, after penetrating 0.95 m.

The last operations with ETB-3 thermal drills in the USSR were carried out at Garabashi Glacier, Elbrus, Central Caucasus in May–June 1988, where two holes were drilled to the depths of 76.7 and 76.0 m (Bazhev et al. 1988). Both holes reached bedrock. Because of limited power (~2 kW), the rate of penetration was half of the normally attainable rate with ETB-3 drills. Drilling operations were complicated by water freezing on the hole walls at 2–4 m, where the minimum hole temperature (−6 °C) was observed.

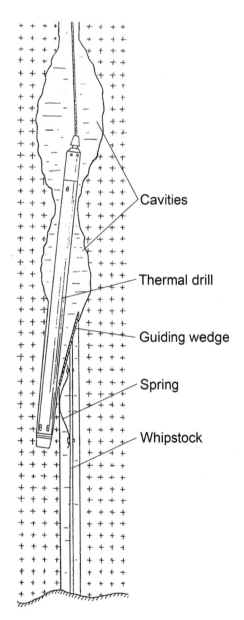

Fig. 2.74 Deviational drilling with an ETB-3 drill (Zagorodnov 1988)

Drillings with ETB-3 in high-mountain glaciers

In the beginning of the 1990s, a new area for the application of the ETB-3 drill was discovered. Two high-elevation glaciers were successfully cored to the bottom. In 1992, at Guliya Ice Cap, China (6200 m asl), a hole was drilled to a depth of 197.5 m using a PICO-4″ shallow drill, where the core began to fracture as a result of the bubble pressure within the ice (Koci and Zagorodnov 1994). From this depth down to the bottom at 307.5 m where a hard object was encountered, an ETB-3 drill was used. Cores were retrieved in continuous lengths of 1–1.2 m. Core quality was good but the cores were small in diameter because debris within the ice reduced the penetration rate, causing excess heat to melt the core and the hole walls. A total of 150 L of ethanol were used to drill the entire 113 m of core, and the hole remained open throughout the five drilling days.

In 1993, at Huascaran, Peru (6048 m asl), two ice cores were obtained using a hand auger from the surface to the ice–firn transition layer at a depth of 30 m and an ETB-3 drill down to the bottom of the holes, which were 160 and 166 m from the surface (Zagorodnov et al. 1998).

Canadian version of the ETB-3 drill

Holdsworth (1984) constructed a thermal drill based on the principles of the ETB-3 drill. The idea was to use two types of drills with the same drill rig, namely to use an electromechanical drill mainly in firn and replace it with a thermal drill below the firn–ice transition layer simply by disconnecting the electromechanical drill. The hermetically sealed heaters of the coring head had a diameter of less than 1.5 mm and power densities as high as 100 W/cm^2 (Koci 1984). They were brazed to a stainless-steel ring and provided a penetration rate greater than 4 m/h with a power input of approximately 1200 W. The core diameter obtained with this drill was 85 mm, whereas the diameter of the melted borehole was approximately 100 mm. Unfortunately, the test results of this drill in the field were not published.

ETB-5 thermal drill

A new electro-thermal drill ETB-5 was developed by the Morev group from the AARI for drilling in cold ice with temperatures lower than −33 °C (Bogorodsky et al. 1984) (Fig. 2.71b). The difference between the ETB-3 and ETB-5 drills is that latter contains two tanks for storing antifreeze. Both of these chambers are equipped with pistons and are connected by a rod. At the surface, both chambers are filled with an antifreeze solution with the required concentration, while the pistons are set at their lowest position. This way, the concentration of antifreeze in the borehole is doubled compared with that obtained with the ETB-3 drill. During drilling, the core lifts the pistons and, as a result, the solution is expelled and moves downwards, where it mixes with the meltwater. At the same time, the low-concentration antifreeze solution is sucked into the upper tank through inlet valves and is then removed from the borehole. It is assumed that, in this way, antifreeze solution of the required concentration is distributed more evenly along the length of the drill. The drill is suspended in the borehole via a single-conductor armored cable (KG-1-40-180) with a diameter of 8.6 mm. The core barrel could be made 1.5 to 3 m long.

The ETB-5 drill was firstly tested at the Vavilov ice cap, Russian Arctic during April–May 1979 (Morev and Pukhov 1981; Morev et al. 1981). A hole was drilled down to 105 m. The average rate of penetration was in the range of 3.5–4.0 m/h at a drill head power of approximately 3.2 kW. The core diameter was 75–78 mm, and the core length per run was ∼2 m. During the drilling process, a slush layer that was 3–5 m thick appeared at the upper part of the fluid column and led to difficulties, particularly during the lowering and raising of the drill in the hole. To reduce the formation of this slush, the weak ethanol solution in the upper part of the fluid column was partially changed to a stronger ethanol solution at the end of each shift. Further drilling in this hole was carried out using an ETB-3 thermal drill.

Then, the ETB-5 drill was tested at Pobeda base, at the 43-km mark of the Mirny–Vostok route in Antarctica. A special mobile drilling shelter was built (Fig. 2.75). In May–June 1980 (25th SAE), a hole was drilled down to 387 m, with an ice core production rate of 8.8 m/day (Kudryashov et al. 1983; Savatyugin and Preobrazhenskaya 2000).

In early 1981 (26th SAE), the drilling shelter was moved to Komsomolskaya Station, 870 km from Mirny. Drilling with the ETB-5 drill was not very successful and, at 73 m, the drill became stuck and was lost (Kudryashov et al. 1983). During March–October 1981, testing of the ETB-5 drill was continued and a few shallow holes were drilled near the runway at Mirny (Dyurgerov et al. 1987; Savatyugin and Preobrazhenskaya 2000).

From November 1981 to February 1982, drilling operations at Komsomolskaya Station were continued and a 800.6-m-deep hole was drilled (Manevskyi et al. 1983; Morev et al. 1988a, 1990). Drilling with the ETB-5 drill started from a depth of 120 m with the following mean parameters: a core length of 0.67 m; a core diameter of 79 mm; a power consumption of 2.5 kW; and a penetration rate of 3.6–3.8 m/h. In total, nearly 1200 runs were done with a mean core production of 10 m/day. The hole was then filled with an ethanol–water solution to a depth of 230 m, and the drilling site was secured and prepared for the next summer season. The temperature in the hole increased from −53.5 °C at a depth of 5 m to −49.8 °C at 800 m. The base was reopened in December 1982 and drilling with the ETB-5 drill was continued after a waiting period of 11 months. However, at a depth of 871.5 m, the drill quickly froze in the ice and became stuck because of an electrical failure within the drill. The 16.5-mm cable ruptured near the surface during a recovery attempt and 870 m of cable were left in the hole at a depth of 110 m. This hole was the deepest one drilled using an ethanol–water solution in coldest ice.

m-ATED and ETED thermal drills

In 1990s, V. S. Zagorodnov, Ohio State University, USA redesigned and improved the ETB-3 thermal drill and referred to it as the "modified Antifreeze Thermal Electric Drill (m-ATED)" (Zagorodnov et al. 1998). The operating principle of the m-ATED does not differ from that of the ETB-3 drill. However, some important modifications make it possible to enlarge the clearance between the wall and the drill, reduce the weight of the drill, and increase diameter of the ice cores.

The body of the m-ATED consists of a thin-walled (1.6 mm) stainless–steel tube (Fig. 2.76). An ethanol–water solution flows to the kerf through eight 6-mm-diameter tubes attached to the outside of the core barrel. This arrangement practically doubles the clearance between the drill and the borehole wall. A wedging mechanism keeps the piston fixed in a low position while lowering the drill in the hole. The piston is set free when the drill touches the bottom of the hole. The coring head consists of a stainless-steel base ring and two identical electrical cable heaters (the OD of the sheath is 1.62 mm; 40 Ω) coiled into a toroidal shape with an OD of 12 mm. The drill with its 2-m-long core barrel weighs 25 kg.

During the spring of 1997, two holes were drilled in the central part of Vetrennyi Ice Cap, Franz Josef Land using the m-ATED (Zagorodnov et al. 1998; Arkhipov et al. 2001). The first hole was drilled to a depth of 314.8 m. Normally, the core diameter was 97 ± 2 mm and its length was 2 ± 0.05 m. The average penetration rate was only 2.67 m/h because the flow of the ethanol–water solution passing through the drill head carried heat away from the kerf. The ethanol–water solution often froze inside the circulation tubes when the drill was raised to the surface and was exposed to colder temperatures. That is why before lowering the drill into the borehole, it was heated with a propane torch and the ice plugs were melted. The average core production rate was 36 m/day under one-shift operation. Borehole inclination was 11° at a depth of 75 m and 4°

Fig. 2.75 Schematic of the drilling shelter at Pobeda base, East Antarctica (Savatyugin and Preobrazhenskaya 2000)

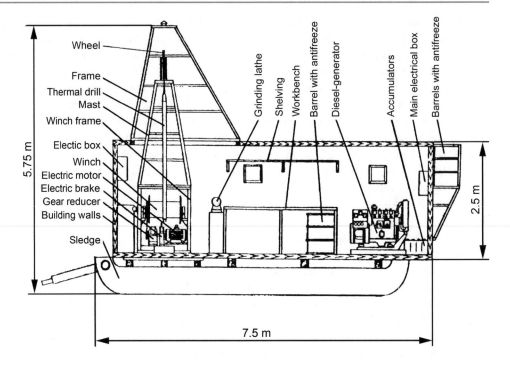

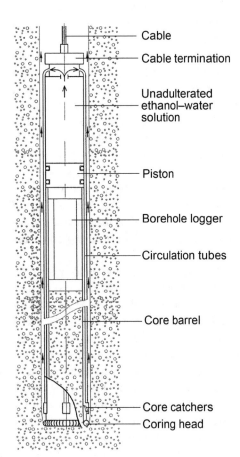

Fig. 2.76 m-ATED thermal drill (Zagorodnov et al. 1998)

at a depth of 315 m. A second hole was drilled 1.5 m away from the first hole and was terminated at a depth of 36 m.

In subsequent years, the antifreeze drill was again modified and renamed as the "Ethanol Thermal Electric Drill (ETED)" (Zagorodnov et al. 2005). A new core barrel was made out of a thick-walled (4.3 mm) stainless-steel tube (Fig. 2.77). This increased the weight of the drill to 48 kg and made the drill more robust. Six channels for the flow of the ethanol–water solution were machined into the core barrel and were covered with segments of stainless-steel square tubing with a wall thickness of 1.2 mm. The coring head consisted of one coiled tubular heater with a working power of 2.25 kW at 450 V DC.

A hand-operated valve was added to the circulation system of the ETED. The function of this valve was to disconnect the inner space of the core barrel (the container for the ethanol–water solution) from the channels while filling the drill with the antifreeze solution, enabling the drill to be filled in a horizontal position while removing the core. The valve was then returned to the drilling position when hanging vertically before descent.

In May–June 2002, an ETED was used for coring from a depth of 180 m, where electromechanical drilling had stopped, down to bedrock at 460 m at the saddle of Mt. Bona and Mt. Churchill in the Wrangell–St. Elias range, Alaska, USA (4200 m asl; temperature at a depth of 10 m: −24 °C). A few coring-head burnouts were experienced in the borehole.

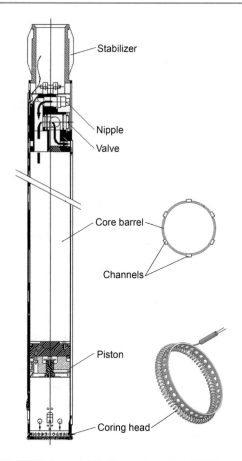

Fig. 2.77 ETED thermal drill (Zagorodnov et al. 2005)

Usually, the coring heads drilled 40–80 m or more of ice before requiring replacement. The ice-core production rate was very steady down to the bottom of the borehole and was ~2 m/h with cores that were 1.8–2 m in length. During the process of drilling, a few core catchers broke. To solve this problem, a neck was melted at the bottom of each ice core. To melt these necks, the drill was held for 40s slightly above the bottom after penetration and before breaking the core. At a depth of 380 m, the thermal head burnt out and the drill got stuck (Zagorodnov, personal communication, 2017). To recover stuck drill, a short ETED was used to deploy glycol down the borehole through a column of approximately 200 m of the ethanol–water solution. In order to release glycol from the drill, a short extension rod was mounted on the piston. When the drill reached the top of the stuck drill, the rod pushed the piston and glycol was expelled. Close to the bottom, a few (5–10 mm thick) layers containing small (4–6 mm) pebbles were penetrated without a noticeable reduction in penetration rate.

In June 2003, the ETED was tried from penetrating into warm ice when approaching the base of the ice sheet at a depth of 3001 m at NorthGRIP base, Greenland. The hole was filled with a hydrophobic drilling fluid Exxsol™ D60 solvent mixed with HCFC 141b densifier. Unfortunately, all

six high-resistance (up to 100 Ω) coring heads with a shielded electric return burnt out at the moment of touching the bottom. The heads had been preliminarily checked at pressures up to 60 MPa for 48 h and were considered to be able to run at up to 800 V. A possible explanation for the failure of the coring heads is dielectric breakdown due to the changes in heater geometry and the opening of micro fractures in the shield under the high hydrostatic and superimposed contact pressures.

The ETED was slightly redesigned for using it without any ethanol–water solution in temperate glaciers (Zagorodnov et al. 2005). The piston and manifold were removed from the drill, which significantly increased the drill's lowering rate and the drilling-run penetration depth. In 2003, a hole drilled using a shallow electromechanical drill at Quelccaya Ice Cap, Peru was continued from 124 m to a final depth of 168 m with this thermal drill (Fig. 2.78). The presence of capillary water in the firn complicated the drilling operations. A small amount of meltwater constantly ran into the borehole. From 100 to 124 m, the core length recovered via electromechanical drilling decreased to 0.1 m per run, and that was a reason for switching to thermal drilling.

Fig. 2.78 Drilling operations at Quelccaya Ice Cap, Peru, 2003 (*Credit* V. Zagorodnov)

Fig. 2.79 Inside the drilling tent, Bruce Plateau, Antarctica, January 2010 (*Credit*: V. Zagorodnov)

During the 2009–2010 season, the ETED was used to drill a hole from 178.5 m (down to this depth, a BPRC electromechanical shallow drill had been used) to bedrock at 447.65 m within the Larsen Ice Shelf System, Antarctica (LARISSA) project on the Bruce Plateau (Fig. 2.79) (Zagorodnov et al. 2012). During the operations, the level of the ethanol–water solution used was lowered (via bailing) to a depth of 244 m.

At a depth of 384 m, the heating element of the drill head burned out and the drill was frozen at the bottom (The Stuck Ice Drill Saga 2010). After five days of preparation and experiments, the drill was recovered in the same way as in the Bona-Churchill drilling operations of 2002. The up and down speed of the drill while travelling in the borehole showed that the diameter of the borehole was slightly constricted at a depth of approximately 288 m. A pump was installed into the bailer and then the weak ethanol–water solution was enriched with pure ethanol at specified depth intervals. Then, approximately 9 L of glycol was delivered to the top of the drill in two different runs. Subsequently, the

Fig. 2.80 Removing an ice core from the recovered ETED drill, January 2010 (The Stuck Ice Drill Saga 2010)

drill cable was placed under tension and, after waiting for approximately 36 h, the ice around the drill had dissolved enough to release it. The drill was raised to the surface with a 0.66-m-long core that had been drilled before the drill became trapped (Fig. 2.80). When drilling was completed, a minimum temperature of −15.8 °C was measured at a depth of 173 m.

2.3.6 Summary

Aside from the benefits of having increased maximum-depth capabilities, this type of thermal drills shows limitations (such as being slow and power-hungry) similar to those of dry-hole thermal drills, including poor ice-core quality due to the thermoelastic stress caused by drilling–melting (Nagornov et al. 1994). For many years, the dispute between the adherents of using hydrophobic and hydrophilic drilling fluids did not stop. The truth is somewhere in between: both of them are reliable under certain specific conditions.

Hydrophobic drilling fluids were usually based on petroleum fuels DF-A, Jet A1, JP-8, and TS-1, which are all kerosenes (Talalay et al. 2014b). Kerosenes have a density of 800–850 kg/m^3 at −30 °C, which is lower compared with the 917–924 kg/m^3 for ice. Therefore, to compensate fully for the overburden ice pressure, they were made denser by mixing them with fluorocarbons that had a density significantly higher than the density of ice. However, all these densifiers are harmful and environmentally dangerous substances, and by now most countries have prohibited their use.

Ethanol is much more desirable than kerosene-based drilling fluids with respect to human toxicity and environmental pollution. However, choosing the right ethanol concentration is the tricky part. If there is too much antifreeze, the core will be attacked, especially at grain boundaries; if there is too little, there is the danger of the drill freezing in. In order to avoid the formation of hard slush in the borehole, concentrations higher than the equilibrium concentration of ethanol must be used in the bottom part of the borehole, which results in the partial dissolution of the ice core and the wall (Zagorodnov et al. 2005). A positive effect of using an excessive ethanol concentration is the increased borehole diameter, which results in a higher lowering rate.

The equilibrium concentration of ethanol aqueous solutions depends on temperature. Therefore, as borehole temperature changes owing to convection and heat from the drill, cable, and logging tools, frozen water precipitates from the aqueous solution, and slush is formed in the borehole. For the purposes of deep drilling in very cold ice (< −30 °C) and especially given that the borehole may be required to stay accessible for many years, experience has demonstrated that hydrophilic fluids are not suitable.

The maximal depth drilled in the cold ice of central Antarctica using thermal drills with kerosene-based drilling fluid reached 2755.3 m at Vostok Station, whereas the deepest hole drilled using ethanol was only 871.5 m deep at Komsomolskaya Station. On the other hand, antifreeze drilling was successfully used to recover thousands of meters of cores in temperate, polythermal, and moderately cold glaciers in Antarctic ice shelves, the Arctic, and high-mountain regions for shallow and intermediate depths. A combination of dry-hole electromechanical drilling techniques and antifreeze drilling provided an effective approach for acquiring intermediate depth ice cores in high-elevation glaciers.

Some theoretical predictions were carried out in order to find the optimal shape of the solid thermal coring heads and the most effective drill load range (Kudryashov et al. 1973b, 1991); however, results often contradict experiments. The following types of resistive heaters were used: (1) cartridge heaters (Ueda and Garfield 1968), (2) open nichrome bare wire (Forage 4000 1986), (3) isolated nichrome wire packaged into a sealed ring body (Bogorodsky and Morev 1984), and (4) sheathed tubular heating elements cast into a copper or aluminum body (Kudryashov 1989) or coiled into a toroidal shape (Zagorodnov et al. 1998). It was observed that cartridge heaters had several undesirable features in practice, such as the frequent failure of sealed electrical connections, the multiplicity of vulnerable electrical lines, and servicing difficulties. Open nichrome wires tend to burn out when immersed in water and, hence, should be installed in a sealed chamber. The most reliable thermal heads used coiled tubular heating elements.

Theoretically, the rate of penetration linearly increases with power density, which is given by the input power per unit cross-section area of the thermal head. To drill faster, one needs to follow a simple rule: the thermal head should have as small a cross-section as possible and the highest allowable power. Typically, the power density of thermal heads was in the range of 50–104 W/cm^2 with two exceptions. The power density of the thermal head designed in the LGGE was ~175 W/cm^2 owing to its small size (with an OD and an ID of 145 and 125 mm, respectively) and its high rated power (7.5 kW). The power density of thermal head of the TBS-112VCh drill was also quite high at 140 W/cm^2, allowing it to drill with a penetration rate of up to 4.2 m/h in cold ice at Vostok station.

Axial load should be applied vertically on the thermal head to squeeze out meltwater from the head's face and minimize the heat transfer losses in the meltwater layer between the head and bottom of the hole. Shkurko (personal communication, 2004) observed that increasing the specific load from 0.5 to 2 kN/m^2 can considerably accelerate the melting rate. However, at loads higher than 2 kN/m^2, the melting rate become stable even at loads as high as 10 kN/m^2.

Even though ice-core cable-suspended electromechanical drills with low-temperature fluids have replaced thermal coring drills in most applications, there are still a few thermodynamic conditions in glaciers (near-temperate and polythermal glaciers) in which thermal drills are still more effective than electromechanical drills. If one plans to renew thermal drilling with hydrophobic drilling fluids, research should be done to study the compatibility of heating elements with new drilling fluids, such as synthetic ESTISOL^{TM} esters or dimethylsiloxane oils.

References

Abyzov SS, Bobin NE, Kudryashov BB et al (1988) Mikrobiologicheskiye issledovaniya ledyanogo kerna iz glubokikh skvazhin lednika Tsentral'noi Antarktidy [Microbiological study of ice core from deep boreholes in Central Antarctic ice sheet]. Informatsionny Byulleten' Sovetkoj Antarkticheskoj Ekspeditsii [Soviet Antarctic Expedition Inform Bull] 111:86–91 (in Russian)

Abyzov SS, Mitskevich IN, Ivanov MV (2007) Above the Lake Vostok. Climate change and polar research. In: Abyzov S, Perovich D (eds) Luso-American development foundation. Textype—Artes Gráficas, Lda., Lisbon, pp 11–39

Anonymous Author (1986) Sovetskie glyatsiologicheskie issledovaniya na Vostochnom ledyanom pole v 1984–1985 gg. [Soviet glaciological investigations on Nordaustlandet in 1984–1985]. Akademiia nauk SSSR. Institut geografii. Materialy gliatsiologicheskikh issledovanii [Academy of Sciences of the USSR. Institute of Geography. Data of Glaciological Studies] 56:10–26. (Text in Russian)

Arkhipov SM, Vaykmyae RA, YeV Vasilenko et al (1987) Soviet glaciological investigations on Austfonna, Nordaustlandet, Svalbard in 1984–1985. Polar Geogr Geol 11(1):25–49

Arkhipov SM, Mikhalenko VN, Tompson LG et al (2001) Stratigrafiya deyatel'nogo sloya lednikovogo kupola Vetrennyi na o. Graham-Bell, Zemlya Ftrantsa-Iosifa [Stratigraphy of the active layer of Windy Ice Cap on Graham Bell Island, Frantz Josef Land]. Rossiiskaya Akademiya nauk. Institut geografii. Materialy gliatsiologicheskikh issledovanii [Russian Academy of Sciences. Institute of Geography. Data of Glaciological Studies] 90:169–186. (Text in Russian)

Augustin L, Donnou D, Rado C et al (1988) Thermal ice core drill 4000//Ice Core Drilling. In: Rado C, Beaudoing D (eds) Proceedings of the third international workshop on ice drilling technology, 10–14 October, 1988. Grenoble, France. Laboratoire de Glaciologie et Geophysique de l'Environnement, Grenoble, pp 59–65

Barkov NI (1960) Elektrobur dlya bureniya skvazhin vo l'du [Electric drill for coring in ice]. USSR Pat. 127,629

Barkov NI (1963) Burenie glubokikh skvazhin v lednikovom pokrove metodom protaivaniya [Drilling of deep holes in the ice by the thermal method]. Informatsionny Byulleten' Sovetkoj Antarkticheskoj Ekspeditsii [Soviet Antarctic Expedition Inform Bull] 40:39–42 (in Russian)

Barkov NI (1970) Predvaritel'nye rezul'taty bureniya lednikovogo pokrova na stantsii Vostok [Preliminary results of ice drilling at Vostok Station]. Informatsionny Byulleten' Sovetkoj Antarkticheskoj Ekspeditsii [Soviet Antarctic Expedition Inform Bull] 80:24–29 (in Russian)

Barkov NI, Bobin NE, Stepanov GK (1973) Burenie skvazhiny v lednikovom pokrove Antarktidy na stantsii Vostok v 1970 g. [Drilling in the Antarctic ice sheet at Vostok Station in 1970].

Informatsionny Byulleten' Sovetkoj Antarkticheskoj Ekspeditsii [Soviet Antarctic Expedition Inform Bull] 85:22–28 (in Russian)

Bazhev AB, Zagorodnov VS, Rototaeva OV (1988) Burovye raboty v oblasti pitaniya lednika Garabashi na El'bruse [Drilling operations in the ice-feeding region of Garabashi Glacier at Elbrus]. Akademiia nauk SSSR. Institut geografii. Materialy gliatsiologicheskikh issledovanii [Academy of Sciences of the USSR. Institute of Geography. Data of Glaciological Studies] 64:11–12. (in Russian)

Bird IG (1976) Thermal ice drilling: Australian developments and experience. In: Splettstoesser JF (ed) Ice-core drilling. Proceedings of the symposium, 28–30 August, 1974. University of Nebraska, Lincoln, USA. University of Nebraska Press, Lincoln, pp 1–18

Bird IG, Ballantyne J (1971) The design and application of a thermal ice drill: technical note 3. Department of Supply, Antarctic Division, Melbourne

Bobin NE, Fisenko VF (1974) Opyt termobureniya skvazhin s otborom kerna v pohodnykh usloviyah [Experimental thermal core drilling on a traverse]. Informatsionny Byulleten' Sovetkoj Antarkticheskoj Ekspeditsii [Soviet Antarctic Expedition Inform Bull] 88:74–76 (in Russian)

Bobin NE, Moiseev BS, Zemtsov AA (1986) Rezul'taty bureniya skvazhiny s primeniem nizkotemperaturnoi zalivochnoi zhidkosti na base Gornaya [Results of drilling with low-temperature fluid at the Base Gornaya]. Dvadtsat' chetvertaia Sovetskaya antarkticheskaya ekspeditsia. Zimovochnyie issledovania 1978–1980 gg.: Obshchee opisanie i nauchnye rezu-l'taty [Twenty fourth Soviet Antarctic Expedition. Over-wintering research 1978–1980: general description and scientific results]. Trudy Sovetskoi antarkticheskoi ekspeditsii. [Transactions of Soviet Antarctic Expedition] 81:93–101. (in Russian)

Bobin NE, Kudryashov BB, Pashkevitch VM et al (1994) Equipment and methods of microbiological sampling from deep levels of ice in central Antarctica. Mem Natl Inst Polar Res Spec Issue 49:184–191

Bobin NE, Kudryashov BB, Pashkevitch VM et al (1998) Tekhnika i tekhnologiya asepticheskogo otbora prob dlya mikrobiologicheskikh issledovanyi tolshchi lednika [Technique and methods of aseptic sampling for microbiological study of ice sheets]. Antarktika; doklady komissii [The Antarctic. The Committee Reports] 34:79–88. (in Russian)

Bogorodsky VV, Morev VA (1984) Equipment and technology for core drilling in moderately cold ice.In: Holdsworth G, Kuivinen KC, Rand JH (eds) Proceedings of the second international workshop/symposium on ice drilling technology, 30–31 August, 1982. Calgary, Alberta, Canada. USA CRREL Spec. Rep. 84–34, pp 129–132

Bogorodsky VV, Morev VA, Pukhov VA et al (1984) New equipment and technology for deep core drilling in cold glaciers. In: Holdsworth G, Kuivinen KC, Rand JH (eds) Proceedings of the second international workshop/symposium on ice drilling technology, 30–31 August, 1982. Calgary, Alberta, Canada. USA CRREL Spec. Rep. 84–34, pp 139–140

Burn-Nunes LJ, Vallelonga P, Loss RD et al (2011) Seasonal variability in the input of lead, barium and indium to Law Dome, Antarctica. Geochim Cosmochim Ac 75:1–20

Chistyakov VK, Skurko AM, Zemtsov AA et al (1988) Eksperimental'niye burobie raboty na Severnoi Zemle v 1975–1985 gg. [Experimental drilling operations at Severnaya Zemlya in 1975–1985]. Geograficheskie i glyatsiologicheskie issledovaniya v polarnikh stranakh [Geographical and Glaciological Investigations in Polar Regions]. Leningrad, Gidrometroizdat, pp 33–42. (in Russian)

Deep Ice Coring Project at Dome Fuji, East Antarctica (1992) Prospect of National Institute of Polar Research. Tokyo, Japan

Dmitriev DN, Vostretsov RN (1979) Opredelenie gorizontal'noi skorosti techenia l'da v raione stantsii Vostok do glubiny 800 m [Estimation of horizontal flow speed of ice in the region of Vostok Station to the depth of 800 m]. Informatsionny Byulleten' Sovetkoj Antarkticheskoj Ekspeditsii [Soviet Antarctic Expedition Inform Bull] 99:32–34

Dmitriev DN, Vostretsov RN, Petukhov IA (1978) Deformatsiya stenok glubokoi skvazhiny v Antartkticheskom lednikovom pokrove na stantsii Vostok [Deformation of deep borehole walls in the Antarctic ice sheet at Vostok Station]. Informatsionny Byulleten' Sovetkoj Antarkticheskoj Ekspeditsii [Soviet Antarctic Expedition Inform Bull] 98:53–57 (in Russian)

Dmitriev AN, Zubkov VM, Krasilev AV et al (1995) Rezul'tati bureniya skvazhini 5G na stantsii Vostok v 1991 g. [Results of 5G hole drilling at Vostok Station in 1991]. Akademiia nauk SSSR. Institut geografii. Materialy gliatsiologicheskikh issledovanii [Academy of Sciences of the USSR. Institute of Geography. Data of Glaciological Studies] 79:174–176. (in Russian)

Donnou D, Gillet F, Manouvrier A et al (1984) Deep core drilling: Electro-mechanical or thermal drill? In: Holdsworth G, Kuivinen KC, Rand JH (eds) Proceedings of the second international workshop/symposium on ice drilling technology, 30–31 August, 1982. Calgary, Alberta, Canada. USA CRREL Spec. Rep. 84–34, pp 81–84

Donnou D, Augustin L, Manouvrier A et al (1988) Setting up a deep ice core drilling facility and preliminary tests Terre Adelie—Antartica//Ice Core Drilling. In: Rado C, Beaudoing D (eds) Proceedings of the third international workshop on ice drilling technology, 10–14 October, 1988. Grenoble, France. Laboratoire de Glaciologie et Geophysique de l'Environnement, Grenoble, pp 66–69

Drilling ice cores (n.d.). National Ice Core Laboratory. About ice cores. Retrieved January 29, 2017 from http://icecores.org/icecores/drilling.shtml

Dyurgerov MB, Korolev PA, Manevsky LN et al (1987) Issledovaniya srednemnogoletney akkumulyatsii atmosfernikh osadkov v raione observatotii Mirny [Investigation of the average precipitation in the region of Mirny Station]. Informatsionny Byulleten' Sovetkoj Antarkticheskoj Ekspeditsii [Soviet Antarctic Expedition Inform Bull] 109:51–57 (in Russian)

Etheridge DM (1989) Dynamics of the Law Dome Ice Cap, Antarctica, as found from bore-hole measurements. Ann Glaciol 12:46–50

Etheridge DM, Wookey CW (1988) Ice core drilling at a high accumulation area of Law Dome, Antarctica, 1987. In: Rado C, Beaudoing D (eds) Ice core drilling. In: Proceedings of the third international workshop on ice drilling technology, 10–14 October, 1988, Grenoble, France. Laboratoire de Glaciologie et Geophysique de l'Environnement, Grenoble, pp 3–5

Fisenko VF, Bobin NE, Stepanov GK et al (1974) Oslozhneniia i avarii pri glubokom burenii-protaivanii, ikh likvidatsiia i preduprezhdenie [Complications and accidents during deep thermal drilling: their elimination and prevention]. Antarktika. Doklady Komissii [The Antarctic. The Committee Reports] 13:161–166. (in Russian)

Forage 4000 (1986) Rapport technique. Laboratoire De Glaciologie et Géophysique de l'Environnement. Centre National de la Recherche Scientifique, Saint-Martin-d'Heres, France

Giles JD (1994) Washington USGS thermal drilling project on the South Cascade Glacier 1994. Polar Ice Coring Office, University of Alaska Fairbanks, USA. PICO Report OR-94–04

Giles J (2004) Russian bid to drill Antarctic lake gets chilly response. Nature 430:494

Gillet F, Donnou D, Ricou G (1976). A new electrothermal drill for coring in ice. In: Splettstoesser JF (ed) Ice-core drilling. Proceedings of the symposium, 28–30 August, 1974. University of Nebraska, Lincoln, USA. University of Nebraska Press, Lincoln, pp 19–27

Goetz J (2015) Collaborative research: the Greenland firn aquifer impacts on ice sheet hydrology: characterizing volume, flow, and discharge (1417987). End-of-season project report. IDDO, University of Wisconsin-Madison, USA

Hamley TC, Morgan VI, Thwaites RJ et al (1986) An ice-core drilling site at Law Dome Summit, Wilkes Land, Antarctica. ANARE Res. Notes 37

Hansen BL, Langway Jr CC (1966) Deep core drilling in ice and core analysis at Camp Century, Greenland 1961–1966. Antarct. J. U.S. 1 (5):207–208

Hantz D, Lliboutry L (1983) Water ways, ice permeability at depth, and water pressures at Glacier D'Argentiere. French Alps J Glaciol 29 (102):227–239

Holdsworth G (1984) The Canadian Rufli-Rand electro-mechanical core drill and reaming devices. In: Holdsworth G, Kuivinen KC, Rand JH (eds) Proceedings of the second international workshop/symposium on ice drilling technology, 30–31 August, 1982. Calgary, Alberta, Canada. USA CRREL Spec. Rep. 84–34, pp 21–32

Jouzel J, Lorius C, Raynaud D (2013) The White Planet: the evolution and future of our frozen world. Princeton University, New Jersey, USA

Kislov BV, Manevsky LN, Savatiugin LM (1983) Gliatsiologicheskie issledovaniia v raione stantsii Novolazarevskoi [Glaciological studies in the Novolazarevskaya Station area]. Sovetskaia antarkticheskaia ekspeditsiia. Trudy [Soviet Antarctic Expedition. Trans] 76:125–130. (in Russian)

Klein ES, Nolan M, McConnell J et al (2016) McCall Glacier record of Arctic climate change: Interpreting a northern Alaska ice core with regional water isotopes. Quat Sci Rev 131:274–284

Klement'ev OL, Korotkov IM, Nikolaev VI (1988) Glyatsiologicheskie issledovaniya v 1987–1988 gg. na lednikovykh kupolakh Severnoi Zemli [Glaciological investigations on ice caps of Severnaya Zemlya in 1987–1988]. Akademiia nauk SSSR. Institut geografii. Materialy gliatsiologicheskikh issledovanii [Academy of Sciences of the USSR. Institute of Geography. Data of Glaciological Studies] 63, pp 25–26. (in Russian)

Koci BR (1984) New horizons in drill development. In: Holdsworth G, Kuivinen KC, Rand JH (eds) Proceedings of the second international workshop/symposium on ice drilling technology, 30–31 August, 1982, Calgary, Alberta, Canada. USA CRREL Spec. Rep. 84–34, pp 51–54

Koci B (1985) Ice-core drilling at 5700 m powered by a solar voltaic array. J Glaciol 31(109):360–361

Koci B (2002) A review of high-altitude drilling. Mem Natl Inst Polar Res Spec Issue 56:1–4

Koci B, Zagorodnov V (1994) The Guliya Ice Cap, China: Retrieval and return of a 308-m ice core from 6200-m altitude. Mem Natl Inst Polar Res Spec Issue 49:371–376

Koerner RM, Kane HS (1967) Glaciological studies at Plateau Station. Antarct. J. U.S. 2(4):122–123

Korotkevich YeS, Kudryashov BB (1976) Ice sheet drilling by Soviet Antarctic Expeditions. In: Splettstoesser JF (ed) Ice-core drilling. Proceedings of the symposium 28–30 August, 1974, University of Nebraska, Lincoln, USA. University of Nebraska Press, Lincoln, pp 63–70

Korotkevich ES, Savatiugin LM, Morev VA (1978) Skvoznoe burenie shel'fovogo lednika v raione stantsii Novolazarevskoi [Drilling through shelf ice near Novolazarevskaya Station]. Informatsionny Byulleten' Sovetkoj Antarkticheskoj Ekspeditsii [Soviet Antarctic Expedition Inform Bull] 98:49–52 (in Russian)

Kotlyakov VM (ed) (1985) Glyatsiologiya Shpitsbergena [Glaciology of Spitsbergen]. Nauka, Moscow (in Russian)

Kovalenko VI, Moiseev BS, Zagrivnyi EA (1981) Burenie-protaivanie skvazhiny na stantsii Vostok-1 [Thermal core drilling at Vostok-1 station]. Sovetskaia antarkticheskaia ekspeditsiia. Trudy [Soviet Antarctic Expedition. Trans] 73, pp 112–116. (in Russian)

Kudryashov BB (1989) Soviet experience of deep drilling in Antarctic. In: Bandopadhyay S, Skudrzyk FJ (eds) Proceedings first

international symposium mining in the Arctic, Fairbanks, July 17–19, 1989. Rotterdam, A.A. Balkema. Netherlands, pp 113–122

Kudryashov BB, Fisenko VF, Stepanov GK, et al (1973a) Opyt bureniia ledyanogo pokrova Antarktidy [Experience drilling into the Antarctic ice sheet]. Antarktika. Doklady Komissii [The Antarctic. The Committee Reports] 12:145–152. (in Russian)

Kudryashov BB, Bobin NE, Sliusarev NI et al (1973b) Teoriia i praktika bureniia-protaivaniia v Antarktide [Theory and practice of thermal drilling in Antarctica]. Akademiia nauk SSSR. Institut geografii. Materialy gliatsiologicheskikh issledovanii [Academy of Sciences of the USSR. Institute of Geography. Data of Glaciological Studies] 22:71–77. (in Russian)

Kudryashov BB, Chistyakov VK, Vartykyan VG (1975) Rezul'taty issledovanii i razrabotok po glubokomu bureniyu-protaivaniyu vo l'dah Antarktidy [Investigation and designing results of deep drilling-melting in ice of Antarctica]. Fyzicheskie protsessy gornogo proizvodstva. Mezhvuzovsky sbornik, Vipusk 1 [Physical Processes in Mining. Inter-institution Collection of Articles, Issue 1] Leningrad Mining Institute, Leningrad, USSR, pp 111–116. (in Russian)

Kudryashov BB, Chistyakov VK, Bobin NE (1977a) Burenie skvazhin teplovim sposobom v lednikovom pokrove Antarktidy [Bore-hole thermal drilling in Antarctic Ice Sheet]. Tekhnika i tekhnologia geologorazvedochnikh rabot. Organizatsia proizvodstva: Obzor [Technique and technology of prospecting. Works managing: Review] Moscow, Vsesouznyi Nauchno-Issledovatel'skii Institut Ekonomiki Mineral'nogo Sir'ia i Gelogorazvedochnikh rabot (VIEMS) [All-Union Research Institute of Minerals Economy and Prospecting]. (in Russian)

Kudryashov BB, Vartykyan VG, Chistyakov VK (1977b) Termoburovoy snaryad dlya polucheniya orientirovannogo ledyanogo kerna. (Thermodrill for recovering an oriented core). Informatsionny Byulleten' Sovetkoj Antarkticheskoj Ekspeditsii [Soviet Antarctic Expedition Inform Bull] 94:60–62 (in Russian)

Kudryashov BB, Abyzov SS, Bobin NE (1978) Otbor prob dlia mikrobiologicheskikh issledovanii glubokikh gorizontov lednikovoi tolshchi na stantsii Vostok [Sampling for microbiological studies of deep glacier layers at Vostok Station]. Informatsionny Byulleten' Sovetkoj Antarkticheskoj Ekspeditsii [Soviet Antarctic Expedition Inform Bull] 98:58–62 (in Russian)

Kudryashov BB, Chistyakov VK, Morev VA (1983) Burenie lednikovogo pokrova Antarktidy teplovym sposobom [Thermal drilling in Antarctic Ice Sheet]. In: Korotkevich ES, Dubrovin LI (eds) 25 let Sovetskoi antark-ticheskoi ekspeditsii [25 years of Soviet Antarctic Expeditions]. Leningrad, Gidrometeoizdat, USSR, pp 138–149. (Text in Russian)

Kudryashov BB, Chistyakov VK, Zagrivnyii EA et al (1984a) Preliminary results of deep drilling at Vostok Station, Antarctica, 1981–82. In: Holdsworth G, Kuivinen KC, Rand JH (eds.) Proceedings of the second international workshop/symposium on ice drilling technology, 30–31 August, 1982, Calgary, Alberta, Canada. USA CRREL Spec. Rep. 84–34, pp 123–124

Kudryashov BB, Chistyakov VK, Bobin NE (1984b) Problema bureniia glubokikh skvazhin v tsentral'nykh raionakh Antarktidy [Problems of drilling deep wells in central parts of Antarctica]. Akademiia nauk SSSR. Institut geografii. Materialy gliatsiologicheskikh issledovanii [Academy of Sciences of the USSR. Institute of Geography. Data of Glaciological Studies] 51:168–172. (in Russian)

Kudryashov BB, Chistyakov VK, Litvinenko VS (1991) Burenie skvazhin v usloviiakh izmeneniia agregatnogo sostoianiia gornykh porod [Drilling wells under changes in the aggregate state of rocks]. Nedra, Leningrad (in Russian)

Kudryashov BB, Vasiliev NI, Talalay PG (1994) KEMS-112 electromechanical ice core drill. Mem Natl Inst Polar Res Spec Issue 49:138–152

Kudryashov BB, Krasilev AV, Talalay PG et al (1998a) Drilling equipment and technology for deep ice coring in Antarctica In: Hall J (ed) Proceedings of the seventh symposium on Antarctic Logistics and Operations, 6–7 August, 1996. Cambridge, United Kingdom, British Antarctic Survey, Cambridge, UK, pp 205–212

Kudryashov BB, Chistyakov VK, Vasiliev NI et al (1998b) Burenie i issledovanie glubokoi skvazhiny na stantsii Vostok [Deep ice drilling and logging at Vostok station] Antarktika; Doklady komissii [The Antarctic. The Committee Reports] 34:73–78. (in Russian)

Kudryashov BB, Vasiliev NI, Vostretsov RN et al (2002) Deep ice coring at Vostok Station (East Antarctica) by an electromechanical drill. Mem Natl Inst Polar Res Spec Issue 56:91–102

Kyne J (2013) Drilling into water. Retrieved 18 March, 2017 from https://earthobservatory.nasa.gov/blogs/fromthefield/category/greenland-aquifer-expedition/page/8/

Langway CC (2008) The history of early polar ice cores. USA CRREL Tech. Rep. ERDC/CRREL TR-08-1

Long Range Drilling Technology Plan (2016) Prepared by the IDDO in collaboration with the Ice Drilling Program Office, University of Wisconsin-Madison, USA. Available on-line at http://icedrill.org/documents/view.shtml?id=272

Lorious C, Donnou D (1978a) Campagne en Antarctique, Novembre 1977–Fevrier 1978. Centre National de la Recherche Scientifique, Laboratoire de Glaciologie, Courrier 30:6–17

Lorious C, Donnou D (1978b) A 905-meter deep core drilling at dome C (East Antarctica) and related surface programs. Antarct. J. U.S. 13 (4):50–51

Manevskyi LN, Morev VA, Nikiforov AG et al (1983) Eksperimental'noe burenie na stantsii Komsomol'skoi [Experimental drilling at Komsomolskaya Station]. Informatsionny Byulleten' Sovetkoj Antarkticheskoj Ekspeditsii [Soviet Antarctic Expedition Inform Bull] 103:71–73 (in Russian)

Marec G, Maitre M, Pinglot F et al (1988) Telemetering and remote control circuits for a 4000 m thermal drill. In: Rado C, Beaudoing D (eds) Ice core drilling. Proceedings of the third international workshop on ice drilling technology, Grenoble, France, 10–14 October, 1988. Laboratoire de Glaciologie et Geophysique de l'Environnement, Grenoble, pp 72–85

Miège C (2015) 18 Days on the ice. Retrieved 18 March, 2017 from https://earthobservatory.nasa.gov/blogs/fromthefield/category/greenland-aquifer-expedition/page/2/

Miège C, Forster RR, Brucker L et al (2016) Spatial extent and temporal variability of Greenland firn aquifers detected by ground and airborne radars. J. Geophys. Res. Earth Surf. 121:2381–2398

Miller O (2016) So what did 6 scientists do for 3 weeks on the ice sheet? Retrieved 18 March, 2017 from https://earthobservatory.nasa.gov/blogs/fromthefield/category/greenland-aquifer-expedition/

Miller OL, Solomon DK, Miège C et al (2017) Hydraulic conductivity of a firn aquifer in Southeast Greenland. Front Earth Sci 5 Article 38:1–13

Morgan VI, McCray AP, Wehrle E (1984). Ice drilling at Cape Folger, Antarctica. In: Holdsworth G, Kuivinen KC, Rand JH (eds) Proceedings of the second internationa workshop/symposium on ice drilling technology, 30–31 August, 1982. Calgary, Alberta, Canada. USA CRREL Spec. Rep. 84–34, pp 85–86

Morgan V, Wehrle E, Fleming A et al (1994) Technical aspects of deep ice drilling on Law Dome. Mem Natl Inst Polar Res Spec Issue 49:78–86

Morev VA (1966) Opyty po bureniiu l'da elektroteplovym sposobom v Mirnom [Experiments in electrothermal ice drilling at Mirnyii]. Informatsionny Byulleten' Sovetkoj Antarkticheskoj Ekspeditsii [Soviet Antarctic Expedition Inform Bull] 56:52–56 (in Russian)

Morev VA (1972) Ob effektivnosti i ekonomichnosti elektrotermoburovykh snariadov pri burenii materikovogo l'da [Efficiency and economics of electrothermal devices during inland ice drilling].

Sovetskaia antarkticheskaia ekspeditsiia, Trudy [Soviet Antarctic Expedition. Transactions] 55:158–165. (in Russian)

Morev VA (1976) Elektrotermobury dlia bureniia skvazhin v lednikovom pokrove [Electric thermal drills for glacier core drilling]. Akademiia nauk SSSR. Institut geografii. Materialy gliatsiologicheskikh issledovanii [Academy of Sciences of the USSR. Institute of Geography. Data of Glaciological Studies] 28:118–120. (in Russian)

Morev VA, Shamont'ev VA (1970) Eksperimental'noye burenie lednikovogo pokrova [Experimental Ice Drilling]. Informatsionny Byulleten' Sovetkoj Antarkticheskoj Ekspeditsii [Soviet Antarctic Expedition Inform Bull] 78:102–104 (in Russian)

Morev VA, Raikovsky YuV (1979) Burenie Antarkticheskogo lednikovogo pokrova v raione stantsii Novolazarevskaia [Drilling of the Antarctic Ice Sheet in the Novolazarevskaya Station area]. Akademiia nauk SSSR. Institut geografii. Materialy gliatsiologicheskikh issledovanii [Academy of Sciences of the USSR. Institute of Geography. Data of Glaciological Studies] 37:198–200. (in Russian)

Morev VA, Pukhov VA (1981) Eksperimental'nye raboty po bureniiu kholodnykh pokrovnykh lednikov termoburovymi snariadami AANII [Using AANII thermodrills in experimental drilling of cold ice sheets]. Trudy Arkticheskogo i Antarkticheskogo nauchno-issledovatel'skogo instituta [Trans Arctic Antarctic Res Inst] 367:64–68 (in Russian.)

Morev VA, Pukhov VA, Yakovlev VM (1981) Burenie skvazhiny na lednike Vavilova, Severnaya Zemlya [Core drilling through Vavilov Glacier, Severnaya Zemlya]. Akademiya nauk SSSR. Institut geografii. Materialy gliatsiologicheskikh issledovanii [Academy of Sciences of the USSR. Institute of Geography. Data of Glaciological Studies] 40:154–157. (in Russian)

Morev VA, Pukhov VA, Yakovlev VM et al (1984) Equipment and technology for drilling in temperate glaciers. In: Holdsworth G, Kuivinen KC, Rand JH (Eds) Proceedings of the second international workshop/symposium on ice drilling technology, 30–31 August, 1982. Calgary, Alberta, Canada. USA CRREL Spec. Rep. 84–34, pp 125–127

Morev VA, Manevskiy LN, Yakovlev VM et al (1988a) Drilling with ethanol-based antifreeze in Antarctica In: Rado C, Beaudoing D (eds) Ice core drilling. Proceedings of the third international workshop on ice drilling technology, Grenoble, France, 10–14 October, 1988. Laboratoire de Glaciologie et Geophysique de l'Environnement, Grenoble, pp 110–113

Morev VA, Klement'ev OL, Manevskii LN et al (1988b) Glyatsio-burobye raboty na lednike Vavilova v 1979–1985 gg. [Ice drilling on Vavilov Glacier in 1979–1985]. Geograficheskie i glyatsiologicheskie issledovaniya v polarnikh stranakh [Geographical and glaciological investigations in polar regions] Leningrad, Gidrometroizdat, USSR, pp 25–32. (in Russian)

Morev VA, Manevskyi LN, Yakovlev VM et al (1990) Opyt bureniia skvazhin s zalivkoi antifriznoi zhidkost'iu na osnove etanola v Antarktike [Experience in drilling bore-holes filled with an ethanol-based antifreeze fluid in Antarctica]. Akademiia nauk SSSR. Institut geografii. Materialy gliatsiologicheskikh issledovanii [Academy of Sciences of the USSR. Institute of Geography. Data of Glaciological Studies] 68:181–184. (in Russian)

Nagornov OV, Zagorodnov VS, Kelley JJ (1994) Effect of a heated drilling bit and borehole liquid on thermoelastic stresses in an ice core. Mem Natl Inst Polar Res Spec Issue 49:314–326

Narita H, Fujii Y, Nakayama Y et al (1994) Thermal ice core drilling to 700 m depth at Mizuho Station, East Antarctica. Mem Natl Inst Polar Res Spec Issue 49:172–183

Naruse R, Okuhira F, Ohmae H et al (1988) Closure rate of a 700 m deep bore hole at Mizuho Station, East Antarctica. Ann Glaciol 11:100–103

Neff P (2010) Ice core paleoclimate records from Combatant Col. British Columbia. In Depth 5(2):1–2

Neff PD, Steig EJ, Clark DH et al (2012) Ice-core net snow accumulation and seasonal snow chemistry at a temperate-glacier site: Mount Waddington, southwest British Columbia. Canada J Glaciol 58(212):1165–1175

Nolan M (2012) McCall Glacier Blog May 2012. Retrieved 18 March, 2017 from http://www.drmattnolan.org/personal/mccall_may2012/index.htm

Paterson WSB (1967) A temperature profile through the Meighen Ice Cap, Canada. In: Proceedings of 14th general assembly, International Union of Geodesy and Geophysics, 25 September–7 October, 1967. Berne, Switzerland, pp 440–449

Paterson WSB (1976) Thermal core drilling in ice caps in Arctic Canada. In: Splettstoesser JF (ed) Ice-core drilling. Proceedings of the symposium 28–30 August, 1974. University of Nebraska, Lincoln, USA. University of Nebraska Press, Lincoln, pp 29–36

Perrin J (1988) Deep ice core drilling equipment depth measurement and drilling process. In: Rado C, Beaudoing D (eds.) Ice core drilling. Proceedings of the third international workshop on ice drilling technology. Grenoble, France, 10–14 October, 1988. Laboratoire de Glaciologie et Géophysique de l'Environnement, Grenoble, pp 70–71

Raikovskyi YuV, Samoilov OYu, Pron NP et al (1990) Gliatsiologicheskie issledovaniia na shel'fovom lednike Eimeri v 1987–1989 gg. [Glaciological investigations of the Amery Ice Shelf in 1987–1989]. Akademiia nauk SSSR. Institut geografii. Materialy gliatsiologicheskikh issledovanii [Academy of Sciences of the USSR. Institute of Geography. Data of Glaciological Studies] 68:114. (in Russian)

Rand J (1980) 1979 Greenland Ice Sheet Program, Phase 1: Casing operation. USA CRREL Spec. Rep. 80–24

Raymond CF, Harrison WD (1975) Some observations on the behavior of the liquid and gas phases in temperate glacier ice. J Glaciol 14 (71):213–234

Ryumin AK, Nozdryukhin VK, Emel'yanov YuN et al (1974) Stroenie lednika Abramova po dannym radiolokatsionnogo zondirovaniya [Structure of Abramov Glacier according with radar sounding]. Trudy Sredneaziatskogo regional'nogo nauchno-issledovatel'skogo gidrometeorologicheskogo instituta (SARNIGMI) [Transactions of Middle-East Regional Research Hydro-Meteorological Institute] 14 (95):27–35. (in Russian)

Savatyugin LM (1980) Gliatsiologicheskie issledovaniia na shel'fovom lednike Shekltona (ianvar'-aprel' 1978 g.) [Glaciological work on the Shackleton Ice Shelf (Jan.-Apr. 1978)]. Informatsionny Byulleten' Sovetkoj Antarkticheskoj Ekspeditsii [Soviet Antarctic Expedition Inform Bull] 100:114–118 (in Russian)

Savatyugin LM (2001) Rossyiskie issledovania v Antarktike. (Tridtsat' pervaya SAE—Sorokovaya RAE). Russian Investigations in Antarctica. (Thirty first SAE—Fortieth RAE), vol. 3. Gydrometeoizdat, St.-Petersburg. (in Russian)

Savatyugin LM, Zagorodnov VS (1987) Glyatsiologicheskie issledovaniya na lednikovom kupole Akademii Nauk [Glaciological investigations on Akademiya Nauk Glacier]. Akademiia nauk SSSR. Institut geografii. Materialy gliatsiologicheskikh issledovanii [Academy of Sciences of the USSR. Institute of Geography. Data of Glaciological Studies] 61:228. (in Russian)

Savatyugin LM, Preobrazhenskaya MA (2000) Rossyiskie issledovania v Antarktike. (Dvatsat' pervaya–Tridtsataya Sovetskaya Antarkticheskaya Ekspeditsya). Russian Investigations in Antarctica. (Twenty first–Thirtieth Soviet Antarctic Expedition), vol. 2. Gydrometeoizdat, St.-Petersburg. (in Russian)

Schwarzacher W, Untersteiner N (1953) Zum Problem der Bänderung des Gletschereises. Österreichische Akademie der Wissenschaften, Mathematisch-naturwissenschaftliche Klasse Nr. 4:52–58

Schwikowski M, Jenk TM, Stampfli D et al (2014) A new thermal drilling system for high-altitude or temperate glaciers. Ann Glaciol 55(68):131–136

Sekurov AV (1967) Osobennosti razrabotki elektrotermoburovogo kompleksa dlia burenia l'da i rezul'taty ego ispitanii v Mirnom v 1965/66 g. Informatsionny Byulleten' Sovetkoj Antarkticheskoj Ekspeditsii [Soviet Antarctic Expedition Inform Bull] 60:59–62 (in Russian)

Sekurov AV, Frolov AI, Leve RR (1965) Ustroistvo dlya elektrotermicheskogo bureniya skvazhin vo l'du [Device for electric thermal drilling in ice]. USSR Pat. 176,217

Shreve RL, Kamb WB (1964) Portable thermal core drill for temperate glaciers. J Glaciol 5(37):113–117

Sukhanov LA, Morev VA, Zotikov IA (1974) Potativnye ledovye elektrobury [Portable thermoelectric ice drills]. Akademiia nauk SSSR. Institut geografii. Materialy gliatsiologicheskikh issledovanii [Academy of Sciences of the USSR. Institute of Geography. Data of Glaciological Studies] 23:234–238. (in Russian)

Suzuki Y (1976). Deep core drilling by Japanese Antarctic Research Expeditions. In: Splettstoesser JF (ed) Ice-core drilling. Proceedings of the symposium, 28–30 August, 1974. University of Nebraska, Lincoln, USA. University of Nebraska Press, Lincoln, pp 155–166

Suzuki Y, Takizawa T (1978) Outline of the drilling operation at Mizuho station. Mem Natl Inst Polar Res Spec Issue 10:1–24

Talalay PG (2016) Mechanical ice drilling technology. Geological Publishing House, Beijing and Springer Science + Business Media, Singapore

Talalay PG, Gundestrup NS (2002) Hole fluids for deep ice core drilling. Mem Natl Inst Polar Res Spec Issue 56:148–170

Talalay PG, Fan X, Xu H et al (2014a) Drilling fluid technology in ice sheets: Hydrostatic pressure and borehole closure considerations. Cold Reg Sci Tech 98:47–54

Talalay P, Hu Z, Xu H et al (2014b) Environmental considerations of low-temperature drilling fluids. Ann Glaciol 55(65):31–40

Talalay P, Liu B, Yang Y et al (2018) Electric thermal drills for open-hole coring in ice. Polar Sci 17:13–22

Taylor PL (1976) Solid-nose and coring thermal drills for temperate ice. In: Splettstoesser JF (ed) Ice-core drilling. Proceedings of the symposium, 28–30 August, 1974. University of Nebraska, Lincoln, USA. University of Nebraska Press, Lincoln, pp 167–177

Tchistiakov VK, Kracilev A, Lipenkov VYa et al (1994) Behavior of a deep hole drilled in ice at Vostok Station. Mem Natl Inst Polar Res Spec (49):247–255

The Stuck Ice Drill Saga (2010). Project on the Larsen Ice Shelf System, Antarctica. Ice Core Drilling Team. Posted on February 1, 2010. Retrieved 10 February, 2018 from http://research.bpcrc.osu.edu/LARISSA/tag/field-camp/

Theodorsson P (1976) Thermal and mechanical drilling in temperate ice in Icelandic glaciers. In: Splettstoesser JF (ed) Ice-core drilling. Proceedings of the Symposium, 28–30 August, 1974. University of Nebraska, Lincoln, USA. University of Nebraska Press, Lincoln, pp 179–189

Ueda HT, Hansen BL (1967) Installation of deep core drilling equipment at Byrd Station (1966–1967). Antarct. J. U.S. 2(4):120–121

Ueda HT, Garfield DE (1968) Drilling through the Greenland Ice Sheet. Hanover, USA CRREL Special Report 126

Ueda HT, Garfield DE (1969) The USA CRREL drill for thermal coring in ice. J Glaciol 8(53):311–314

Ueda HT, Talalay PG (2007) Fifty years of Soviet and Russian drilling activity in Polar and Non-Polar ice. A chronological history. USA CRREL Tech. Rep. ERDC/CRREL TR-07-20

Vartykian VG, Kovalenko VI, Moiseev BS (1977) Opyt iskrivleniia skvazhin v usloviiakh Antarktidy [Experience of bore-hole deviation in Antarctica]. Informatsionny Byulleten' Sovetkoj Antarkticheskoj Ekspeditsii [Soviet Antarctic Expedition Inform Bull] 96:24–25 (in Russian)

Vasiliev NI, Zhigalyev SP, Zubkov VM, et al. (1993) Burenie neglubokikh skvazhin v lednikakh [Shallow drilling in glaciers].

Rossiiskaya Akademiya nauk. Institut geografii. Materialy gliatsi-ologicheskikh issledovanii [Russian Academy of Sciences of the USSR. Institute of Geography. Data of Glaciological Studies] 77:74–77. (in Russian)

Vasiliev NI, Talalay PG, Bobin NE et al (2007) Deep drilling at Vostok station, Antarctica: History and recent events. Ann Glaciol 47:10–23

Vedachalam N, Vadivelan A, Umapathy A et al (2017) Concept and testing of a remotely operated vehicle-mountable inductive elec-trothermal polar under-ice corer. Mar Technol Soc J 51(6):33–43

What Can Ice Reveal About Fire? (2010) National Science Foundation. Where discoveries begin, Press Release 10–228. Retrieved 11 February, 2017 from https://www.nsf.gov/news/news_images.jsp? cntn_id=118162&org=NSF

Zacny K, Paulsen G, Bar-Cohen Y et al (2016) Drilling and breaking ice. In: Bar-Cohen Y (ed) Low temperature materials and mech-anisms. CRC Press, pp 271–347

Zagorodnov VS (1981) Issledovanie stroeniia i temperaturnogo rezhima shpitsbergenskikh lednikov s pomoshch'iu termobureniia [Using thermal drills in studying temperature regime of Spitsbergen glaciers]. Akademiia nauk SSSR. Institut geografii. Materialy gliatsiologicheskikh issledovanii [Academy of Sciences of the USSR. Institute of Geography. Data of Glaciological Studies] 41:196–199. (in Russian)

Zagorodnov VS (1988) Antifreeze-thermodrilling of cores in arctic sheet glaciers. In: Rado C, Beaudoing D (eds) Ice core drilling. Proceed-ings of the third international workshop on ice drilling technology, Grenoble, France, 10–14 October, 1988. Laboratoire de Glaciologie et Geophysique de l'Environnement, Grenoble, pp 97–109

Zagorodnov VS (1989) Antifriz-termicheskoe kernovoe burenie ark-ticheskikh pokrovnykh lednikov [Antifreeze-thermal core drilling in arctic ice sheets]. Akademiia nauk SSSR. Institut geografii. Materialy gliatsiologicheskikh issledovanii [Academy of Sciences of the USSR. Institute of Geography. Data of Glaciological Studies] 66:143–149. (in Russian)

Zagorodnov V, Thompson LG (2014) Thermal electric ice-core drills: history and new design options for intermediate-depth drilling. Ann Glaciol 55(68):322–330

Zagorodnov VS, Zotikov IA (1981) Kernovoe burenie na Shpitsber-gene [Core drilling at Spitsbergen]. Akademiia nauk SSSR. Institut geografii. Materialy gliatsiologicheskikh issledovanii [Academy of Sciences of the USSR. Institute of Geography. Data of Glaciolog-ical Studies] 40:157–163. (in Russian)

Zagorodnov VS, Zotikov IA, Barbash, VR et al (1976) O termoburenii na lednike Obrucheva [Thermal drilling on the Obruchev glacier]. Akademiia nauk SSSR. Institut geografii. Materialy gliatsiologich-eskikh issledovanii [Academy of Sciences of the USSR. Institute of Geography. Data of Glaciological Studies] 28:112–118. (in Russian)

Zagorodnov VS, Samoilov OIu, Raikovsii YuV et al (1984) Glubinnoe stroenie lednikovogo plato Lomonosova na o. Zap. Shpitsbergen [Deep structure of the glacial Lomonosov Plateau on western Spitsbergen]. Akademiia nauk SSSR. Institut geografii. Materialy gliatsiologicheskikh issledovanii [Academy of Sciences of the USSR. Institute of Geography. Data of Glaciological Studies] 50:119–126. (in Russian)

Zagorodnov VS, Arkhipov SM, Macheret YY (1985) Reconstructions of ice-formation conditions on a subpolar glacier from core analyses [Rekonstruktsiia uslovii l'doobrazovaniia na subpoliarnom lednike po rezul'tatam issledovanii kerna]. Akademiia nauk SSSR. Institut geografii. Materialy gliatsiologicheskikh issledovanii [Academy of Sciences of the USSR. Institute of Geography. Data of Glaciolog-ical Studies] 53:36–44. (in Russian)

Zagorodnov VS, Thompson LG, Kelley JJ et al (1998) Antifreeze thermal ice core drilling: an effective approach to the acquisition of ice cores. Cold Reg Sci Tech 28(3):189–202

Zagorodnov V, Thompson LG, Ginot P et al (2005) Intermediate depth ice coring of high altitude and polar glaciers with a light-weight drilling system. J Glaciol 51(174):491–501

Zagorodnov V, Nagornov O, Scambos TA et al (2012) Borehole temperatures reveal details of 20th century warming at Bruce Plateau, Antarctic Peninsula. Cryosphere 6:675–686

Zagrivnyi EA, Moiseev BS (1988) Oslozhnenia i metodi ikh ustranenia pri burenii glubokoi skvazhini na stantsii Vostok [Complications and methods of their removal at deep drilling at Vostok station]. Zapiski Leningradskogo Gornogo Instituta [Trans Leningrad Min-ing Inst 116:87–93 (in Russian)

Zagrivnyi EA, Zemtsov AA, Vostretsov RN et al (1980) Eksperimen-tal'noe burenie skvazhiny, zalitoi nezamerzaiushchei zhidkost'iu [Experimental drilling of the hole filled by non-freezing liquid]. Informatsionny Byulleten' Sovetkoj Antarkticheskoj Ekspeditsii [Soviet Antarctic Expedition Inform Bull] 100:119–123 (in Russian)

Zagrivnyi EA, Zemtsov AA, Kononov YuB et al (1981) Opyt burenia-plavlenia skvazhin, zalitikh nezamerzayushchey zhidkostiu, v Antarktike i Arktike [Experience of drilling by melting of holes filled by non-freezing liquid in Antarctica and Arctic]. Zapiski Leningradskogo Gornogo Instituta [Trans of Leningrad Mining Inst] 86:79–83 (in Russian)

Zagrivnyi EA, Moiseev BS, Skurko AM (1985) Rezul'tati polevikh ispitanii vysokochastotnogo termoburovogo kompleksa TBS-112VCh pri burenii glubokoi skvazhini v nizkotempera-turnom lednikovom pokrove (stantsia Vostok) [Results of field tests of high frequency thermal drill TBS-112VCh at drilling of deep hole in low-temperature ice sheet (station Vostok)]. Zapiski Leningradskogo Gornogo Instituta [Trans of Leningrad Mining Inst] 105:103–107 (in Russian)

Zemtsov AA, Men'shikov NG (1988) Kompleks tekhnicheskikh sredstv dlya otbora prob na uglerodnyi analiz iz ledovikh tolshch [Ice sheet sampling devices for carbon analyzing]. Zapiski Leningradskogo Gornogo Instituta [Trans of Leningrad Mining Inst] 116:78–81. (in Russian.)

Zotikov IA (1979) Antifreeze-thermodrilling for core through the central part of the Ross Ice Shelf (J-9 Camp), Antarctica. USA CRREL Rep. CR 79–24

Zotikov IA, Zagorodnov VS, Raikovskii YuV (1979). Core drilling through Ross Ice Shelf. Antarct. J. U.S. 14(5):63–64

Zotikov IA, Zagorodnov VS, Raikovskii YuV et al (1981) Kernovoe burenie na shel'fovom lednike Rossa [Core drilling on the Ross Ice Shelf]. Informatsionny Byulleten' Sovetkoj Antarkticheskoj Eks-peditsii [Soviet Antarctic Expedition Inform Bull] 102:68–74 (in Russian)

Hot-Water Ice Drills

3

Abstract

Hot-water drills provide fastest penetration in glaciers and, nowadays, are actively used for the observation of ocean cavities under ice shelves, the retrieval of sub-ice seabed samples, the study of internal ice structures, video imaging, temperature logging, measurements of deformation within ice, the determination of basal sliding velocity, clean accessing to subglacial lakes. During drilling, hot water is pumped at high pressure through a drill hose to a nozzle that jets hot water to melt the ice. The water from the nozzle uses the melted hole as the return conduit and then, at the surface, it usually reuses by the hot-water drill.

Keywords

Thermomechanical jetting action • High-temperature water • Nozzle • High-pressure hose • High-pressure pump

Hot-water drilling is one of the fastest methods to drill boreholes in glaciers. A hot-water drilling system with an open circuit was successfully used for the first time in 1956 for soundings by Forages et Cie. on the Saleina glacier, Alps (Iken et al. 1977). Unfortunately, details of this drilling project were not published. It is on record that hot-water jets were used for tunneling in the Mer de Glace glacier located also on the northern slopes of the Mont Blanc massif, Alps (Reynaud and Courdouan 1962). In this project, a hot-water spray with temperature from 20 to 30 °C was jetted on the tunnel face via 17 horizontal nozzles under a pressure of 0.4 MPa. The water flow rate was 1.7 L/min. The average driving progress was 5 m per day.

The Berendon Glacier in British Columbia, Canada provides an interesting example of large-scale glacier ice removal via hot-water melting, even though it is not directly related to ice-drilling technology. Following a series of positive-surface mass balance years in the 1960s, it was hypothesized that the Berendon Glacier would advance enough to bury the adjacent Granduc copper-concentrating terminal by the mid-1990s (Fisher and Jones 1971). This copper terminal had been constructed only 20 m away from the glacier terminus, presumably during a time when glaciers were less widely recognized as dynamic features of the landscape (Eyles and Rogerson 1977). With the hope of counteracting an impending glacier advance, the Granduc Operating Company discharged wastewater at 30 °C directly onto the surface of Berendon Glacier at a rate of 13,500 L/min, year-round for 5 years. The wastewater stream quickly incised through the ice down to the glacier bed, bisecting the terminal ~ 200 m of the glacier tongue. By the end of this 5-year period, the wastewater stream had become the effective terminus of the glacier, with 30-m-thick ice flowing at 28 m/year into the stream, where it was undercut by, and calved into, the stream. Although the anticipated glacier advance did not occur, the geometry of the Berendon Glacier was permanently modified by this intentional glacier thermokarst.

In the 1970s, hot-water ice-drilling technology was introduced in glacial researches by French and Swiss scientists (Gillet 1975; Iken et al. 1977). Nowadays, hot-water drill systems are actively used for the observation of ocean cavities under ice shelves, the retrieval of sub-ice seabed samples, the study of internal ice structures, video imaging, temperature logging, measurements of deformation within ice, the determination of basal sliding velocity, clean accessing to subglacial lakes, and many other scientific objectives (Makinson and Anker 2014; Rack et al. 2014; Tsutaki and Sugiyama 2009; and others). Even the coring of ice is also possible via hot-water drilling (Browning et al. 1979; Engelhardt et al. 2000). Multiple boreholes with depths well over 1000 m have been achieved in one season.

Numerous hot-water drills have been constructed by combining water pumps and heating units with individual specifications matched to achieve certain drilling rates over anticipated ranges of borehole depths, borehole widths, and ice temperatures. Modern ice hot-water drill systems can be categorized by depth as: (1) near-surface drilling for depths up to 50–60 m for the installation of ablation stakes, temperature measurements, accessing lake/ocean waters, and seismic surveys; (2) shallow drilling at depths of up to 300–400 m for monitoring glacier dynamics, basal sliding, englacial water pressure, etc.; (3) intermediate drilling to depths of up to 1500 m for the study of marginal parts of the Antarctic and Greenland ice sheets, for creating access holes through ice shelves, and for subglacial lake exploration; (4) deep drilling up to depths of 3500 m for subglacial exploration and other scientific issues (e.g., the installation of neutrino detectors, as in the IceCube Project). In terms of borehole diameter, hot-water drills can produce small-diameter holes (up to 100–150 mm), intermediate-diameter holes (up to 300–350 mm), and large-diameter holes (up to 600–1000 mm).

At first view, hot-water drill systems appear quite simple. During the initial stage of drilling operations, a certain volume of water should be prepared to start drilling. Water is pumped at high pressure through a drill hose to a nozzle that jets hot water to melt the ice. The hose and nozzle are lowered slowly to form a straight hole because gravity is used as the steering mechanism. The water from the melted hole is used as the return conduit. An additional pump (either submersible or at the surface) is in most cases installed to pump water to a surface water tank. This water is then reused by the hot-water drill.

It was found that hot-water drills can easily do certain types of jobs that would be very difficult for mechanical drills. For example, during tests sessions in the Arctic in 1981, a hole was drilled through the ice, freeing a cable that had been frozen in place. This allowed a suspended instrument to be recovered (Verrall and Baade 1984). Hot-water drill systems also can be used to recover stuck electromechanical drills (Kuivinen et al. 1980). Another job that could not have been easily done with a mechanical drill was the freeing of wooden posts in the ice and lowering them to a greater depth (Verrall and Baade 1984). These posts, which suspended a small building on the ice, had effectively been lifted by a summer's melt; the water had run down cracks and fissures leaving the posts and the building was suspended higher in the air by approximately half a meter. With a hot-water drill, the posts were easily melted out and set lower in the ice.

3.1 Hot-Water Systems for Near-Surface Drilling in Glaciers and Making Access Holes Through Sea and Lake Ice

Near-surface and sea/lake ice drilling to depths of up to 50–60 m is done using typically small sets of equipment that can be towed by people or using snowmobiles. They are aimed at making holes for the installation of ablation stakes,

temperature measurements at the bottom of active layers, accessing lake and ocean water, and carrying out seismic surveys. These drills can also effectively measure ice thickness, which is required for many polar engineering applications. Examples include ice road construction, transiting across ice sheets, profiling man-made spray ice pads, barriers, and natural ice rubble piles, and defining thick ice floes and ridge geometry (the thickness of these features may exceed 30 m).

Another important issue related to hot-water jets is cutting sea and lake ice for opening wide channels of water in fields of ice and cutting away ice from boats and docksides. Hot-water ice cutters were extensively used for making open channels in first-year and multi-year Arctic sea ice (Mellor 1986; Morev et al. 1986; and others). Even though hot-water-jet ice cutters are quite similar to hot-water drills, this technology is beyond the scope of this book; however, ice drills for making large holes ("groove-melting" devices) in sea ice are reviewed at the end of this section.

Some methods for making sea ice access holes are used for military purposes, e.g., by submarines for breaking the ice and get to the surface or for sub-ice missile launching, and the details of such units are confidential and their specifications cannot be obtained. For example, the materials of two Workshops on Ice Penetration Technology (Hanover, USA, June 12 and 13, 1984 and Monterey, USA, June 16–19, 1986) are still under limited distribution and are not included in the following review.

3.1.1 PICO Shallow Hot-Water Drill

The shallow hot-water drill system of the PICO, University of Nebraska–Lincoln, consisted entirely of off-the-shelf components (Kuivinen et al. 1980; Koci 1984). Three Malsbary oil-fired water heaters (two Model 21-H and one Model 221-H) were used in varying combinations to melt the required snow in a 2250-L reservoir tank and to heat the meltwater for drilling. To use the system at high altitudes, the oil burner nozzles were changed, reducing the fuel consumption rate from 21 to 7 L/h. The heated water in the tank was pumped through a Synflex hose.

At Dome C, Antarctica from late December 1979 through early January 1980, the PICO hot-water drill was used to make shot holes of depths between 15 and 60 m. A total of 37 holes were drilled at 17 sites over a 30-km area, providing 1085 m of holes in ten days. The water pools in these holes were at approximately half the firn–ice transition depth, so many of the holes at Dome C were unusable below 40–45 m because there was no way to place the explosive charges before refreezing occurred.

It was found that ∼1135 L of water were needed to drill a 60-m-deep hole. It was possible to deliver water to the drill head at 95 °C (7 °C above the boiling point) because of the backpressure at the nozzle. Water delivery was at a rate of 32 L/min with 160 kW used to heat the water. Drilling rates ranged from 360 m/h for the top 10 m of the firn to 120 m/h to reach 60 m with a minimum hole diameter of 76 mm. The cycle time for each 60-m hole was approximately 2.5 h.

Upon completion of the drilling operations in support of the seismic program, the hot-water drill was used to melt free the Swiss shallow drill that had gotten stuck at a depth of 65 m during the previous season. The drill was recovered intact and was retrograded to the USA for overhauling.

Later on, the PICO modified the drill's system (Kuivinen and Koci 1984). Two 80-kW oil-fired burners were used to generate water from snow on the surface and to heat this water to 90 °C. The drill hose used was a standard 20-mm reinforced hose surrounded by 20 pairs of #24 wires and jacketed with a Kevlar strength member. The flow rate was 30 L/min. A thermal gradient of 12 °C/100 m was observed between the heater outlet and the nozzle during drilling operations. Tanks similar to fuel bladders with a hole in the top proved very satisfactory for water storage. Tank capacities were 2 and 4 m^3.

During the 1983–1984 field season, this system was used to drill an array of seismic shot holes at Upstream B on the Siple Coast. All drilling equipment was moved between sites using two snowmobiles. An array of seventeen 80-mm-diameter holes, each one with a depth of 23 m, was completed in one working day. Hole depth was limited because water pooled at this depth. The water required to drill each shot hole was in the range of 600 L.

One of the holes was drilled deeper to test the feasibility of using PICO hot-water drill to place thermistor strings at various depths on the ice. The drilling of this hole was carried out using four 80-kW heaters, two of which were used to generate water on the surface and two to heat the water while drilling. The drill hose with thermistors attached was left downhole. All the thermistors in the string survived the freezing-in process.

During 1985–1986 season, the PICO hot-water drill system was again used on the Siple Coast to drill 96 shot holes to a depth of 15 m (Koci and Kuivinen 1986). During the 1987–1988 field season, the drilling of the shot holes continued at Downstream B (Boller and Sonderup 1988). A Tucker Sno-Cat was used to tow the pallet of drilling equipment, which consisted of two 80-kW heaters, a high-pressure pump, a hose reel, a generator, a 550-L storage tank, and several drums of antifreeze and fuel. The two heaters were first used to generate water on the surface in the storage tank. Then, they were used to heat the water to 95 °C on its way down the drilling hose. A flow rate of 44 L/min provided a 100-mm-diameter hole with a depth of 17 m in 6–8 min.

Drilling was conducted along a grid that consisted of a straight line of 100 holes (360 m between holes), with four crosslines of 10 holes each, for a total of 140 holes. Besides this grid, the crew created an additional small grid that consisted of 20 holes in one line, thus bringing the total number of shot holes drilled during that season to 160. Once a steady routine was established, a complete cycle (i.e., melting the required amount of water, drilling the shot hole, packing up to move, and moving 360 m down the line) could be completed in 40 min.

3.1.2 DREP Hot-Water Shallow Drills

A simple hot-water ice drill capable of making holes 200 mm in diameter through ice at least 50 m thick was designed at the Defence Research Establishment Pacific (DREP), Esquimalt, Canada (Verrall and Baade 1984). The drill consisted of a water heater, a water pump, a hose to feed the water down the hole, and another hose to recover it (Fig. 3.1). The snow melter could be heated separately. Water could be routed either to the ice hole or to a snow melter, and it could be picked up from either of the two.

The water heater used was a commercial APEX steam generator. It consisted of a coil of ½″ steel tubing, which acted as the heat exchanger, and an oil-fired burner unit. It put 68.9 kW of heat into the water when a fuel nozzle rated at 13.2 L/h was used. The total weight of the unit was ~150 kg.

The snow melter consisted of two concentric barrels. The snow was melted in the central drum, and the flames from the burner rose up through the gap between the drums. Water was removed through a plumbing fitting and a valve

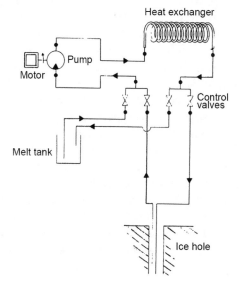

Fig. 3.1 Schematic of the DREP hot-water drill (Verrall and Baade 1984)

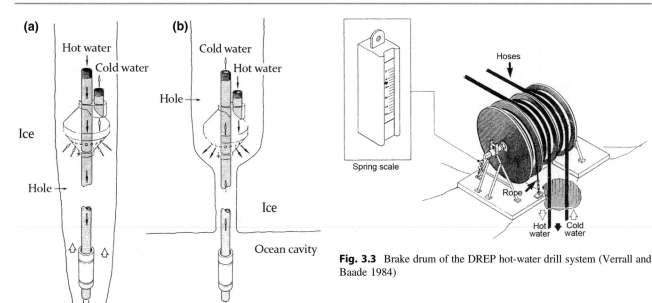

Fig. 3.2 Schematic of the DREP drill stem: **a** usual operation of the drill; **b** water flow reversal once the pipe has dropped below the bottom of the ice (Verrall and Baade 1984)

Fig. 3.3 Brake drum of the DREP hot-water drill system (Verrall and Baade 1984)

at the bottom. The fuel burner was the same unit as the one that was used on the main heat exchanger.

The drilling head consisted of a 1.5-m-long 1″ plumbing pipe with a 20-kg weight attached to it (Fig. 3.2). Screwed to the top of the pipe was an attachment that picked up the return water and fed it to the return hose. In addition, it prevented the pipe from descending until the ice supporting it had melted; thus, it ensured a minimum diameter for the hole. Because a fairly large hole was desired, a jet nozzle and its associated high-pressure pump were not required. Once the pipe had dropped below the bottom of the ice shelf, the water flow was reversed. Hot water was ejected from the conical head, and cold water was picked up by the pipe.

The upward force was produced by a brake drum (Fig. 3.3). This tension, which was less than the weight of the drill head, helped to keep the hole plumb. Both hoses were wrapped around the drum at least one and quarter turns so that there would be no slippage. The braking action was produced by a length of rope wrapped around the drum and attached to the frame. A spring scale indicated the braking force exerted on the drum. The braking force was quite constant because the spring in the scale stabilized the operation. If the rope grabbed a little and increased the tension, the spring lengthened, and this extra length caused the rope to slacken off and reduce the tension to its normal value. Usually, a force of 70–90 N was used. The drum was

also used as a capstan to bring up the pipe and hoses. The brake rope was removed and an electric drive was attached over the square shaft of the drum. The hoses were then pulled out onto the ice with the capstan providing most of the lifting power.

A Mann West 1″ gear pump was used to pump the water out of the hole through the heat exchanger and down the hole again. It was powered by a 470-W electric motor, and the electric power was supplied by a 5-kW Briggs and Stratton motor-generator. This generator also provided power to run the fuel pump on the heater.

The hoses were Gates EPDM hoses (ID/OD: 25.4 mm/36.1 mm). On the surface, the hoses were coiled in insulated boxes: one box for the feed line and another one for the return hose. All attachments to the hoses were made with quick disconnects.

During testing on the Ellesmere Island Ice Shelves, Canadian Arctic in April 1982, a hole was sunk at a rate of 11 m/h. This hole was drilled such that, in the upper 8 m, it was 23 cm in diameter with a variation of 1 cm. Problems with the equipment prevented information about the rest of the hole from being logged. Presumably, the hole decreased in diameter at the bottom. The most serious problem was a crack that developed in the tubing of the heat exchanger. It was probably caused by the freezing of a slug of water left in the tube, although the tubes had been carefully steamed out. Then the drill was used to make holes through the 45-m thick Ward Hunt Ice Shelf and to cut large holes in much thinner ice (Verrall 2001). This type of drill was also very useful for recovering equipment that is suspended by a frozen-in cable. Occasionally, too, it was used to rescue mechanical augers that have become stuck in the ice.

3.1.3 Soviet/Russian Portable Hot-Water Drills

In the beginning of 1980s, a portable hot-water ice drill was designed in the Institute of Geography of Siberia and the Far East (IGSFE), Irkutsk, USSR (Kravchenko 1984). Two boiler modifications were available. The first option with open heating included a metal tank that was heated by two blow torches (Fig. 3.4). A hand air pump created a small overpressure inside the tank ($\sim$50 kPa) so that hot water could delivered into a rubber hose and a drill stem with a nozzle at the end. The boiler was mounted on a small sled; the system weighed 32 kg. This method was effective if the water tank was quite small (20 L) for limited hole lengths. The time for heating up to a working water temperature of 95 °C was approximately 25 min, and the water consumption was 1.1 L/m of the hole length.

In the second option, three blow torches heated gas tubes that passed through the water tank with volume of 70 L. The working temperature inside the tank was near 125 °C, which allowed it to deliver hot water without requiring overheating pressure (<200 kPa). To control the inside pressure, the tank was equipped with a monometer. This system was more effective but heavier. The heating time was 40 min and the water consumption was 0.8 L/m of the hole length. The system weighed 106 kg.

Both systems were used at ice field site Sinyi Kamen' (Blue Stone), Western Sayans, Russia during three field seasons. The maximum drilling depth achieved was 10 m; the average penetration rate was 66 m/h with an open heating tank and 90 m/h with closed heating tank. The diameter of the holes (near the surface) was 25 mm.

A similar portable hot-water drill system was designed in the AARI (Morev et al. 1984). It consisted of a water tank mounted on a sled and was protected via heat-insulating housing, inside of which a gasoline blow torch was mounted (Fig. 3.5). Water was heated in the tank and pumped via a flexible hose to the drill nozzle. On the return stroke of the pump, water in the hole was transferred back to the tank for recycling. To protect the pump and drill nozzle from clogging during drilling in debris-laden ice, two filters were used. The weight of the unit was 30 kg. The drill was tested on the Pamir Glaciers and on various ice mounds in Eastern Siberia. The maximum drilling depth achieved was approximately 20 m, with a penetration rate of $\sim$60 m/h at water flow velocity in the nozzle in the range of 7–10 m/s.

In 1985, the Sakhalin Research, Design, and Survey and Institute of Oil and Gas (SakhalinNIPImorneft) designed and tested another portable hot-water drill system (Fig. 3.6) (Astaf'ev et al. 1997). Two blow torches warmed water in a 40-L water tank to a temperature of 80–90 °C. A hand pump delivered hot water from the water tank through rubber hoses into 40-mm-diameter insulated aluminum pipes with threaded joints, each 1.5 m long. This drill was used for thickness measurements of sea ridges in Sea of Okhotsk, Russia.

In 2001, a new hot-water drill with computer recording of the drilling parameters was designed in the AARI (Mironov et al. 2003). The drill consisted of (Fig. 3.7): (1) an 85-kW water heater operating on diesel fuel; (2) a 2-kW power generator; (3) 0.6-kW and 1.1-kW power supply pumps; (4) an 80-m-long supply hose; (5) a drill stem with a nozzle; (6) a control box for computer recording of the drilling parameters; and (7) a tank with antifreeze.

During transportation, the pumps and hoses were filled with antifreeze, which was replaced by hot water when drilling started. Preparation of the system for drilling took $\sim$30 min. Seawater was pumped via electrical pumps through the water heater and then to the control box via the supply hose. The temperature of the hot water was controlled automatically and was typically in the range of 70–80 °C. The water pressure at the water heater outlet was $\sim$1 MPa.

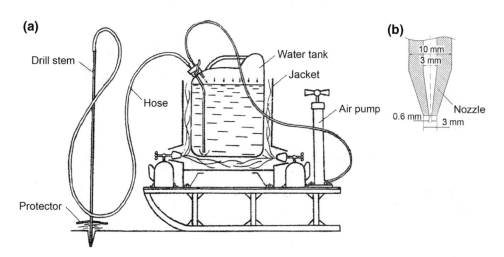

Fig. 3.4 **a** IGSFE portable hot-water ice drill and **b** its nozzle design (Kravchenko 1984)

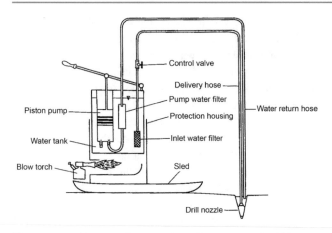

Fig. 3.5 Portable AARI hot-water ice drill (Morev et al. 1984)

depth and, thus, to determine ice type and voids with a vertical size larger than 4 cm. This drill was used to drill thousands of holes in the northwestern Caspian Sea, in Baydaratskaya Bay of Kara Sea, and at the shelf sea of Sakhalin Island (Smirnov and Mironov 2010).

3.1.4 Unmanned Autonomous Hot-Water Ice Drill of the University of New Hampshire

An unmanned autonomous hot-water ice drill was developed in the Marine Systems Engineering Laboratory, University of New Hampshire for drilling holes in sea ice in inaccessible regions of the Arctic to facilitate the collection of oceanographic data (Beverly 1982, 1983; Beverly and Westneat 1982a, b). Researchers intended to deliver the system to the site via airdropping. The drill was designed to make 150-mm-diameter access holes through 10–15 m of ice cover. After penetrating and reaching beneath the ice, it served as a weight to pull as much as 1000 m of instrumentation cable through the hole. The freezing of the hole after drilling was acceptable.

This drill recirculated hot water via a three-stage centrifugal pump through a heating chamber and jetted it out through a nozzle at the drill tip (Fig. 3.8). To recirculate water, the drill had to submerge itself in its own melted water, and this was accomplished by the use of a parabolic-shaped aluminum hot point, which initiated the drilling process. The hot point was heated using nine

The control box included a transformer for drill moving, water temperature and pressure sensors, the electronics block, and the computer. Hot water from the control box was flowed through the drilling hose to a 24-mm-diameter thermal drill. Drilling was carried out until the drill got in contact with sediments or accessed the water. Upon the end of the drilling operations, water was pumped out of the system and it was filled with antifreeze. Three people operated the system.

The presence of ice with different densities and porosities in stamukhas and ice ridges led to significant changes in penetration rate, which varied from 120 to 1200 m/h. In addition, the drill suddenly dropped in cavities and voids. The drill allowed operators to register penetration rate versus

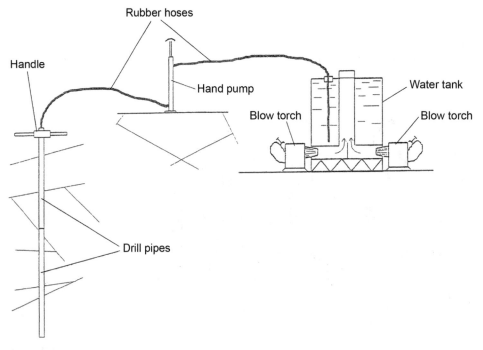

Fig. 3.6 SakhalinNIPImorneft portable hot-water ice drill (Astaf'ev et al. 1997)

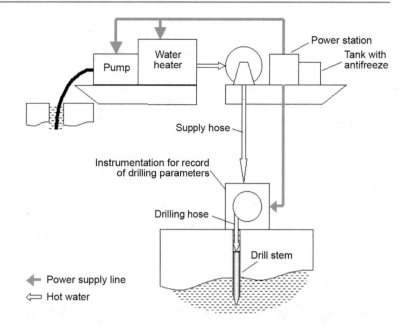

Fig. 3.7 AARI hot-water ice drill with computer recording of drilling parameters (Mironov et al. 2003)

cartridge heaters, each rated at 0.78 kW at 300 V, arranged in banks of three and controlled through three thermistors equally spaced within. A 15.9-mm-diameter hole was placed in the center of the hot point for the hot water jet.

The drill melted into the ice until water reached two sensors located 100 mm above the pump intake. When these sensors were triggered, power was diverted from the hot point to the water-heating chamber and the pump. The

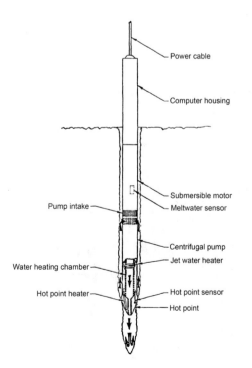

Fig. 3.8 Schematic of an unmanned autonomous hot-water ice drill (Beverly 1982)

0.53-m-long water-heating chamber contained 20 cartridge heaters arranged radially within a tube and was divided into four banks for control. Each heater was rated at 0.62 kW, yielding 3.1 kW per bank and 12.5 kW in total. Two out of every five heaters within each bank were equipped with thermistors to monitor the surface temperature of the heater.

A modified commercial three-stage centrifugal pump driven by a 1.1-kW, 240-V-AC submersible motor recirculated the meltwater at rate of 180 L/min. The drill was controlled by a microprocessor located in the water-tight chamber at the top of the drill. Sudden loss of meltwater due to the encountering of voids and cracks in the ice could be sensed and the drill could be reinitialized to provide meltwater. The power cable served as a strength member for deploying the drill as it proceeded through the ice. It supplied 300 V at a maximum current of 42 A for the heaters, 6 V for the computer, and 240 V AC for the motor, and also provided a communication line. The capsule that remained on the surface of the ice after the operation was completed contained a satellite communication system, a computer for formatting the sensor data, a 30-kW h battery supply, and an inverter to generate the AC power for the motor. The guidance of the drill was controlled via cable tension coupled with a low center of gravity and a high center of buoyancy to keep the drill vertical. The drill weighed ~45 kg.

Eight test runs with the drill prototype were conducted at the USA CRREL ice well facility during 1982, which encompassed both the initialization phase (hot-point drilling) to establish meltwater and water-jet drilling. The ice well was 0.9 m in diameter and 60 m deep and was maintained at −25 °C. The drill prototype typically produced holes with an average diameter of 165 mm. During the initialization phase, the drill proceeded with a powered-on hot point at 5.8 kW at an average penetration rate of 2.1 m/h down to a depth of 0.83 m, where sensors detected that the drill was sufficiently submerged into meltwater. Then, the internal centrifugal pump and the water heaters (at a full power of 12.5 kW) were activated. During this phase, the drill proceeded at an average penetration rate of 4.4 m/h. Unfortunately, all drilled holes were inclined by 4.5–5°, and it was possible to drill only to depths of 5–6 m. The physical limitations of the diameter of the ice well prevented the continuation of the drilling process to either prove or disprove that the drill could continue at that angle or whether the angle would increase enough to cause penetration problems. There is no information available about this project after these tests.

3.1.5 Exxon Hot-Water Ice Drill

To drill vertical holes of various diameters through consolidated and unconsolidated 30-m-thick ice, Exxon Production Research Co. designed and extensively used a hot-water drill capable of functioning in the Arctic at temperatures as low as −45 °C (Poplin et al. 1987). The drilling system was composed of five modules (Fig. 3.9): (1) a pump-and-motor module, (2) a water-heater module, (3) a reserve tank, (4) a hose-and-probe module, and (5) a sled. The assemblies could be transported separately or as a complete system. The typical onsite setup time was 15 min.

The pump-and-motor and water-heater modules were protected by fiberglass-reinforced cases. The pump was a positive-displacement high-pressure piston model capable of pumping water at a rate of 19 L/min at pressures of up to 14 MPa. The pump was driven through a gear-reduction unit by a 3.7-kW four-cycle gasoline engine. The assembly also included an 18-kg propane bottle that contained fuel for the water heater. The exhaust heat of the pump engine was used to prevent the propane from solidifying. The water-heater section was capable of increasing the temperature of a water flow of 11.4 L/min by 38 °C. After setting them, the burners were controlled automatically. Water for start-up was contained in a 57-L insulated reservoir. A separate tank held antifreeze, which was used for system purging and shutdown.

Most of the warmed water left the heater section through a feed line to the probe (drill stem). The remaining heated water was carried from the heater section via two auxiliary lines. Flow from one of the auxiliary lines entered the reservoir barrel to maintain a water temperature of approximately 24 °C. The quantity of heated water returned to the reservoir barrel was controlled by an operator. Under normal operating conditions, a small quantity (0.4 L/min) of heated water was also returned to the seawater-intake hole to help prevent the formation of frazil ice, which could restrict the intake flow.

Hot water from the heat exchanger was supplied through a feed line with an ID of 9.5 mm to a 37-m-long hose

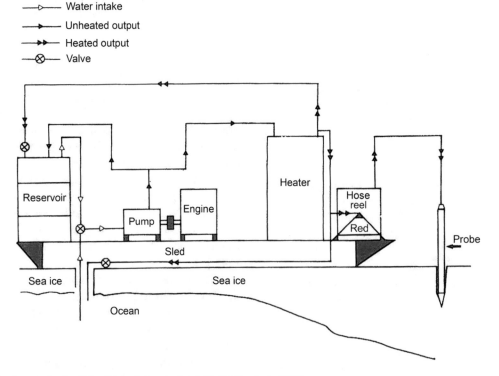

Fig. 3.9 Principal components of the Exxon hot-water ice drill (Poplin et al. 1987)

calibrated in 0.3-m increments and then to the probe. The feed line could be omitted for drilling in proximity to the drill or could be added in lengths of 30 m to drill at a distance (a feed line that was 91 m long was used successfully, reducing the number of required relocations). The calibrated hose was wound on a reel capable of storing 100 m of hose.

The 1.5-m-long probe consisted of three parts: a stainless-steel tip, a brass nozzle, and a brass casing attached to the nozzle. Probe casings with ODs ranging from 22 to 152 mm could be used depending on the required hole size. Hole diameters ranged from 25 to 300 mm. The nozzle had a parabolic shape with a single central orifice (1.7–2.9 mm in diameter), which yielded the highest penetration rates. Multiple nozzle orifices with the outflow directed at angles of up to 30° from the vertical were used successfully with a 51-mm-diameter probe to drill holes up to 100 mm in diameter.

The entire drill system was fixed onto a fiberglass toboggan or a wooden sled. The total shipping weight of the drill was 176 kg with the toboggan and 196 kg with the wooden sled. Under field conditions, the total weight of the drill with the sled (including a propane supply with a capacity of 18 kg and 57 L of water) was 285 kg.

In several 1983–1987 field programs, a cumulative ice thickness of over 24 km was drilled in the Arctic. The 18-kg propane bottle was sufficient to operate the drill for up to 4 h. The pump was normally operated at ~1 MPa. The thickness of the drilled ice was estimated using depth marks on the calibrated hose and the rapid changes in penetration rate when the probe entered the sea. The ability to detect penetration through the bottom of the ice could be verified to within 0.15 m using a lead line. This degree of accuracy was acceptable for most thick-ice feature profiling applications. The snow/ice interface was also easily detected via changes in the probe's penetration rate. Snow depth was subtracted

from the total thickness to calculate the ice thickness. For water depths less than ~60 m, the probe and hose were also used to determine the depth to the sea floor.

In one of the field programs, the performance of the hot-water drill was evaluated on a large, thick multi-year ridge drifting in the nearshore of the Alaskan Beaufort Sea. Ice thickness data were collected along the sail crest and two profile lines perpendicular to the ridge. A total of 61 ice thickness measurements totaling 1300 m were collected over six hours of drilling. The maximum ice thickness measured along the profile was 30.2 m. Two operators were required: one to operate and monitor the drill and one to feed the probe into the ice. Two probes with diameters of 25 and 32 mm (with a 2.8-mm-diameter orifice) were most often used to measure ice thickness. The use of these probes resulted in average penetration rates of 258 and 180 m/h, respectively, in sea ice.

3.1.6 CRREL Portable Hot-Water Drilling System

The CRREL portable hot-water drilling system was developed for conducting detailed thickness surveys of multi-year sea ice (Tucker and Govoni 1987). During drilling, seawater is pumped from a source hole in the ice, heated, and sent directly to the drill probe (Fig. 3.10). The gate valves were used to switch to closed-loop operation, in which glycol from the reservoir was circulated continuously through the system. Closed loop operation was normally used to start and warm up the system each day prior to drilling. For shutdown periods longer than a few minutes, the entire system, including the intake line and the drill's probe and hose, were filled with glycol. The glycol was then purged back into the reservoir prior to drilling or used to drill the initial seawater source hole. While drilling, a small volume

Fig. 3.10 Principal components of the CRREL hot-water ice drill (Tucker and Govoni 1987)

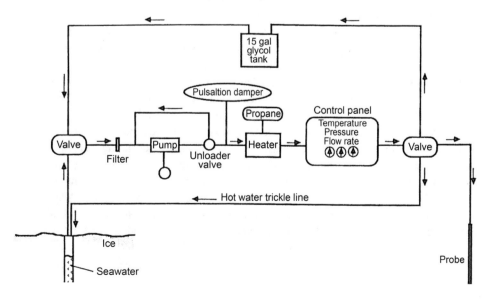

of hot water ($\sim$1 L/min) was directed into the seawater source hole through a trickle line to prevent frazil ice from forming and plugging the water intake line or pump prior to reaching the heating unit.

The heater used was a Spartan QRI propane-fired burner. It was capable of causing a 75 °C increase in water temperature at a flow rate of 11.4 L/min. Water temperature was regulated by a valve that controlled the fuel flow and was located on the heater control panel. This heating unit required a 12-V battery to provide a pulsed ignition spark in lieu of a pilot flame. Water was pumped through the heater and to the drill probe with a 11.4-L/min twin-piston pump capable of reaching a pressure of 3.45 MPa. The pump was driven with a 1.5-kW Honda four-cycle gasoline engine and required a pulsation damper to prevent excessive vibrations.

System pressure and flow rate were controlled by an unloader valve located in the loop circuit around the pump. Hot water was pumped through 60 m of a 12.7-mm high-pressure hydraulic hose to the drill probe. This length of hose was necessary to avoid the continuous relocation of the drilling system when conducting thickness surveys. The probe itself was simply a 1.2-m-long 9.5-mm steel pipe fitted with a straight-jet fog nozzle having an orifice diameter of 3.05 mm. These components, plus a 57-L glycol reservoir, an 18.1-kg propane cylinder, fittings, hoses, valves, and other components, were mounted on a 1.8-m double-runner steel sled. Quick-disconnect hose fittings allowed the set-up or take-down time to be $\sim$30 min. The weight of the entire system with both the glycol reservoir and the propane cylinder full was $\sim$250 kg.

In April and May 1986, this drilling system was used in the field to assess thickness variations on multi-year floes. A typical layout on floe consisted of a 15 × 15 point grid with a 10-m spacing. With a 60-m-long hose, the drill system was relocated only a few times on grids of this size. Typical operating parameters were a water pressure of 0.8 MPa, a flow rate of 9.4 L/min, and water temperatures ranging between 60 and 82 °C. Operating the propane heater at lower fuel flow rates resulted in lower water temperatures (60–72 °C) but made the propane last significantly longer. The 18.1-kg cylinder was normally used for approximately 2–3 h (ambient temperatures were as low as −29 °C) and was then removed and placed into a separate heated box for warming while another cylinder was installed on the drill. After warming, the original cylinder could be used for approximately one additional hour. The typical mode of operation in the ice-thickness measurement program involved three people. The penetration rates were in the range of 120–180 m/h depending on the temperature and pressure of the heated water and ambient environmental conditions (air temperature, wind speed, and ice temperature).

Later, the system was modified by changing the gas boiler to an oil-fired heater, Mi-T-M model HM 500, with a rated output of 443.1 MJ (Govoni and Tucker 1989). This unit was capable of delivering an increase in water temperature of 75 °C at a flow rate of 15.9 L/min. Water temperature was regulated via a thermostat that switched the burner on and off as required. The heating coil, which was made of 13-mm-diameter continuous-length stainless-steel tubing, could withstand pressures of 13.8 MPa. A 120-V, 2.3-kW generator was used to provide power for the fuel pump, the blower, and the flame igniter of the heater. Only 1.5 kW were required for the burner; however, more power was required for starting the burner on cold mornings.

Water was pumped through the drill system using a triplex ceramic plunger pump. This pump was capable of generating 10.3 MPa of pressure at 15.1 L/min. A 3.7-kW Honda gasoline engine powered the pump through gear-reduction unit. The pump was normally operated at flow rate of 15.1 L/min and at a pressure of $\sim$2.2 MPa. A total of 84 m of hose with an ID and an OD of 9.5 mm and 16.4 mm, respectively, was kept on a hose reel to allow for drilling at considerable distances from the unit.

The thermal probe was also redesigned. It consisted of two parts: a brass parabolic-shaped nozzle that was 170 mm long with an orifice size of 3.05 mm, and a brass casing that was 32 mm in diameter and 0.83 m long attached to the nozzle. The total weight of the nozzle and casing was 3.6 kg. The casing added the necessary weight required to force the nozzle to penetrate the ice and maintained the vertical orientation of the probe. The entire drill system, with all its components, hoses, and fittings, was installed on a 2-m-long aluminum-framed sled. The sled had aluminum runners faced with a strip of Teflon to decrease the sliding resistance on ice or snow. The penetration rates attained with the oil-fired drill were significantly higher at 180–270 m/h.

3.1.7   Kovacs Small Hot-Water Drill

Kovacs Enterprise designed a portable hot-water drill that drew power from a 2–3-kW electric generator (Thermal drilling equipment n.d.). This system's boiler and pump unit were fitted into a latched shipping-field container measuring only 0.36 m × 0.53 m × 0.58 m (Fig. 3.11). The box and boiler-pump unit weighed $\sim$91 kg. The pump was rated at 8 L/min and the boiler could heat water to 80–90 °C depending on flow rate and supply water temperature. A high-pressure hose was also provided, which was specifically rated for use with hot water at 121 °C continuous and 135 °C intermittent flows. Two options were available: using a hose with an ID of either 9.5 mm or 12.7 mm. The former weighed 0.33 kg/m and the latter weighed 0.48 kg/m. The hose was attached to a stainless-steel drill stem with changeable nozzles.

Fig. 3.11 Kovacs small hot-water drill (Thermal drilling equipment n.d.)

3.1.8 Shallow Hot-Water Drills for Seismic Surveys

The primary purpose of these drills is to create shallow shot holes down to depths of 30–35 m for conducting seismic surveys. All of them are lightweight and portable.

A simple hot-water drill was designed in Pennsylvania State University (Penn State) in the beginning of the 2000s. This drill system was very light and transportable: all equipment, which comprised a hand-operated winch, a boiler, and an aluminum tripod (if available), was installed on a polyethylene Siglin cargo sled (Fig. 3.12a) or a Nansen sled (Fig. 3.12b) to be towed by a snowmobile. The boiler could heat water up to 93 °C; the drilled holes had a diameter of 50–60 mm, which was enough to load a charge (weighing typically 0.2–5 kg). The penetration rates varied in the range of 25–50 m/h depending on density of the drilled snow/firn.

The BAS shallow hot-water drill was one of the smallest in this family (Fig. 3.13). The thermal power of the drill was ∼40 kW. It was highly-portable, but its maximum drilling depth was limited to 20 m.

The ANDRILL shallow hot-water drill system was more bulky and heavier. It was aimed at making 75-mm-diameter shot holes down to depths of ∼35 m (Field Update: Dec. 12 2010). Snow was melted in a ∼0.25-m³ insulated tank to provide the water supply for the drill. Then, the water went through a hose to a set of the pressure washers on the next sled. The hot water continued through a manifold to a hose reel mounted together with a tower on a third sled. The hose was connected to a drill stem with a nozzle. One person controlled the speed at which the hose unwound, while another manually guided the hose into the hole to make sure that the nozzle melted a vertical hole (Fig. 3.14). The process of melting each hole used up all of the water in the storage tank. This means that the snow had to be continuously replaced in the tank via shoveling and that water had to be heated and recirculated to keep the temperature and volume of the water in the tank high enough to maintain snow melting. The heaters and the pumps could be adjusted to recirculate hot water into the storage tank.

The IDDO small hot-water drill system used a commercially available set of pressure washers and could drill up to 120–150 m/h. It needed four sleds for its melt tank, a drum with fuel, pumping/heating devices, and a tower with a hose reel. The shipping weight of the IDDO drill was ∼1500 kg. The diameter of the drilled holes was ∼60 mm. The fuel consumption for drilling 30-m holes was 4.5 L of AN8 fuel and 2.2 L of gasoline. During the 2011–2012 field season, the IDDO provided a small hot-water drill for two seismic studies on Whillans Ice Stream (Fig. 3.15a). The investigators, using the drill, drilled over 280 shot holes and collected

Fig. 3.12 Penn State shallow-depth hot-water drill installed on: **a** a Siglin cargo sled (*Credit* K. Riverman); **b** a Nansen sled (*Photo* A. Muto), Antarctica (Penn State News 2013, 2017)

Fig. 3.13 BAS shallow-depth hot-water drill, Pine Island Glacier, Antarctica, in the season of 2006–2007 (*Photo* R. Stilwell, British Antarctic Survey)

Fig. 3.14 ANDRILL hot-water drill system, Coulman High, December 2010 (Field Update: Dec. 12 2010)

seismic data from more than 9 km. Another seismic survey was carried out on Beardmore Glacier, Antarctica, during the 2012–2013 and 2013–2014 field seasons (Fig. 3.15b).

3.1.9 IDDO Sediment-Laden Drill

The new Sediment Laden Lake Ice Drill (SLLID), which was designed and built by the IDDO in 2017, is a 28-kW hot-water drill useful for creating holes with very little initial water, freeing instrument cables from ice, and accessing equipment deployed beneath ice (Sediment Laden Lake Ice Drill n.d.). It is expected that this drill will be effective for drilling through sediment-laden ice, especially for long-term ecological studies in environmentally sensitive areas, such as the Dry Valleys in Antarctica. It includes an HL2-35 Espar Hydronic heater (with an operating temperature range of 85–118 °C) and a 6000SC Flowtronic pump. Permanent indicators are included on the hose, which indicate depth and distance to the tip. The diameter of holes needed will vary, most likely being in the range of 130–250 mm. The maximum drilling depth with this system is 6 m, and its shipping weight is $\sim$200 kg.

3.1.10 Ice Drills for Making Large Holes in Sea Ice

Oceanic research in arctic seas often requires access to the sea through thick sea ice. In the Arctic Ocean, the average ice thickness is $\sim$3 m; first-year ice can reach thicknesses of 1–1.5 m, whereas 10-year-old ice in the central Arctic Ocean may be 7 m thick. To lower down different oceanographic instruments, holes must be somewhat over 1 m². This size

(a)

(b)

Fig. 3.15 IDDO small hot-water drill: **a** sled with pumping/heating devices, Whillans Ice Stream, Antarctica, 2011–2012 season (*Photo* H. Horgan from Antarctic glaciology n.d.); **b** start of drilling operations

in Beardmore Glacier, Antarctica, 2012–2013 summer field season (*Photo* M. Conway from Small Hot Water Drill n.d.)

allows almost all types of cable-supported oceanographic instrumentation to be passed through ice canopies. One of the easiest ways to create such large holes is the coring approach, wherein an annular groove is melted through sea ice.

All these types of ice drills use the thermal energy of hot water to melt a groove in ice and are able to make holes of almost any desired shape (round, oval, rectangular, etc.) depending on having suitable cutter head designs and multiple nozzles. In the strict sense, some of the devices reviewed below do not belong to the hot-water drilling systems class because hot water or another liquid is pumped into the cutting ring in closed-loop mode. However, all of them were combined together because they employ a "groove-melting" method with a heat-transfer liquid.

3.1.10.1   University of Washington Ice Drills for Making Large holes in Sea Ice

"Groove-melting" devices of several sizes were built and successfully employed by the University of Washington, USA in the 1970s and involved cold-starting systems for working in arctic seas at ambient temperatures down to −35 °C (Francois and Harrison 1975; Francois 1977). The width of the groove had to be large enough to allow the core to be pushed down or lifted up without sidewall interference, implying core length-to-diameter ratios of 5 or 6 and a diameter of 1 m. It was estimated that groove widths of approximately 7–10 cm would be required. For a 1-m square hole, this is equivalent to an energy requirement of ∼40 kW h every meter depth.

The system consisted of a heat exchanger package, a cutting head, fuel supply (a propane tank), and a source of electrical power. Water was heated in the heat exchanger package and delivered to the cutting head, where it melted a groove in the sea ice. The cooled water delivered and the meltwater were returned to the heat exchanger for reheating and recirculation to the cutting head in a continuous process. The suction pump used (rated at ∼40 L/min and driven by a 250-W electric motor) was identical to that in the water delivery circuit. The main water heater was a standard commercial "hydronic boiler" rated at a 101-kW input on propane.

The cutting heads could be made in many sizes and shapes within the following constraints. Using a pump with a capacity of 40 L/min and a cutting-head tube size of 25 mm, the total orifice area should be 160 mm^2, with the orifices spaced from 13 to 19 mm. A practical maximum length for the cutting head perimeter is 7.3 m. Five operational systems were built and tested in the field. Two of them are shown in Fig. 3.16. The first operational unit, with an output of 23 kW, made as large as 0.71-m-diameter holes in Greenland Sea ice up to 5.5 m thick at a penetration rate of 1.5 m/h (Fig. 3.16a, b). Another unit, with an output of 64 kW (Fig. 3.16c, d), made an ice hydrohole sized

1.22 m × 3.66 m through sea ice that was ∼8.5 m thick at Ice Island T-3. In the latter case, a 1.22-m^2 cutter head was used to make three adjacent cuts with a rate of 0.73 m/h to achieve the desired hole dimensions. Each hole required one normal working day and two men to melt the core and remove the ice core columns.

Various tests showed that minimizing the water level above the suction was desirable and that a multiple-orifice suction manifold minimized core erosion compared with single-point suction intake. The largest unit thus far had an output of 95 kW. It consisted of two packages, a heat exchanger (238 kg), and a support box, which included fuel, a generator, hoses, the cutting head, etc. (225 kg). All these units were transportable via helicopter or sled.

3.1.10.2   CRREL Hot-Water Ring for Large-Diameter Holes

The CRREL portable hot-water drilling system can be used either with a normal drill stem or a 43-cm-diameter ring used for making larger holes, which might be used to install instrument packages into ice or water (Fig. 3.17) (Tucker and Govoni 1987). The ring can be substituted for the drill stem in less than a minute. The penetration rate was greatly enhanced by perforating the bottom side of the ring with 0.08-mm-diameter holes with a spacing of 1.27 cm. Drilling effectiveness was also increased by drawing the intake water from the meltwater just above the ring. This closed loop operation, combined with shorter hose lengths, results in water temperatures 15–20 °C higher than that of the water exiting the normal drilling probe. The penetration rate of the ring was ∼2.5 m/h. The ice cores produced by the ring could be removed in smaller pieces after cutting them with a saw or an ice chisel or could pulled out intact using a winch and tripod arrangement.

3.1.10.3   Thermal Hole Opener of Oceanographic Service, Inc.

A thermal hole opener capable of producing 0.76 m 1.52 m holes through ice thicker than 3 m was developed at Oceanographic Service, Inc. during the winter of 1984–1985 (Hansen 1987). In a strict sense, hole opener does not belong to the hot-water drilling systems class but has similar with jetting cutters design. This closed loop system consisted of four major components: a liquid heater, a pump, risers, and a cutting frame (Fig. 3.18a). Lightweight (70 kg) propane-fired boiler delivered >35 kW of heating power to the circulating 50% glycol–water solution, which prevented the freezing-up of the system at working temperatures well below −40 °C. The gas control valve required either AC or DC power. The boiler had temperature and pressure gauges, with the temperature being controlled by a manual thermostat.

(a)   **(b)**

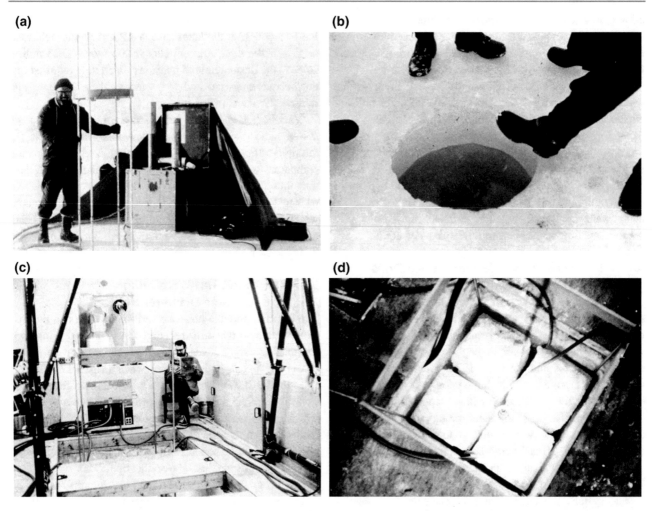

(c)   **(d)**

Fig. 3.16 University of Washington ice drills for making large holes in sea ice: **a** field operations with a prototype unit using a 25-cm square cutting head; **b** 71-cm-diameter hole made with the prototype unit in Greenland Sea; **c** testing of a modified unit (an hydrohole was made inside an hydrohut through removable floor panels); **d** ice columns formed by melting with a 1.22-m² head (Francois and Harrison 1975)

Fig. 3.17 Drilling ring for large-diameter holes (Tucker and Govoni 1987)

A gasoline-powered water pump was used on the first unit. The latest version included a 0.55-kW electric centrifugal pump powered by a small generator. This self-primed pump circulated the antifreeze through the boiler to the cutting frame and back to the pump at a rate between 150 and 225 L/min and a system pressure of ∼140 kPa. The boiler, the propane bottle, and the pump were mounted on a sled.

The cutting frame was connected to the pump and the boiler with a high-temperature polyurethane hose with an ID of 19 mm, which was rated at 107 °C and remained flexible at −40 °C. The cutting frame was made of copper tubing with silver-soldered elbows and tee pieces and an ID of 13 mm. Depending on the size of the frame, two to six hose connections were made to keep the temperature near 80 °C throughout the frame. Special flow mixers in the cutting

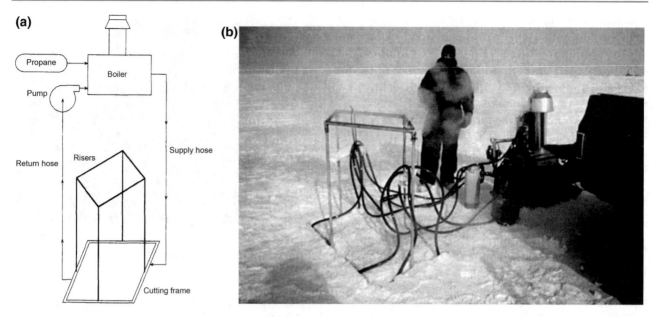

Fig. 3.18 Thermal hole opener: **a** schematic diagram; **b** field testing in Alaskan Beaufort Sea, April 1985 (Hansen 1987)

frame encouraged turbulent flow and enhanced the heat exchange to the ice. Risers were connected to the frame to ensure a vertical penetration of the ice sheet and to simplify the recovery of the frame. Ice penetration depth was limited only by the lengths of the hose and the riser.

Because the cutting frame melted only a 25-mm-wide slot, a method to remove the heavy free ice block (>2.7 t for a 3.5-m^3 block of sea ice) was also designed. The block was removed by installing a toggle and wire through a small-diameter hole (50 mm) in the central part of the block and running the wire through a tripod erected over the cut block. The block was winched up until a portion 1–2 m long was above the ice surface. Using a chainsaw, vertical cuts were made in the ice block, and then a hammer and an ice wedge were used to shear away smaller ice blocks.

The first model of the hole opener was designed specifically to deploy a small, tethered, remotely operated vehicle through sea ice in Beaufort Sea in April 1985 (Fig. 3.18b). The unit was tested nearshore on smooth grounded sea ice near Deadhorse, Alaska and was then used at three separate sites, each ~80 km offshore, where it yielded cutting rates in the range of 1.2–3.0 m/h. The system was easily set up at the sites in ~0.5 h. Typically, the system was operated by two persons and could be transported via snowmobile and sled or using a medium-size helicopter.

The same model of hole opener was used in February 1986 to deploy divers through the sea ice surrounding an Arctic drilling platform located in Beaufort Sea. Two holes were cut through rafted first-year ice near the side of the platform in order to install ice pressure sensors below the water line on the platform's hull.

A second unit was built for the Naval Underwater Systems Center as a tool for deploying sound-source and hydrophone arrays under the sea. The cutting rates of the second model (0.3–0.9 m/h) were lower than those observed for the first model but, nonetheless, it was successful at cutting a 0.9-m square hole in hard multi-year ice that was 4.9 m thick.

Owing to uneven heat transfer in the cutting frame and the non-uniformity of ice, differential cutting was observed, with the cutting frame and risers beginning to tilt. However, the vertical risers prevented exaggerated tilting of the cutting frame, resulting in only partial contact with the ice and a slower cutting rate. To correct these situations, the operator leveled the frame by manually lifting and pressing down on the respective risers until they were again vertical. The ideal operating temperature and pressure for the circulating fluid were 80 °C and 140 kPa, respectively.

3.1.10.4   DRDC Drills for Making Large Holes in Sea Ice

Defence Research and Development Canada (DRDC) was established in 2000 and continued conducting research in the Canadian High Arctic and perfecting the equipment and methods for making large holes in sea ice (Verrall 2001). The first heat commercial exchanger, a steam-jenny, that was used by DREP, predecessor of DRDC, had low efficiency (see Sect. 3.1.2). The present version of the hot-water drill uses an ordinary furnace burner Riello Mectron 15 M to produce the heat. Water is circulated through the heat exchanger by a gear pump driven by a 0.37-kW electric motor. As the hot water comes out of the heat exchanger, it is directed in one of two directions by a Y-valve. One route leads the water into a 45-L tank built right into the framework of the hot-water drill. This tank is used for melting snow and for storing a reserve of hot water for filling the long hoses, priming the pump, etc. The other route feeds the water to the long hoses and drill fittings that actually do the hole melting. The hot water was pumped through a metal pipe with holes along its entire length. The pipe can then be used to slice into the ice. The hot water from the pipe continually filled the grooves between the ice and kept the cut from re-freezing. The pipes can be fashioned to make either straight cuts or circular holes.

In 1996, a 2 m × 13 m ice-hole was cut in Joliffe Bay near Canadian Forces Station Alert, Ellesmere Island, Canada, to launch the Theseus autonomous underwater vehicle (AUV). Series of access holes were drilled through 1.8-m thick ice in the Beaufort Sea, north of Alaska, in 2007 and in the Lincoln Sea, north of Canada's Ellesmere Island, in 2008 to deploy a small (3-m long, 20-cm in diameter), man-handleable Gavia AUV (Doble et al. 2009). The holes were prepared by a hot-water drill feeding a 1-m diameter circular ring or using a 1-m long straight cutter. In 2010, an 8.5 m × 2.5 m hole was cut in 2-m thick fast ice near Borden Island to deploy Explorer AUV under ice (Fig. 3.19) (Kaminski et al. 2010). The ice blocks taken from the hole were 2.5 m × 1.0 m × 2.0 m and were removed using a Bobcat loader. Recently, archaeologists use BRDC hot-water drill (T-shaped copper pipe with small holes on its underside) to make a triangular dive hole through 2-m sea ice to explore the wrecks of two ships Erebus and Terror

Fig. 3.20 Hot-water drill developed by DRDC to cut large ice holes in Arctic sea ice (The Franklin Expedition 2017)

happened in 1845 in Canada's Arctic (Fig. 3.20) (The Franklin Expedition 2017).

3.1.11 Summary

Hot-water drills have the potential for considerably higher penetration rates than electrically heated thermal drills because the turbulent action of the hot-water jet produces a high heat transfer coefficient and efficient transfer of heat to the ice. This thermomechanical jetting action allows the probe to penetrate the ice with high efficiency. Higher water velocities increase the turbulent transfer of heat to the ice. An additional benefit of using high water velocities may be the mechanical erosion of the ice. The velocity of the water is controlled via a combination of flow rate, nozzle orifice diameter, and the pressure that is allowed to build behind the nozzle. Penetration rate also increases with the temperature of the water. With the complete utilization of the thermal energy (discharge water is cooled to the freezing point), the specific energy of this process is the latent heat plus the sensible heat, i.e., ~ 290 MJ/m^3 (Mellor 1986).

Small hot-water drills can drill holes 50–100 mm in diameter at penetration rates of 180 m/h or more using a single nozzle and can operate at air temperatures down to −45 °C. Certainly, to use drills in such cold environments, well-timed methods should be considered to remove water from the system and protect it from water refreezing. One of the most important considerations in the design of these hot-water drilling systems are the weight of their components, their system size, and the fuel type used. The size and weight of these systems were of primary concern. The drill should have the stand-alone capability of operating at small field camps at remote sites with no heavy equipment. Typically, the design goals required sled-mounted system that could be easily moved around the ice by operators or using a

Fig. 3.19 DRDC cutting ice-hole in 2-m thick fast ice near Borden Island, Northern Canada (Kaminski et al. 2010)

snowmobile. The fuel used had to be easily ignitable and burn efficiently in extremely cold environments.

Two primary types of heaters were used in these systems: oil-fired and propane heaters. Both are "instant" heaters, having a very hot flame that heats the water circulating in coils around it. The key advantage of propane heaters is that they weigh much less than fuel oil burners. However, the heating capacity of propane heaters is less, and propane cylinders add significant weight to the drilling system. The availability and handling of propane are also substantial considerations. Propane cylinders have to be refilled or replaced every several hours (depending on cylinder size, consumption rate, and ambient temperature).

An advantage of propane burners, however, is that they require little or no external power. A 12-V battery is generally an option for models requiring a power source for the igniter. Fuel oil burners, on the other hand, require an AC current for the burner pump and blower, and thus a generator is a necessary part of drilling systems that incorporate this type of burner. One of their major advantages is that fuel oil, which can be arctic diesel or jet-turbine fuel, is readily available in most polar locations. That is perhaps the main reason why oil-fired units are far superior than propane-fired units.

Many of the latest portable hot-water drills used commercially available pressure washers manufactured by Kärcher GMBH, Germany or Hotsy, the North American brand of Kärcher. These units generally require little maintenance. Even though they are quite heavy, bulky, and not dismountable, their use can save the time required for the design and testing of original systems.

Hot-water drills have demonstrated their capability of making large holes in sea ice of great thickness. These drills can use open-loop (essentially hot-water drills) or closed-loop circulation systems. In any case, water jets or a tubing cutting head cut the ice by transferring heat from the circulated fluid to the ice, melting a slot down through the ice sheet and leaving a block of ice floating in the hole. The maximum melting depth is limited only by the thermal drill's auxiliary equipment, such as the length of the feeding hoses and guide pipes.

Once an ice block is cut, the cutting head must be removed to eliminate the possibility of damaging it when removing the block. In some instances, the removal of the tubing cutting head from the narrow slot is rather difficult, especially in deep cuts (>3 m) in which the ice block pinches the risers and meltwater refreezes, thus impeding the pulling of the cutting head. If this happens, the cutting head can be also recovered after the ice block is removed. The best practice is to remove the melted water from the groove to prevent the refreezing of the groove behind the melting front and sidewall buildup, which would reduce the clearance required for the passage of the block and the withdrawal of

the melting head. Another problem that needs to be solved is the recovery of heavy ice blocks from the hole. For this purpose, different kind of hoists can be used.

3.2 Shallow Hot-Water Ice-Drilling Systems

These systems allow for drilling holes in firn and ice down to depths of 300–400 m, sometimes even slightly deeper, for bedrock topography studies, monitoring glacier dynamics, the study of interglacial or subglacial waterways and water pressure, installing subglacial borehole instrumentation, and accessing ice shelf cavities (Table 3.1). Most of them are quite potable and can be transported to the drilling sites using snowmobiles, helicopters, or light aircrafts. Some of these systems are light enough to be carried by people. Even though shallow hot-water ice-drilling systems are more power-consuming and heavier than electric-heated thermal drills, they are preferable in many applications because they are many times faster and are able to penetrate dirty ice.

3.2.1 LGGE Shallow Hot-Water Drills

The first LGGE (at this time, it was named as the Laboratoire de Glaciologie, Grenoble, France) portable hot-water drilling system was likely the first of such kind of hot-water drills (Gillet 1975). A motor-driven pump pumped water at a rate of 16–32 L/min into a boiler consisting of a coil and a propane burner that consumed 2.5 kg of gas per hour (which corresponded to a heating power of ~ 28 kW). The water was heated in this way up to ~ 150 °C and pumped via a PVC hose with an ID and an OD of 20 and 26 mm, respectively, that weighed 0.32 kg/m to a downhole drill–nozzle made from a rigid tube. While coming back up to the surface (except in the rare cases in which the water drained into the glacier), the meltwater was taken back by the pump (Fig. 3.21). The weight of the boiler and burner was 20 kg, that of the pump was 32 kg and that of the drill–nozzle was 10 kg.

During the summer of 1972, this system was used to drill a series of holes with depths from 50 to 120 m on Glacier de St-Sorlin, Alps at 2700 m asl. On average, one tank of propane with 13 kg of gas was consumed for every 55 m drilled. The penetration rate for the first few meters varied from 15–20 to 20–30 m/h. The speed decreased slowly as the drill–nozzle went deeper. With a 40-mm-diameter drill–nozzle, hole diameter reached 70 mm at a depth of several meters, whereas it varied at the surface and was, on average, in the range of 120–150 mm. A depth of 100 m was reached in less than 5 h.

In 1979–1980, a hot-water drilling project was carried out by the LGGE on the glacier d'Argentiere in the French Alps

Table 3.1 Parameters of selected shallow hot-water ice drilling systems

Institution or drill name	Years	Drilling site(s)	Flow rate/ (L min^{-1})	Max depth/m	Borehole diameter/mm	ROP in ice/ (m h^{-1})	References
Laboratoire de Glaciologie, Grenoble, France	1972	Glacier de St-Sorlin, Alps	16–32	120	70	15–30	Gillet (1975)
ETH Zurich, Switzerland	1972–1976*	Swiss glaciers	12–12.5	430	~45 mm near the bottom	58–91	Iken et al. (1977)
University of British Columbia, Canada	1976–1981	Hazard Glacier, Canada	12	220	60–80	50–120	Napoléoni and Clarke (1978)
LGGE, Grenoble, France	1979–1980	Glacier d'Argentiere, French Alps	12.5	~250	NA	NA	Hantz and Lliboutry (1983)
US Geological Survey	1976–1983	US glaciers	14.4	~400	NA	84 in average	Taylor (1984)
BAS, United Kingdom	1978–1989	George VI Ice Shelf, Antarctica	18	~300	>30	NA	Makinson (1993)
Electrochaude drill	1982*–1986	Alps, Spitsbergen	NA	140	NA	Up to 14	Rado et al. (1987)
Geological Survey of Greenland	1987–1988	West Greenland	18	390	NA	125–200	Olesen (1989)
Aberystwyth University, UK	1992–up to now	Alps, polar ice caps, Antarctica, Greenland, Tibet	13	142	NA	30–120	Hubbard and Glasser (2005)
HSD, National Energy Authority, Iceland	2005	Langjökull ice cap, Iceland	7.5	110	NA	50	Thorsteinsson et al. (2008)
ANDRILL Hot Water Drill	2006–2007	McMurdo Sound, Antarctica	120	97	>600	NA	Falconer et al. (2007)
ILTS, Hokkaido University, Japan	2007–2012	Alps, Patagonia, Antarctica	26.6 (2 washers)	516	100–150	48–50	Sugiyama et al. (2010)
ARA hot-water drill system	2010–2013	South Pole, Antarctica	55	>200	>150	20	J. Cherwinka, personal communication, 2012, 2013

Note NA—data are not available; *—author's estimation

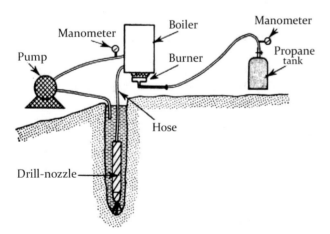

Fig. 3.21 Schematic of the French shallow-depth hot-water ice drill (Gillet 1975)

(Hantz and Lliboutry 1983). A total of 31 boreholes with a maximal depth of ~250 m were drilled in the ablation area at ~2400 m asl, most of them reaching bedrock. A hot-water cleaner for building facades Kärcher HDS 800 was used: it supplied 12.5 L/min of water at 50–80 °C at a pressure of up to 7.5 MPa. Together with a braided hose that was 400 m long (over four segments), a winch, a 5-m-long rigid hose, and a 4-kVA generator set (for the pump of the HDS 800 and the motor of the winch), the device weighed ~700 kg (without fuel). The average fuel consumption was 7 L/h.

Later on, the LGGE developed the small-diameter Electrochaude (translated from French, meaning "electro-hot") drill, which was easily transportable and capable of drilling through glacier ice in spite of there being fine rock debris (Rado et al. 1987). The system could not work without water in the borehole and was intended to finish holes started with either steam drills or hot-point thermal drills.

The body of the drill was made from a stainless-steel tube that was 28 mm in diameter and 1.5 m long (Fig. 3.22). The upper part of the drill consisted of submersible electrical connection that electrically and mechanically joined the drill to the cable. Below, the drill housed a filter and an electromagnetic pump. Its coil was supplied with an alternating current and a diode in series acted as a half-wave rectifier. The piston forming the magnetic core moved alternately under the force of magnetic attraction and a pumping spring, forcing the water out of the priming cage at the frequency of the AC power supply.

In the central part of the drill, a refractory ceramic tube that was 12 mm × 8 mm in diameter and 1 m long (Electroquartz SL 60 ZA) was installed, containing a nichrome spiral heating wire (0.7 mm in diameter, 3.4 Ω/m) capable of dissipating 1.6 kW of power. The electrical supply for both the pump and heating was initially divided (the pump coil was supplied with 240 V). At the bottom of the probe, there was a brass nozzle that had a calibrated hole. In the first version of the probe, the nozzle had a short parabolic shape and water flow was adjusted by a needle valve. In a later version, the nozzle had a long and smooth parabolic taper.

After being pumped into the central tube, water was warmed by the heating element and exited as a steady jet. At the bottom, the water flow made a U-turn and was drawn in 20 cm above the nozzle through openings in the outer tube. Then, the water moved up along the ceramic tube, recovering any heat lost by the tube and the pump. With part of the solid impurities already removed from the water via settling, the filter took out the remaining fines. The probe recovered only a portion of the debris at the bottom of the hole. This amount depended on the debris concentration in the ice, but it did not exceed 1 cm^3 after 100 m of drilling. This debris settled in the space below the water-inlet holes and was removed at the end of the drilling operations.

The cable was hand-held during drilling so that the probe was suspended just above the bottom, and this was done to a depth of 150 m. On the surface, an adjustable transformer was used to increase the voltage to reduce line losses. Power was supplied by a 3.2-kVA generator.

More than 1200 m of holes were drilled in the Blanc, Bossons, Argentiere, Mer de Glace, and Mont de Lans glaciers. Some problems were encountered owing to power shortages in the drill. All holes were started with steam or hot-point thermal drills. At Mer de Glace, the maximum drilling rate of 14 m/h was achieved with a 1.2-mm hole in the tip. The water-flow temperature was approximately 30–40 °C.

This drill was used on the Austfonna ice cap, Nordaustlandet, Spitsbergen, in May 1986. A tank was fixed to the top of the probe to inject antifreeze into the meltwater (the ice temperature reached −10 °C). Ethylene glycol created many problems owing to its high viscosity at these low temperatures. In spite of these problems and the time spent filling the tank every 4 m of drilling, one hole was drilled to a depth of 140 m and three holes were drilled to a depth of 40 m for temperature measurements.

3.2.2 ETH Hot-Water Drill System

In 1972, H. Rothlisberger and A. Iken from the ETH (Eidgenössische Technische Hochschule Zürich—Swiss Federal Institute of Technology Zurich) developed an original shallow hot-water drill (Iken et al. 1977). The designed system consisted of the following units (Fig. 3.23): (1) a centrifugal pump with a gasoline engine (65 kg; gasoline consumption of 3 L/h); (2) a heating unit consisting of an adjustable diesel oil burner and a water-circulation coil mounted on a sledge-shaped frame (80 kg, diesel oil consumption of 6–7 L/h); (3) a small generator for the operation

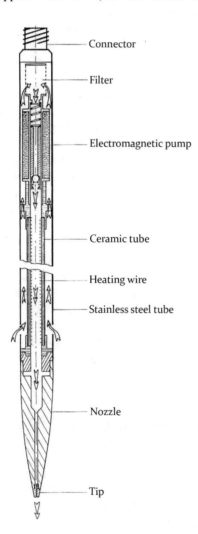

Fig. 3.22 LGGE Electrochaude drill (Rado et al. 1987)

Fig. 3.23 Pump and heating unit beside the melt stream on Gorner-gletscher, July 1974 (*Photo* H. Röthlisberger from Iken et al. (1977))

of the oil burner (300 W); (4) a drilling hose consisting of a pressure-resistant thick-walled hose and a thin PVC-sleeve (hose's ID/OD: 12 mm/20.5 mm; sleeve's ID/OD: 29.5 mm/32 mm; weight of the hose with the sleeve: 0.45 kg/m); (5) a 2-m-long double-walled guiding tube filled with lead in its front part.

Local surface meltwater was used as the circulating fluid. A few openings in the sleeve allowed the water to fill the space between the hose and the sleeve when the hose was submerged in the borehole. This water layer was more or less stagnant and represented an additional insulation layer that did not add weight to the equipment. The water used the space on the outside of the sleeve for ascending from the nozzle back up to the glacier surface. The guiding tube ended in a nozzle that could be changed (ID of 2.5–4 mm). For controlling verticality, a mercury switch with an annular shape could be mounted on the tube. If the drilling direction was deviating from the vertical by more than a given tolerance, which was indicated by the mercury switch, verticality could be restored by pulling up the hose by ~ 2 m and then slowly lowering it again.

This drill was used extensively for borehole studies in the Swiss glaciers to depths as high as 430 m, which was encountered in the Gornergletscher. The hot-water flow rate was 12–12.5 L/min, and the water temperature at the heater exit was kept in the range of 60–80 °C. The hole diameters

obtained with a nozzle with an OD of 3 mm were ~ 45 mm near the bottom and ~ 170 mm near the surface. Drilling rates decreased with depth. Whereas the first 100 m were drilled in 1.1 h with an average drilling rate of 91 m/h, the drilling rate decreased to 58 m/h between depths of 217 and 275 m.

H. Blatter adapted the ETH drilling system to use propane heaters (Müller 1976). In this system, the water, heated by two propane burners and pumped with a pressure of ~ 2 MPa through a rubber hose insulated with PVC, was released through a 4-mm nozzle at the end of the 2-m-long drill head, which was stabilized with ~ 3 kg of lead grains. During the summer of 1974, three boreholes with a maximum depth of 280 m were drilled in the cold ice of White Glacier, Axel Heiberg Island, Canadian Arctic Archipelago with penetration rates ranging from 25 to 50 m/h. From 1974 to 1981, a total of 32 boreholes were drilled on White Glacier using this system to the maximal depth of 380 m and the vertical ice-temperature profiles were measured (Blatter 1987). It was difficult, or even impossible, to judge whether the drill had reached the glacier bed because large rocks in the ice were likely to obstruct the drill. At some sites in the lower ablation zone, the water level in the holes, which were usually filled with water, dropped abruptly at the end of drilling operations.

In the summer of 1982, an electrical profile was measured at Grubengletscher in 16 boreholes drilled using the ETH system to depths between 25 and 85 m (Haeberli and Fisch 1984). A simple electrical measurement indicated where and whether the drill had reached the glacier bed. Borehole water froze within approximately one day, except when its level was more than approximately 20–30 m below the surface, presumably in the transition zone between cold surface ice and deeper temperate ice.

During the snow-melting season of 1980, 1982, 1985, and 1994, a few tens of boreholes were drilled with the ETH drill system for making basal-water pressure recordings down to the bed of Findelengletscher, a temperate valley glacier in the Swiss Alps (Iken and Bindschadler 1986; Iken and Truffer 1997). In the studied area, the glacier was up to 180 m thick. Later, the ETH hot-water drill system was modified for drilling intermediate-depth boreholes (see Sect. 3.3.3).

3.2.3 Hot-Water Drill of University of British Columbia, Canada

The Department of Geophysics and Astronomy at University of British Columbia (Vancouver, Canada) designed a hot-water drill with a propane water heater and a gasoline-driven pump (Napoléoni and Clarke 1978). The drill was designed to reach depths of 300 m in cold ice and for making holes with a minimum diameter of 30 mm (Fig. 3.24).

Water was supplied by a self-priming BVG4 centrifugal pump driven by a 1.5-kW Briggs & Stratton gasoline engine. When the engine ran at 2700 rpm, it provided 15 L/min under a pressure of 0.2 MPa. At the end of the suction hose, a fine strainer was installed because silt was harmful to the small high-pressure gasoline-driven C5330HR pump that was used to pump water into the hole. A high-pressure piston pump was belt-driven by a 3.7-kW Briggs & Stratton gasoline engine. It was equipped with a surge tank, a manometer, and a relief valve and could provide 12 L/min under a pressure of 3 MPa.

A commercially available Culvert King propane heater was used. This heater consisted of a 3-m copper pipe coiled in the shape of a long truncated cone. A large burner blew a flame inside the cone, thus heating the water flowing through the pipe. At sea level, water flowing at 12 L/min could be heated from 0 to 45 °C with a propane consumption of 5 kg/h. In practice, the temperature of the heated water dropped to 32 °C at an altitude of 1800 m asl and to 28 °C for a 300-m gain in elevation.

A Conquest Agricultural hose with an ID and an OD of 13 mm and 26 mm, respectively, was used. The working pressure was 5.3 MPa. Three 100-m-long sections (47 kg per 100 m) were stored on light wooden reels and were carried separately. Before drilling, they were connected and rolled on a large metallic reel fitted with a feed-through swivel joint and equipped with a geared crank.

Different nozzle shapes were tested by drilling through large blocks of clear ice. Some had grooved surfaces made to create a more turbulent flow in the hope of increasing the speed at which the ice melted. However, the tests showed that these asperities are useless because they catch against the walls of the hole and slow the advance. The best results were obtained using the long smoothly tapered tip shown in Fig. 3.25.

This drill was used in August 1976 on Hazard Glacier in the Steele Glacier area (Saint Elias Mountains, Yukon

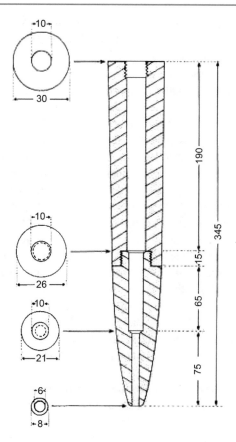

Fig. 3.25 Drilling tip of the Canadian shallow-depth hot-water ice drill; all dimensions are in mm (Napoléoni and Clarke 1978)

Territory, Canada). Drilling sites were located near the upper limit of the ablation zone and their altitudes ranged from 1800 to 2100 m. Near the sites, the glacier was fairly clean, with rocks covering less than 5% of the ice surface. Three holes were drilled in three days by three people. The distance between sites was ∼1 km of fairly smooth ice, and sleds were used to carry the equipment, which, all things included, weighed 400 kg. The heaviest piece was the propane bottle; when full, it weighed 80 kg. Three sites were instrumented with thermistor cables that ran from near the glacier bed to the glacier surface. Temperature measurements taken in July 1977, after the disappearance of the thermal disturbances caused by drilling, indicated that Hazard Glacier is cold, with a typical 10-m ice temperature of approximately −5.5 °C.

Drilling went as fast as 120 m/h near the surface. The deepest hole (220 m) took 4 h to drill and, near its completion, the drilling speed was 50 m/h. The flow was maintained at 12 L/min and the water temperature at the surface was ∼30 °C. Water pressure oscillated between 1.4 and 1.9 MPa with occasional peaks to 2.3 MPa. The circuit was protected by a relief valve set at 2.6 MPa. Because the burner was not automatically regulated, there was the possibility of overheating the coil when the water flow stopped. This risk was avoided by using a flow switch connected to a

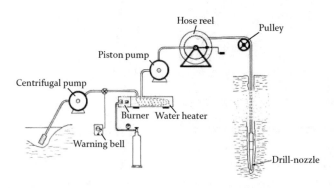

Fig. 3.24 Canadian shallow-depth hot-water ice drill (Napoléoni and Clarke 1978)

warning bell. The consumption of propane was 5 kg/h, even when air temperature dropped to −5 °C. Refreezing of the hole never created any difficulty. After the first 20 m had been drilled, the hose unrolled from the reel under its own weight. Using a pulley to guide the hose into the hole was not essential because, even when left on the glacier surface, the hose slid smoothly into the hole.

No device was used to control verticality. Radar soundings made at the drill sites yielded ice thicknesses that were 10–15% smaller than the actual length of the holes, which may indicate that significant deflection had occurred. It was usually impossible to be certain that the glacier bed had actually been reached and that the nozzle encountered a large rock. In one instance, drilling came to a halt at a depth of 110 m. Water circulation was continued for approximately 20 min, time at which silty water appeared at the surface.

Provided that the hole was kept filled with water, pulling up the hose was fairly easy and it was not necessary to use the reel crank. The hose had to be drained to prevent water from freezing inside during the night. This was done by blowing compressed propane from the fuel bottle through the hose.

In 1980 and 1981, 17 shallow holes (all to depths <70 m) were drilled to the bed of Trapridge Glacier, Yukon Territory, Canada and were instrumented with thermistors (Clarke et al. 1984). These holes were located on a single line crossing the bulge near the glacier centerline. The arrival of the drill at the glacier bed is typically indicated by a sudden reduction in drilling rate followed by a marked increase in the turbidity of the water flushed from the hole. Of the seven holes drilled in 1980, one, where basal ice was at the melting point, connected to the subglacial drainage system. At the moment of connection, water level dropped rapidly from 100% of the ice thickness to 55%. Attempts to fill the hole by pumping water into it at ∼ 12 L/min had no noticeable effect on water level. In summer 1981, a hole drilled at a neighboring site again connected to the subglacial drainage system. On this occasion, the hole drained completely.

3.2.4 USGS Hot-Water Drill Systems

The hot-water drill system designed by the US Geological Survey (USGS) was intended for drilling holes in temperate ice to depths of 400 m (Taylor 1984). The first field tests during the summer of 1976 on South Cascade Glacier and Blue Glacier in Washington, USA involved the use of a single rubber hose with an ID of 9.5 mm, a propane-fired water heater, and surface runoff water. The penetration rates reached ∼ 90 m/h at a depth of 125 m using hot water at a temperature of 38 °C and a flow rate of 8.5 L/min. A drill stem that was 3.7 m long with an ID of 25.4 mm and an OD of 33 mm (40 steel pipe) was used, filled with lead shot around a central tube weighing 20 kg (Fig. 3.26).

Many difficulties were encountered, mainly with the longitudinal stretch of the hose, estimated to be ∼ 10%, which resulted in a jerky, uncontrolled advance of the drill. Later, the hose was changed to a lightweight and flexible

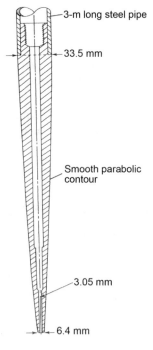

Fig. 3.26 USGS hot-water drill stem and nozzle schematic (Taylor 1984)

high-pressure Synflex 3000-06 hose with high longitudinal stiffness. This hose weighed 0.14 kg/m in air. The hose had a polyurethane cover and a synthetic fiber braid over a nylon core tube. The maximum continuous length was 76 m, which is why pieces were connected with swaged termination fittings.

While drilling during the summer of 1977 on South Cascade Glacier using an oil-fired water heater and an electric pump at 11.4 L/min, 3.4 MPa, and 80 °C, penetration rates of 160 m/h were achieved near the surface, which decreased to 125 m/h at 50 m and to 60–70 m/h at a depth of 210 m near the glacier bed. A strong longitudinal oscillation ("bucking") with a frequency of ~2–3 Hz occurred, which was attributed to the plugging and the release of the nozzle jet against the ice at the bottom of the borehole. The technique that was finally settled on was to lower the drill with a variable-speed winch at a rate slightly less than the free-fall maximum rate. If the lowering rate was increased slightly, the "bucking" oscillation was encountered, and the winch speed could then be reduced.

During the season, 12 additional holes were drilled down to the glacier bed with an average penetration rate of 84 m/h. Fuel consumption was approximately 6.6 L/h. Only one borehole connected with the basal water system (Hodge 1979). At the very end of the field season, approximately 4–6 weeks after drilling, the water level in two of the remaining boreholes was starting to drop, so it is possible that more holes eventually formed a connection. An inclinometer was used in one of the holes and it indicated a deviation of only a few degrees from the vertical; 4° 4′ was maximum, returning to 1° 35′ from the vertical at 200 m.

For the 1979 summer season on South Cascade Glacier, the development of the system had progressed to a convenient sled-mounted unit using an oil-fired commercial water heater and an electrically driven pump producing 14.4 L/min with a pressure of 5.5 MPa at 77 °C. The fuel consumption of the heater was approximately 9.8 L/h. Surface meltwater was used and expended down the hole. The drill system required approximately 4 kW of electrical power, which was obtained from the cabin generator system using a 1.5-km-long power cable and 1-kV step-up and step-down transformers.

The next design modification, which involved using a 3.7-kW gasoline engine-driven pump at 6.9 MPa, was implemented and used on Variegated Glacier, Alaska, where twelve holes were drilled to the bottom at 400 m. The penetration rate at 300 m was approximately 60 m/h, and a total of ~5.5 h of drilling time was required to reach the bed. The drill became stuck several times during retrieval, which was traced to be caused by the tendency of the warm hose to side-melt the hole, producing a "key-hole" shaped cross-section. To remedy this problem, removable spacers 51 mm in diameter were placed on the hose at 5–10-m

intervals; spacer installation and removal was performed just above the borehole mouth using a simple tool. After that, video monitoring of a borehole on South Cascade Glacier showed no noticeable "key-hole" effect.

By the summer of 1980, the equipment and drilling techniques had progressed to a point where operational reliability was achieved. The entire drill equipment and sleds weighed 1200 kg (this weight does not include the fuel and the generator). The drilling system included (Fig. 3.27): (1) a drill-supply polyethylene tank with an open top (170 L); (2) a Cuno 1M1 filter with a G78L2 cartridge (50 μm); (3) a Giant P-41 pump (maximum rating of 17 L/min at 8.3 MPa) driven by a 1.5-kW electric motor; (4) a diesel oil-fired burner and an Alkota 300 heater (rated at 88 kW); (5) a winch drum holding up to 460 m of hose with a Chickshaw hydraulic swivel; (6) a 110-V, 0.55-kW variable winch drive with a two-speed gear (tripping rate could be changed from 0.038 to 110 m/min); (7) a meter wheel; (8) a Synflex 3000-06 hose (ID/OD: 9.5 mm/16.3 mm, rated at a working pressure of 15.5 MPa); (9) a drill stem (1″ NPS pipe, 33 mm in diameter, 3.7 m long, 20 kg); (10) a replaceable nozzle; (11) a compact submersible firn pump (15 L/min at a 6-m head) used also in the melt tank; (12) and a firn drill (3/8″ NPS pipe, 22 mm in diameter, 2 m long) with a Spraco 15A4 spray nozzle.

A snow melter separate from the drill system was used when no surface or reliable firn water supply could be found. The total weight of the sled with the melter circuit equipment was ~700 kg. The snow melter consisted of a sled-mounted polyethylene tank having a volume of ~1.5 m^3 with an open top and a removable cover. Three Spraco 15A4 spray nozzles were fixed on the top of the melter. Hot water was recirculated through a heater (the same as the one used with drill) using a screw Teel 3P569A pump driven by a 370-W motor. Snow was shoveled into the tank. With the tank full of water at the start, a 200-m hole could be drilled without stopping.

The drill supply tank was also used as a snow "mini-melter", using a hand-held firn drill used to make a firn well or to recirculate water back to drill tank when used as a "mini" snow melter. This would accumulate enough meltwater for ~10 min of drilling.

During the summer of 1980, nine holes down to 180–200-m-deep glacier bed at South Cascade Glacier were made for basal water-pressure studies, and ten holes down to depths of 400 m were drilled at Variegated Glacier (Fig. 3.28). A total of ~3.5–4 h were required to drill to the glacier bed at 200 m and return to the surface. Sleds were transported using a small track vehicle.

Experiments at South Cascade Glacier during 1981 were continued in an attempt to drill using recirculated water from the borehole. These efforts were unsuccessful in producing a reliable circuit mainly owing to the complexity of having to

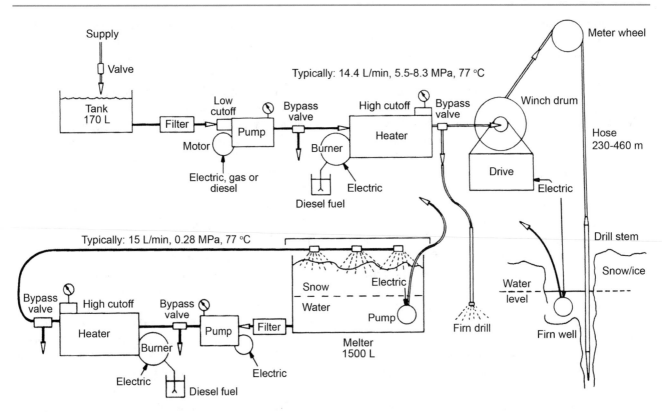

Fig. 3.27 Schematic of the modified version of the USGS hot-water drill system (Taylor 1984)

Fig. 3.28 USGS hot-water drill system at South Cascade Glacier (the hose in the foreground brings water from a firn well near the previously drilled hole) (Taylor 1984)

handle hoses, cables, and pumps in a return hole and difficulties in the cross-connections between adjacent holes. In addition, the ability to recirculate water was found to be very dependent on the depth and permeability of the overlying firn and snow and on the behavior of the water found or placed there. The snow melter was then set up, and fourteen holes were drilled down to the glacier bed in this manner.

A tensiometer for the hose was designed and used to monitor the weight of the drill stem. This alerted the operator within 10–15 s of any drill hang-up, and it was also used to determine contact with the glacier bed or obstructions. Several of the holes drilled during the 1981 season had to be abandoned at mid-depth, where the drill would suddenly stop its advance as if it had hit a large rock and would not continue despite 10–30 min of continued melting. This had happened at least three times in previous years and it had also been noted at Variegated Glacier. Drilling became difficult when the air temperature became less than approximately −5 °C because of the risk of serious damage to the system from freezing water, especially in the high-pressure pump, the heater coils, and the hydraulic swivel on the winch drum. Compressed air was used to blow out the lines at night, but it was not completely reliable for preventing ice plugs in the coils of hose on the drum. Flushing with antifreeze, using approximately 20–40 L of a 25% mixture, provided protection at temperatures of approximately −12 °C. This was added to the drill supply tank when nearly empty, and the system was shut down when the fluid first appeared at the drill stem. Upon startup during the next morning, approximately half of this mixture could be recaptured to be used again at the end of the day.

Drilling at Variegated Glacier was continued by the California Institute of Technology, Pasadena, USA (Caltech) in 1983 to investigate the reasons for the hundredfold speedup in glacier motion in a surge of the kind of those took place in 1982–1983 (Kamb et al. 1985). Five holes were drilled down to the glacier bed (maximum depth of 385 m). A modified USGS hot-water drill system was used for drilling 24 boreholes with maximum depth of 203.3 m down to the bed of South Cascade Glacier to investigate the hydraulics of subglacial water flows in 1986 and 1987 (Fountain 1994).

3.2.5 BAS Shallow Hot-Water Drill

British Antarctic Survey (BAS) has a long history related to the development of hot-water drill. Several near-surface, shallow-depth, and intermediate-depth drilling systems have been developed and were successfully used; however, two projects for deep hot-water ice drilling failed (see Sects. 3.4.3 and 3.4.4). The first BAS hot-water drill was developed in 1978 using the components of a commercially available high-pressure water power washer, which required only minor modifications for field use in the Antarctic (Makinson 1993). A heat exchanger and burner unit provided 80 kW of heating power, and a twin diaphragm pump drew water with a rate of 18 L/min under a maximal pressure of 14 MPa (Fig. 3.29). The pump was driven by a 4.9-kW four-stroke air-cooled petrol engine. The reel used contained 200 m of Dunlop thermoplastic hose with an ID of 1/2″. The working flow rate of the submersible return pump (240 V, 6.5 A) was 40 L/min from a depth of 40 m.

During 1979–1980 field season, the first access hole was drilled through 125 m of near-temperate ice at the northern end of George VI Ice Shelf. First, using a 30-mm nozzle, a pilot hole was drilled through the ice shelf to the ocean cavity below. Then, the hole was enlarged to a depth of ∼10 m below sea level using 70-mm-diameter and 127-mm-diameter nozzles. The submersible pump was lowered down the hole and the storage tank was refilled with sea water. Until this hole refroze, there was unlimited sea-water available at the submersible pump to allow for the drilling of other access holes. Subsequent holes were drilled to the base of the ice shelf using the same 30-mm nozzle and were also enlarged using the 70 and 127-mm nozzles.

The 5-m^3 tank limited drilling to depths of ∼200 m, at which point drilling had to stop to allow for the melting of additional snow. To extend the drilling range of the system, a 12-m^3 tank was obtained. Moreover, to accommodate for the additional length of hose for deeper drilling, an electrically powered winch able to hold 400 m of hose was built. These modifications allowed for the drilling of holes with depths of 250 m using the 127-mm nozzle in 1987–1988 (Fig. 3.30). In the following season, the addition of an 80-kW burner and heat exchanger increased the water temperature at the surface from 60 °C to just over 100 °C, allowing for the drilling of a 300-m-deep hole through the ice shelf with an increased penetration rate.

Along a presumed flow line and at mid-depth (the coldest level in ice shelves), the recorded temperatures were −6 °C at Moore Point (70° 30′S), −4 °C at Carse Point (70° 15′S) and, near the northern ice front (70° 00′S), between −1.6 and −1.8 °C (Paren and Cooper 1988). The drilling program on George VI Ice Shelf was completed in 1989 and attention was then focused on the thicker and colder Ronne Ice Shelf, which required further redesign and development of the hot-water drill (see Sect. 3.3.7).

Fig. 3.29 Schematic of the BAS hot-water drill system used at George VI Ice Shelf (Makinson 1993)

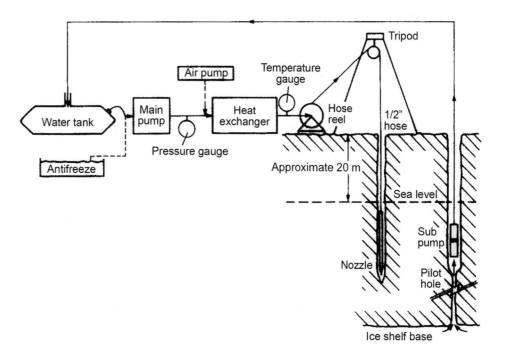

Fig. 3.30 BAS shallow
hot-water drilling system in use,
George VI Ice Shelf, 1988 (*Credit*
K. Makinson)

3.2.6 Sweden Shallow Hot-Water Drill Systems

During the last few decades, intensive hot-water drilling activities were carried out at Storglaciären, a small poly-thermal glacier in arctic Sweden that has the longest research program in the world (since 1946). In 1983 and 1984, a shallow hot-water drill system had been used for drilling a series holes in Storglaciären (Hooke et al. 1987). The deepest hole was drilled with a rubber hose that had an estimated average stretch in the hole of ∼6%. The total amount of hose used for drilling this hole, including the guide tube and nozzle, was 126 m. Based on the hose stretch, the final depth was assumed to be 133–134 m (radar measurements indicated an ice thickness of ∼118 m). The stretching of the hose made it difficult to sense exactly when the drill stopped, but the nozzle was kept at its maximum depth for several minutes; it can be assumed that the hole reached either the glacier bed or, at worst, a large englacial rock quite near the glacier bed. The total length of casing inserted was 131.6 m for deformation measurements.

Approximately 30 holes were drilled in that place between 1983 and 1987 for the recording of water-pressure variations (Hooke et al. 1988). In ∼70% of these, water stood at the glacier surface until the drill tip had penetrated between 40 and 95% of the local ice thickness; the water level then dropped suddenly. However, upon the completion of the holes, the water level in all these cases was only 15–45 m below the surface and remained relatively stable at these levels, even when a water flow was entering the holes from the surface. The other holes did not drain. Drilling operations for water-level observations were continued in 1988–1990 (Hooke and Pohjola 1994).

In 1985, 1987, and 1988, several boreholes were drilled to a maximal depth of 162 m, again for performing deformation measurements (Hooke et al. 1992). Holes were cased with a 40-mm (1985 and 1987) or 50-mm (1988) square aluminum tubing. The diameter of the boreholes was approximately 0.2 m at the surface and probably 0.1 m at a few meters above the glacier bed. At the glacier bed, they may have been somewhat larger, as the drill was normally held there for several minutes. The holes were accessible for ∼2 months after casing was set. In 1989, four holes (with a maximum depth of ∼140 m) were drilled at Storglaciären for making video observations of englacial voids (Pohjola 1994). During the 1991–1994 summer field seasons, a number of holes was drilled each year through the glacier for making water-pressure measurements (Hanson et al. 1998).

During the summers of 2001–2003, 48 holes were drilled at Storglaciären to intercept englacial conduits, and five more were drilled near the grids in support of radar investigations, resulting in 3900 m of ice drilled (Fountain et al. 2005). Approximately 79% of the holes intersected a hydraulically connected englacial feature. Englacial crevasses were found at all depths from near the surface to near the glacier bed. A video camera was used to examine the feature of the local water flow. Surface streams and pools were used as water sources for the drill. The penetration rate was approximately 60 m/h; hole diameter was ∼150 mm at the surface and ∼50 mm near the drill tip at higher depths. While drilling in impermeable ice, the hole was completely filled with water, and excess drill water flowed over the glacier surface. When an efficient connection was made to an englacial drainage passage, the water level in the hole dropped and remained below the surface. The holes were

Fig. 3.31 TRS hot-water drill system in action at Storglaciären, Sweden, July 2013 (Return to the glacier 2013)

Fig. 3.32 Gasoline engine and high-pressure pump (on the right) and heating unit (on the left) (Olesen 1989)

drilled until their first encounter with an efficient englacial hydraulic connection or until the glacier bed was reached.

In 2008, drilling operations at Storglaciären were resumed for downhole radar zero-offset profiling (Gusmeroli et al. 2010). During the summer of 2009, a total of 14 boreholes were drilled to the depths deeper than 80 m to spatially discriminate the water distribution versus depth at different sites on the glacier (Gusmeroli et al. 2009). The maximum depth drilled was 130 m. Typically, these boreholes had diameters of 100–150 mm, which varied with depth.

The last version of the Tarfala Research Station (TRS) hot-water drill system is shown in Fig. 3.31. The drill system had a pump that gathered supra-glacial running water into a diesel-fueled boiler, where the heated water ($\sim$90 °C) was then pressurized and redirected via a system of rubber hoses into a 1.5-m-long metal drill stem with a brass nozzle. The hose was fed manually by an operator that was seating down near or on the ladder-shaped leg of the mast. It would take up to 3 h to drill a 100-m-deep hole, which was the typical depth for boreholes at Storglaciären.

3.2.7 Hot-Water Drilling System of the Geological Survey of Greenland

This hot-water drilling system was designed for bedrock topography studies near Jakobshavn, West Greenland (Olesen 1989). The power and pump unit consisted of a 6-kW four-stroke gasoline engine with a 1:2 reduction gear connected to a high-pressure piston pump through a flexible coupling (Fig. 3.32). At 1750 rpm, the pump delivered 18 L/min of water with a pressure of up to 10 MPa, at which the relief valve was set. The engine was also connected through a belt drive to a small 0.6-kW, 220-V generator that was used for the ignition system in the heater. Fuel

consumption was 2.5 L/h. The whole unit was mounted on a stainless-steel frame and weighed 79 kg.

The heating unit consisted of a water coil with an oil burner (modified for using Jet-A1 fuel). An air blower and an oil pump were driven by a flexible axle, which was connected to the rear end of the generator axle. The water coil was used in a vertical position, making it easy to empty when drilling was completed. The air for the burner was blown in from the bottom of the heater, passing between an inner and an outer mantle up to the burner, which was at the top. With a fuel consumption of 11.6 L/h corresponding to 113 kW and an outlet temperature of 82 °C, the heating efficiency under field conditions was higher than 90%. The heating unit was safeguarded by a safety valve, a water-flow contact, a thermostat, and a high-temperature cutoff, and it was equipped with gauges for temperature and pressure. It was also mounted on a stainless-steel frame and weighed 127 kg.

The ½" hose used came in lengths of 100 m and was fitted with hydraulic couplings. Its working temperature and pressure limits were 121 °C and 13.8 MPa and it weighed 0.21 kg/m. The drill stem was 2 m long with a 25-mm outer diameter and a 10-mm inner tube, with the space between filled with lead in the bottom half. Both tubes were made from stainless steel. The end of the drill stem was fitted with a 180-mm-long tapering bronze tip with interchangeable 25-mm stainless-steel nozzles. The original inclinometer, which was small enough to fit inside a 32-mm tube, was designed and installed on top of the drill stem together with the 10-mm inner tube leading the hot water from the hose to the nozzle. The signal was sent to the surface via a cable running along the hose. The same cable was used to supply the inclinometer with DC voltage from the surface battery.

The drilling set also included a lightweight tripod with a winch and a pulley with a load cell (Fig. 3.33) and a

Fig. 3.33 Lightweight tripod with a winch and a pulley; the reel for the inclinometer cable is located below the winch (Olesen 1989)

low-pressure centrifugal pump that was used when drilling water had to be drawn from far away. The total weight of the complete drilling system with 600 m of high-pressure hose was 473 kg.

During two field seasons in 1987 and 1988, a total of 5657 m were drilled, with the deepest hole being 390 m deep. In 1987, a nozzle with an inner diameter of 2.7 mm was used for all holes, and this resulted in a pump pressure of 3.5 MPa with one length of the 100-m hose. In 1988, drilling operations were conducted using a 2.5-mm nozzle, which required a pump pressure of 5 MPa. During drilling, the hoses were stretched out on the ice and became connected as drilling progressed.

The penetration rate decreased from 300 m/h in firn to 125–200 m/h in clear ice. At depths deeper than 290 m, the penetration rate dropped to ~66 m/h. However, penetration through 5–10-m debris layers, which were very often found between depths of 150 and 250 m, required 1–2 h of drilling.

The load cell and inclinometer (which were both added in the field season of 1988) worked well together, immediately alerting the operator about any slowing of progress or deviations from the vertical. They were most useful when debris layers were encountered because the operator could add pressure on the drill tip by paying out more hose as long as the drill stem remained vertical. This procedure often resulted in a more rapid penetration than when the drill tip had little or no contact with the ice. When a constant pressure had been maintained for at least half an hour and no progress had been made, it was assumed that the bottom had been reached.

In 1990, several boreholes were drilled to the depths of 202–600 m in the Greenland ice sheet margin at Pâkitsoq to measure englacial temperatures and to investigate the subglacial drainage system (Thomsen et al. 1991). All holes reached the bottom of the ice except one site where the ice

thickness was expected to be ~500 m, but drilling was carried out to 600 m without reaching bedrock; this was the deepest that could be reached with the available supply of drill hose. No draining of the water took place in the boreholes.

3.2.8 Icefield Instruments Hot-Water Drilling System

Hot-water drilling in Trapridge Glacier, a small sub-polar glacier in the St. Elias Mountains, Yukon Territory, Canada begun in 1980, and most of these holes are known to have reached the glacier bed at depths less than 80 m (Clarke and Blake 1991). Starting from the end of the 1980s, drilling operations were carried out with the technical support of Icefield Instruments Inc. (Whitehorse, Yukon, Canada), mainly to install temperature sensors and instruments into the glacier substrate. In all years except 1992, holes were drilled to their final diameter in one pass with the drill in roughly 1 h of drilling (Waddington and Clarke 1995). In 1992, however, the thermal efficiency of the drill was impaired and thus the hole obtained after one pass was too small to install the instrument; a second pass was required. The first pass took approximately 1 h, and the second one took approximately 10 min.

The hot-water drill used had a 1.14-mm-diameter jet nozzle with a pressure drop of 7–14 MPa (Fig. 3.34) (Blake et al. 1992). The holes were nominally 50 mm in diameter; however, they were usually larger at the surface. Boreholes froze from the top down because the coldest ice was near the surface (approximately −5 °C). A 48.1-m-deep hole froze ~13.4 h after the cessation of drilling.

3.2.9 Hot-Water Drilling System for Water Well Construction

The first water wells (Rodriguez wells or Rodwells) in ice were built using stream drills (see Sect. 4.2.1). Similar water wells were developed in 1962 at under-ice Camp Tuto, Greenland using hot water heated by the rejected heat from one of two Caterpillar diesel engines driving a 75-kW, 120/240-V generator that supplied for the camp's electrical requirements (Russell 1965). During May–June 1962, the well produced 5.3 m^3 of water per day with a fuel consumption of 137 kg/day.

Rand (1982) proposed using hot-water drills, which would be used to melt a vertical shaft down to the firn–ice transition layer and then maintain an ice cavity (Fig. 3.35). An initial quantity of water was obtained from the surface water-melting system. At depths where firn becomes water-impermeable, water starts to be collected in the hole.

Fig. 3.34 Hot-water ice drilling at Trapridge Glacier, Canada, 1992 (*Credit* E. Blake)

When the water level reached the sensor placed above the submersible pump inlet housed in the drill, the pump would turn on automatically. Once the water began to circulate out of the hole, the initial surface water supply was no longer required. To enlarge the water pool, the water was made to circulate from the water well to the boiler to increase its temperature and then back into the hole. The drill was then lowered more slowly to create a large sub-ice reservoir of water. The amount of water in the reservoir at any time was controlled by the water produced via melting and by the water lost to the surrounding firn/ice via percolation, the water withdrawn for use, and the recirculated warm water.

This concept was realized at Amundsen-Scott station, South Pole, Antarctica where the first experimental Rodriguez well, Rodwell-1, was constructed in the 1990s. The hot-water equipment was originally installed in 1991–1992, housed in two stacked shipping containers that were placed in a pit (Fig. 3.36). The supporting piping systems in the power plant included waste-heat recovery silencers, and the associated heat exchangers, pumps, etc., were installed during the subsequent summer in 1992–1993. The surface container contained the equipment for hot-water drilling and for operating the well; the wellhead was in the lower container (Taylor et al. 1997).

A 30-cm-diameter pilot hole was drilled into the firn to a depth of 60–70 m and accommodated the electrical cable connected to the submersible pump, an emergency heat trace, and two insulated hoses (one for carrying water to the surface and one for returning warm water to the well). Because contamination of the well with fuel was a major concern, a low-temperature EPDM mat surrounded the wellhead and extended out 7 m.

The water well was certified as a source of potable water in early 1994. On February 26, 1994, the pool was approximately 16 m deep and 22 m in diameter, and the well bottom was 101 m below the wellhouse (103.5 m below the snow surface). Before consumption began, however, an electrical fire in the pump cable on March 1, 1994 forced a nine-month shutdown. During the shutdown, a 4-m-thick ice layer formed on the pool surface and 6–11 m of freezeback is repeated freezing occurred on the walls and the bottom. The well was restarted in December 1994 by drilling through the ice layer and recirculating warm water as had been done before. By March 1995, the well had melted below the pre-fire level. Figure 3.37 shows the approximate well geometry before and after the fire and at the time of deployment in December 1995.

This water well supplied potable water from January 1995 onwards. The reservoir reached a relatively stable size of approximately 24 m in diameter and 16 m in depth and contained approximately 5000 m^3 of water. The consumption rate was approximately 2000 m^3/year. The pump drew water from approximately 1 m below the water surface. Nearly 10% of the water was consumed and the rest was heated using waste heat from the station and returned to the well. The flow rate of the returning water was $\sim$60 L/min and it was discharged 3 m below the water surface through a nozzle designed to produce a uniform 90° cone-shaped jet. Peak discharge velocity 1 m away from the nozzle was kept at less than 1 mm/s, and no remnants of the jet should persist at the bottom of the well.

To collect micrometeorites at the bottom of the pool, a second access hole approximately 2 m from the central hole was drilled using a hot-water drill (Fig. 3.38). Water from

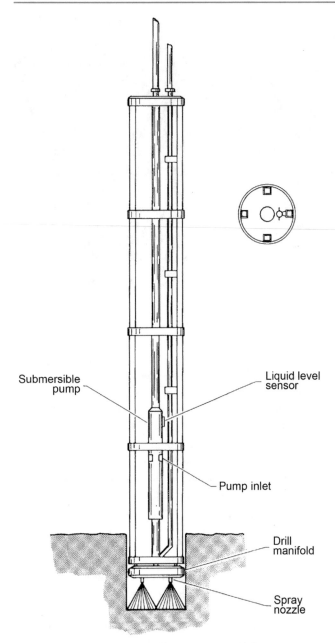

Submersible
pump

Liquid level
sensor

Pump inlet

Drill
manifold

Spray
nozzle

Fig. 3.35 Hot-water drill proposed for the construction of water wells (Rand 1982)

the well was heated in boilers located in the wellhouse and fed into a 30-cm-diameter cylindrical drill via a neoprene rubber hose. This hot water discharged through a 90° conical nozzle to melt the ice as the drill descended. A water pump positioned inside the drill pumped water from downhole to the surface via a return hose. The drill was lowered using a winch and stainless-steel cable. The hoses were attached to the cable with stainless-steel clamps and cloth ties. The cable was guided into the hole using a sheave suspended from a spring scale. The scale allowed operators to determine whether the drill was suspended in air, was on ice, or was in water; this set the penetration rate. Drilling took ∼36 h,

which was longer than expected but yielded a vertical, wavy-walled access hole with a minimum clearance of 30 cm in diameter.

A typical Rodwell lasts for 7–10 years before the well becomes too deep (>150 m) and it becomes too energy intensive to extract water. In 2001–2002, all the available water was pumped out from Rodwell-1 and a new well, Rodwell-2, was started in January 2002 a hundred meters away from Rodwell-1 (Fig. 3.39). The average water consumption amounted to 4500 m³/year (Fig. 3.40). The construction of Rodwell-3 was started in the season of 2006–2007, including startup drilling operations using an electric-heated hot point (see Sect. 1.2.32). However, it was not put in service until 2012–2013. It had been circulating water during the winter of 2012 and it was brought online in January 2013. Meanwhile, Rodwell-2 has been being converted to contain the station's sewage outfall. During exploration, the Rodwell pump froze up several times, and the emergency snow melter had to be put into service.

3.2.10 Hot-Water Drill of Aberystwyth University, UK

A portable hot-water drill capable of melting boreholes up to depths of ∼150 m and based on the commercially available Kärcher pressure washer was designed in Aberystwyth University, UK (Hubbard and Glasser 2005). The chosen machine was fixed within a frame measuring ∼1.2 m 0.8 m × 0.9 m and weighed 185 kg. The unit was composed of a pump, powered by a gasoline (HDS1000BE) or diesel (HDS1000DE) generator, and a diesel burner. Running at a pressure of ∼9 MPa, the Kärcher HDS1000 washer provided a water flow of ∼13 L/min. It was fed by a 4000-L water bladder and a 200-L water butt via a short hose (∼10 m in length), incorporating both inlet and in-line filters with a pore size of ∼250 μm (Fig. 3.41a).

During snow-free summer drilling, the water butt was supplied from supraglacial stream. However, once drilling started, water exiting at the top of the borehole was recirculated via the pump. If surface snow was available on site, a propane-powered tar melter was used to prepare start-up water from snow. Hot water was pumped into a high-pressure hose attached to stainless-steel drill stem that was 1.5–3 m long and 30 mm in diameter and had a changeable nozzle with a single 1-mm-diameter central hole (Fig. 3.41b, c).

Boreholes with this system were drilled for the first time in 1992 in Haut Glacier d'Arolla, a 4-km-long temperate valley glacier with a maximum thickness of ∼180 m located in the Swiss Alps (Hubbard et al. 1995). During each of the summers of the years 1992–1999, an array of boreholes was drilled ∼1.5 km away from the glacier snout and east

Fig. 3.36 Upper container of Rodwell-1 at Amundsen–Scott station, 2001 (*Photo* D. Hrubes from Down the Hole n.d.)

Fig. 3.37 Evolution of Rodwell-1 at Amundsen–Scott station, 1994–1995 (Taylor et al. 1997)

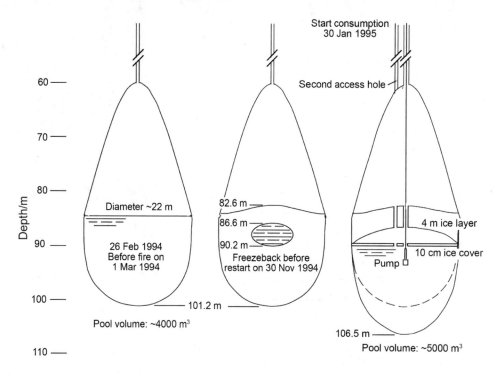

of the glacier's center line (Willis et al. 2003). On average, 25 boreholes were drilled each year, ranging in depth from ~30 m near the margin to larger than 130 m on the center line, and they have been used to study the englacial and subglacial hydrology and geochemistry of this part of the glacier (Copland et al. 1997; Hubbard et al. 1998).

Water left the washer at pressure of ~9 MPa with a thermostat setting of ~80 °C. This combination resulted in penetration rates that fell progressively from ~120 m/h near the ice surface to ~30 m/h at ice depths deeper than 120 m. This decrease in penetration rate primarily resulted from heat losses from the lengthening section of hose suspended

Fig. 3.38 Hot-water drill used to drill the second access hole in Rodwell-1, Amundsen–Scott station (Taylor et al. 1997)

within the water-filled borehole. The hose was fed down the borehole manually by the driller at the optimal rate. This rate of lowering was defined by the requirement of keeping the water jet as close to the bottom of the borehole as possible but without resting any of the weight of the drill stem directly onto the ice. Typical fuel consumption was ∼2 L/h of gasoline and 4 L/h of diesel with the HDS1000BE generator and 6 L/h of diesel with the HDS1000DE generator.

During the last two decades, this basic drill set and its modifications were intensively used in polar and mountain regions for borehole deformation measurements at Glacier de Tsanfleuron, Switzerland (Chandler et al. 2008); investigations on the subglacial hydrology of Kronebreen, which is a fast-flowing tidewater glacier in Svalbard (How et al. 2017); and optical televiewing at the midre Lovénbreen, Svalbard (Fig. 3.42) (Roberson and Hubbard 2010), Roi Baudouin (Hubbard et al. 2012), and Larsen C ice shelves, Antarctica (Hubbard et al. 2016; Ashmore et al. 2017) and in Khumbu Glacier at an altitude of ∼5000 m asl in northeastern Nepal at the foot of Mount Everest (Fig. 3.43) (Miles et al. 2018).

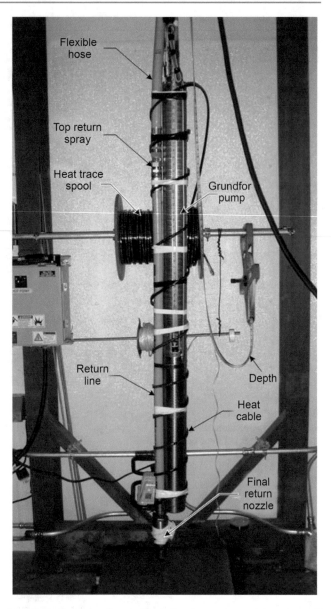

Fig. 3.39 Pump head assembly before it was inserted into the well shaft of Rodwell-2, Amundsen–Scott station (Haehnel and Knuth 2011)

3.2.11   Icelandic Hot-Water Drilling Systems

Icelandic hot-water drilling systems were primary aimed for accessing subglacial lakes beneath the Vatnajökull ice cap, which is the largest and most voluminous ice cap in Iceland. Lake Grímsvötn was first accessed with a hot-water drill in 1990 (Björnsson 1991). In 1991, two hot-water boreholes were drilled through the 250-m-thick layer of ice covering the lake, and the lake's temperature was measured and samples for a geochemical study were obtained (Ágústsdóttir and Brantley 1994).

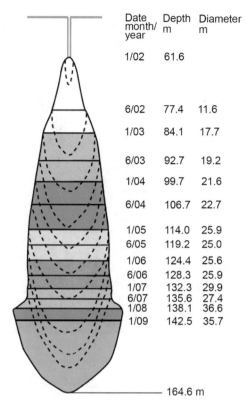

Date month/year	Depth m	Diameter m
1/02	61.6	
6/02	77.4	11.6
1/03	84.1	17.7
6/03	92.7	19.2
1/04	99.7	21.6
6/04	106.7	22.7
1/05	114.0	25.9
6/05	119.2	25.0
1/06	124.4	25.6
6/06	128.3	25.9
1/07	132.3	29.9
6/07	135.6	27.4
1/08	138.1	36.6
1/09	142.5	35.7

164.6 m

Fig. 3.40 Progression of Rodwell-2, Amundsen–Scott station, starting in January 2002 (Haehnel and Knuth 2011)

Another 300-m-deep hole was drilled into Lake Grímsvötn using hot-water drilling equipment provided by the Science Institute, University of Iceland in 2002 (Gaidos et al. 2004). Twin diesel burners heated meltwater to 80–90 °C in the drilling system (Fig. 3.44). Some of this hot water was recirculated into a container of snow to produce additional meltwater, and the remainder was pumped through a hose to the steel drill stem, where it was ejected under pressure from a nozzle with four holes.

The latest hot-water drilling system was designed and built at the Hydrological Service Division (HSD), National Energy Authority, Iceland and incorporated in-line sterilization to minimize the probability of subglacial lake contamination (Figs. 3.45, 3.46a) (Thorsteinsson et al. 2008). Snow was melted in a 600-L plastic tank, in which a heat exchanger was placed (Fig. 3.46b). Glycol was heated in a combustion unit (diesel burner) and was circulated in a closed-loop system between the burner and the heat exchanger. Meltwater was pumped from the plastic tank through twin filters (50 µm) to remove large particulates and then through a Pura UV20-1 UV-sterilization unit before entering a commercially available Kärcher HDS1000DE washer that combined a high-pressure pump and a heater (operating pressure range: 4–21 MPa; maximum temperature: 150 °C).

Water was then pumped into a 600-m-long hose made of synthetic rubber, which was reinforced with two high-tensile steel braids and deployed on a winch. The hose winch was powered by a geared motor and was fitted with a reduction gear that allowed for a minimum rotation speed of 0.055 rpm, corresponding to a penetration rate of 5.4 m/h for the outermost layer of hose on the winch. The maximum attainable winching speed in boreholes was 400 m/h.

Fig. 3.41 Hot-water drill of Aberystwyth University: **a** Kärcher HDS1000 washer (in the background) fed by a 4000-L water bladder and a 200-L water butt; **b** drill stem and hose; **c** nozzle (Hubbard and Glasser 2005)

Fig. 3.42 Hot-water drilling on
Svalbard, 2009 (Climate change
scientists from Aber are heading
to the Himalayas to study the
world's highest glacier 2017)

Fig. 3.43 Hot-water drilling on
Khumbu Glacier, Nepal
Himalaya, April 2017 (Miles
2017)

Water emerged at the front nozzle of the 2-m-long
stainless-steel drill stem. Two versions of the drill stem were
made, with ODs of 32 and 40 mm. During operations, the
hose passed over a top wheel with a diameter of 0.25 m,
fixed at the end of a 2.8-m-long inclined tower mounted at
the end of the drill's housing base (Fig. 3.46c). A depth
counter with a digital display was connected to the top wheel
axle. Two 2-m-long pieces of steel wire connected the drill
housing and a small hydraulic jack unit attached to the
tower, thereby holding the tower in place. A pressure meter

connected to the jack served to measure hose tension. The
drilling system was fitted with water-sampling outlets at
different locations (Fig. 3.46d).

A 2.2-kW gasoline generator supplied electricity for all
units except the Kärcher washer, which run on its built-in
6.6-kW diesel motor. All units except the snow melter were
mounted on a wheeled trailer suitable for glacier travel and
were housed in an aluminum shelter that could be opened up
from three sides. The total weight of the system, including
the trailer, was ~1300 kg.

Fig. 3.44 Hot-water ice drilling into Lake Grímsvötn, Vatnajökull ice cap, Iceland, 2002 (*Credit* T. Thorsteinsson)

Fig. 3.45 Schematic diagram of the HSD hot-water drilling system (modified from Thorsteinsson et al. 2008)

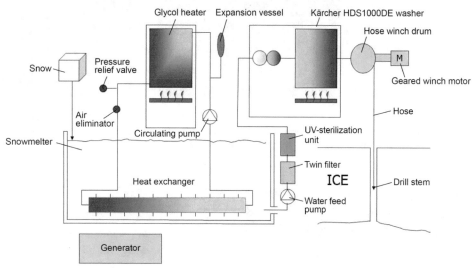

During initial testing, the high-pressure pump of the Kärcher washer was found to deliver 7.5 L/min of meltwater running at full load, but the efficiency of the snow melter turned out to be only 3.3–4.2 L/min. The latter was improved by inserting larger spirals in the combustion unit and covering the 15-m-long hoses connecting the combustion unit and the heat exchanger with insulating material.

This hot-water drill was tested on the Langjökull ice cap in June 2005. For sterilization purposes, all parts of the drilling system between the water feed pump and the drill stem were filled with 35 L of a 95% ethanol solution on the day before departure to the ice cap. The ethanol was emptied from the system into a container before the start of the drilling operations. Two test boreholes were drilled using a 32-mm-diameter drill stem. A nozzle with seven 0.8-mm-diameter holes (Fig. 3.47a) was used in the first attempt and a depth of 80 m was reached in 2.5 h. The drilling direction then appeared to deviate from the vertical and, when the drill stem had been pulled to the surface, one of the holes in the nozzle was found to be blocked.

During the second drilling operation, a drill nozzle with a single 1.5-mm-diameter hole (Fig. 3.47b) was used,

Fig. 3.46 Testing the hot water drilling system on Langjökull ice cap in June 2005: **a** drill system setup: a snow melter, hoses, and a trailer with the drill's parts in the background; **b** heat exchanger in the snow melter; **c** drill ready for operation; **d** outlets for sampling at different locations within the system (Thorsteinsson et al. 2008)

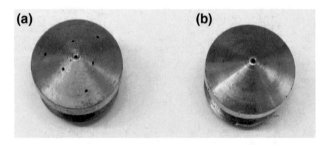

Fig. 3.47 Different drill nozzles: **a** with seven 0.8-mm-diameter holes; **b** with a single 1.5-mm-diameter hole (Thorsteinsson et al. 2008)

resulting in steady penetration to a depth of 110 m with an average penetration rate of 50 m/h and a hot-water consumption of 1.1 L/m. The hose was centered in the hole throughout the drilling process, indicating that no deviation from the vertical had occurred. The snow melter produced water at the rate of 500 L/h. The temperature of the water emerging from the drill stem at surface level was 95 °C. Fuel consumption during drilling was as follows: 10 L/h for the Aquila HB303 diesel burner for glycol heating, 7.5 L/h for the Kärcher diesel burner, 1 L/h for the diesel motor powering the Kärcher high-pressure pump and the burner fan, and 1–1.5 L/h for the gasoline generator.

3.2.12   Shallow Hot-Water Drilling Systems of the University of Wyoming

A series of shallow hot-water drilling systems were developed in the Department of Geology and Geophysics, University of Wyoming, USA. However, detailed descriptions of these systems were not published. It was reported that the first version of these drills was used at Worthington Glacier, a temperate valley glacier located in Alaska, to make sixteen holes, totaling more than 2750 m in length, for making video observations in the beginning of the 1990s (Harper and Humphrey 1995). These holes were 20–30 m apart and 175–200 m deep. Six holes were drilled to the glacier bed; the remaining ten were drilled to ~10 m above the bed. Video observations showed that entrained debris was limited to within ~1 m of the glacier bed.

Then, a new hot-water drill was designed to create straight holes with uniform and smooth walls for uncased inclinometry measurements (Harper et al. 1998, 2001). Straight and vertical holes with an almost constant diameter were created by lowering a heavy (40 kg) drill stem using an electromechanical in-line drive. A load cell on the drill tower monitored the hanging weight of the 2-m-long drill stem with a precision better than 1%, giving constant input to a computer controlling the penetration rate. Hole diameter was restricted to ~100 mm. The long and narrow stem, which had a tapered cross-section along its length going from 34 to 14 mm, caused turbulent mixing not just ahead of the drill nozzle but also within the upwelling water alongside the drill stem. The thermal decay length of the up-welling water was on the order of meters, and hence the long and narrow end of the drill stem gave turbulent eddies space to grow as they moved up the hole.

Afterwards, 31 closely spaced boreholes were drilled at Worthington Glacier in 1997. These boreholes were inspected for smoothness and diameter using a borehole video camera and a digital-recording caliper over a 70-day period. These measurements revealed that the drilling system produced overly wide holes until the drill stem and the high-pressure jet were submerged several meters within the borehole water column. Consequently, inclinometry data could not be collected in the top 5–8 m of the boreholes. Boreholes were stopped ~10 m short of the glacier bed. This was done to prevent the circulation of borehole water arising from potential connections with the basal hydrologic system.

Refreezing of the boreholes was a problem that needed to be addressed after the boreholes were ~20–30 days old. Freezing occurred both near the surface and at higher depths and affected approximately a third of the boreholes with an apparent random distribution. Holes were protected against freezing via re-drilling with a specially designed reamer that focused melting to only the parts of the borehole with a decreased diameter. After reaming, the holes were re-inspected with the video camera to ensure that the hole geometry had not been grossly altered by the reaming process.

Three boreholes were used for measurements of year-long deformations. Each hole was fitted with a wire running along its entire length. The holes were then re-drilled the following year (1998) using a specially designed wire-following drill tip. The inclination of the re-drilled holes was measured, and these measurements represented 372–374 days of deformation.

A network of boreholes (>60) with depths of 140–190 m spanning the length of the 7-km-long Bench Glacier, Alaska, USA was created during the springs of 2002, 2003, and 2006 (Fig. 3.48) (Fudge et al. 2008; Harper et al. 2005, 2010).

Fig. 3.48 Drilling with a hot-water drill on Bench Glacier, Alaska (*Credit* J. Harper from Choi 2010)

When the drill failed to advance near the glacier bed, it was reversed and then made to re-advance repeatedly for 20–30 min in an effort to penetrate englacial and/or subglacial debris. Rarely did the drill advance from the initial point of stoppage and never for more than a few decimeters.

Recently, a more powerful hot-water ice drill has been designed in the University of Wyoming for drilling intermediate-depth holes in Greenland (see Sect. 3.3.11).

3.2.13 ILTS Hot-Water Drilling Systems

To drill shallow holes to depths down to 200 m, the Institute of Low Temperature Science (ILTS), Hokkaido University, Japan designed a hot-water drilling system (Fig. 3.49) based on the commercially available Kärcher HDS1000BE washer in which a gasoline-driven water pump engine (Honda GX390) and a diesel combustion heater were combined (Fig. 3.50a) (Tsutaki and Sugiyama 2007, 2009; Sugiyama et al. 2008). The total weight of the washer was 165 kg. A safety device was installed between the pump and the heater, which turned off the heater when the water pressure dropped below an operating threshold. During drilling tests, the washer generated a hot-water jet with a temperature of 60–76 °C and a flow rate of 15.8–16.6 L/min.

The pump drew water from a 3000-L basin with a diameter of 2 m and a weight of 28 kg (Fig. 3.50b). On site, it could also draw water from natural streams. The drilling hoses, which individually were between 50 and 100 m long and had an ID of 8 mm, were connected with couplings. During drilling, the hose was supported by an aluminum tripod and a pulley equipped with hose-length gauge (Fig. 3.50c). The tripod had a height of 2.2 m and weighed 10 kg.

Two 1.5-m-long stainless-steel pipes with an ID of 19.6 mm and an OD of 27.2 mm were connected with a brass coupling and used as a drill stem with a replaceable nozzle (Fig. 3.50d). If necessary, a conic-spray nozzle (Type K-20) was used to ream the borehole. The total weight of the system was ∼220 kg, plus 30 kg per 100 m of hose.

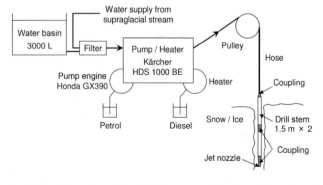

Fig. 3.49 Schematic of the ILTS hot-water drill system (Tsutaki and Sugiyama 2009)

In July 2007, eight boreholes with a total length of 925 m and a depth ranging from 87 to 138 m were drilled in Rhonegletscher, Swiss Alps. The pump and the heater were always located near the mouth of the first borehole, whereas the hose was extended to subsequent holes. Usually, water was pumped into the washer directly from a supraglacial stream. Drilling was carried out using a 2.0-mm or 2.5-mm jet nozzle. The borehole diameters were 100–150 mm near the surface. The 2.0-mm conic-spray nozzle was also used to ream some boreholes. The rate of penetration was in the range of 27–70 m/h; this rate decreased when water drained from the borehole and also near the glacier base. The rates in the farthest holes from the washer (>200 m) were lower than for those near the machine because the temperature of the water decreased with hose length. Drilling was continued until the nozzle reached the glacier bed, which was recognized via changes in the hose tension.

Drilling at Rhonegletscher was carried out again in 2008, year in which 24 boreholes were drilled to depths ranging from 18 to 99 m with a total length of 1118 m. The system consumed 6.9 L/h of diesel for the heater and 1.8 L/h of gasoline for the pump in both the 2007 and 2008 operations.

In subsequent years, the system was modified to drill deeper (Sugiyama et al. 2010). A second washer was added and a simple manifold was constructed to combine hot water from the two machines. The system generated hot water at ∼65 °C at a rate of 26.6 L/min and a pressure of 10 MPa. The length of the drilling hose was extended to 550 m. An electrically powered feeding system was developed to lower down and lift up the hose from the borehole because the hose had become too heavy to be handled by men (Fig. 3.51a). Hot water flowed through high-pressure hoses with an ID of 12.7 mm (Bridgestone WAR08, OD: 19.8 mm) before being emitted as a water jet through a 3-mm-diameter nozzle. The total weight of modified system increased to ∼600 kg.

In February–March 2010, two boreholes were drilled through Glaciar Perito Moreno, Southern Patagonia Icefield, Argentina (Fig. 3.51b). The first hole reached bedrock at 516 m after 21 h, including the time spent for testing and repairing equipment. The total fuel consumption rates were 11.2 L/h of diesel and 4.3 L/h of gasoline. The second borehole was drilled 5 m apart to a depth of 515 m and operations took 10.5 h. The diesel and gasoline consumption rates were 15.7 and 3.9 L/h, respectively. The penetration rates were 48–50 m/h. The boreholes were usually filled with water but drainage also occurred several times during drilling.

Basically the same drilling system was used in January 2012 at Langhovde Glacier, an outlet glacier coming out into Lützow–Holm Bay, East Antarctica, ∼20 km south of the Syowa Japanese research station (Fig. 3.52) (Sugiyama et al. 2014). A third Karcher HDS1000BE washer was added to the drill set. Two boreholes were drilled and both of them

Fig. 3.50 Subsystems of the ILTS hot-water drilling system: **a** high-pressure Kärcher HDS1000BE washer; **b** 3000-L water basin; **c** operator controlling hose feeding; **d** jet nozzle mounted at the tip of drilling stem (Tsutaki and Sugiyama 2009)

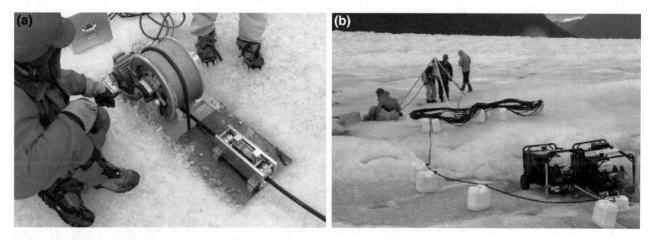

Fig. 3.51 Drilling operations at Glaciar Perito Moreno, Argentina, February–March 2010: **a** feeding system with brake; **b** drilling setup (Sugiyama et al. 2010)

penetrated into an ocean cavity. The mean penetration rate was 40 m/h. The ice and water column thicknesses were found to be 398 and 24 m in the first borehole and 431 and

10 m in the second one, respectively. Judging from the ice surface and bed elevations, the drilling sites were situated within several hundred meters from the grounding line.

Fig. 3.52 Hot-water drilling at
Langhovde Glacier, Antarctica,
January 2012 (Hot water drilling
at Langhovde Glacier, East
Antarctica 2012)

3.2.14 ANDRILL Hot-Water Drill

ANDRILL (ANtarctic geological DRILLing) was a multi-national project established by four nations (Germany, Italy, New Zealand, and the US) to recover geological stratigraphic records from the Antarctic margin (Talalay and Pyne 2017). McMurdo Ice Shelf was chosen as the platform for the initial drilling operations. To select the best drilling site, two hot-water access holes were made in 2003 through the ice shelf and a few 0.8-m-long gravity sediment cores were taken (Falconer et al. 2007).

After that, the innovative Hot Water Drill (HWD) was designed and built in New Zealand, which was able (1) to make access holes with a diameter larger than 600 mm for the deployment of a sea riser; (2) to maintain an open hole around the sea riser during two or more months of drilling; and (3) to provide sea water for the production of drilling fluid necessary for geological coring. It was therefore different from the

hot-water drills built previously and consisted of the following major components: an initial surface water supply (melted snow was used on site); boilers to heat the low-pressure primary heating circuit; a secondary high-pressure circuit with heat exchangers and a high-pressure pump suitable for heated seawater; a submersible pump for water recirculation; and flexible hoses, winches, and jetting tools. The HWD was designed for stationary operations linked to the main geological drill rig platform and was installed in two 20′ modified insulated containers (Fig. 3.53). Moreover, for certain cases, it was also capable of stand-alone operation using additional electrical generators.

The heating plant used slightly modified standard-model water boilers, for a total power output of 820 kW and using a propylene glycol aqueous solution in the low-pressure primary circuit coupled to the shell and the tube heat exchanger separating the secondary circuit (Fig. 3.54). Boilers could run individually or together. Propylene glycol

Fig. 3.53 ANDRILL hot-water
drill setup at McMurdo Ice Shelf,
October 2006 (*Credit* A. Pyne)

Fig. 3.54 Two boilers in the primary loop of the HWD system (*Credit* A. Pyne)

Fig. 3.55 Secondary-loop CAT pump with variable speed drive of the HWD system (*Credit* A. Pyne)

was chosen as the heating fluid because it poses low health risks and has a low flammability rating.

The high-pressure secondary circuit was capable of either open-circuit fresh/seawater drilling or recirculating-mode drilling (using the propylene glycol aqueous solution). The fluid in the secondary loop was circulated by a CAT pump with a flow rate of up to 120 L/min at 8.3 MPa (Fig. 3.55). The design of the heating system consisting of primary and secondary circuits prevented the boilers from being subject to seawater and reduced the likelihood of corrosion, which causes very limited boiler service life. The secondary circuit was subject to seawater at ∼85–90 °C.

A 4000-L snow melting/water supply tank (Fig. 3.56) and a 1000-L glycol/water tank were incorporated into the heating system. Tipping the 4000-L snow-melter tank allowed the container doors to be closed when not being filled with snow. The winch system consisted of two single-sheave capstans, two hose drums (each with 400 m of neutrally buoyant 1″ Kevlar/EDPM hose supplied by IVG, Italy), and two rope winches, each with 400 m of 3/8″ Dyneema braid (Fig. 3.57). The capstans were installed in the cellar area of the drill platform and the hose drums and rope winches were installed in the HWD containers ∼6 m

away. The winches were powered electrically with an electronic feedback control via a PLC, enabling either the operation of a single capstan and a hose drum/rope winch for initial drilling or tandem operation for the twin-hose sea riser over the reamer ring.

Three different downhole tools were used. Initial drilling was carried out with a single jet nozzle that could make holes ∼100 mm in diameter (Fig. 3.58). This was followed by using the reamer tool, which was configured to spray upwards. The third tool was a ring tool 600 mm in diameter with an internal clearance of 450 mm to allow the sea riser to be deployed through it. This tool was capable of both upwards and downwards hot-water cutting and also recirculating the heated glycol aqueous solution. It was used initially to ream the hole at the base of the ice shelf and then periodically as an over-reamer to maintain the sea riser free in the ice shelf hole and the drilling fluid water supply.

To start drilling, hot water was generated by melting snow using the closed loop circulation of the warm propylene glycol aqueous solution. During the next phase, fluid was circulated in a closed loop to recover the water-glycol solution and replace it with fresh water to allow for the melting of snow via the heat exchanger system. Once the

Fig. 3.56 Filling of a 4000-L melter tank with snow (*Credit* A. Pyne)

Fig. 3.57 Dual winch systems for hoses and ropes (*Credit* A. Pyne)

initial drill hole had been created through the ice shelf, a water-filled well was melted into the ice to one side of the drill hole. The well was filled with seawater, which was forced up the drill hole. This was then used as source of water for the HWD system as well as for drilling fluid production.

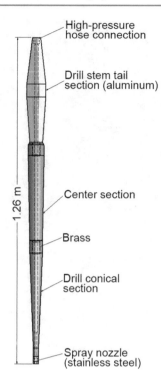

High-pressure
hose connection

Drill stem tail
section (aluminum)

1.26 m

Center section

Brass

Drill conical
section

Spray nozzle
(stainless steel)

Fig. 3.58 HWD drill stem

The HWD was tested on McMurdo Ice Shelf in February 2006, approximately 4.1 km east of the future drill site, and it successfully drilled through 97 m of ice shelf and reamed holes larger than 600 mm in diameter (Fig. 3.59).

In October 2006, the HWD melted a 60-cm-wide, an 82-m-deep access hole through McMurdo Ice Shelf, ~9 km southeast from Scott. The over-reamer tool, used to ream the ice shelf hole once the sea riser was in place, ran down a wire guide on the sea riser (Fig. 3.60). When not in use, the

reamer was parked approximately 2 m below the cellar floor in the cellar well. Hot-water reaming of the hole was carried out at scheduled times throughout the drilling operations (usually every 7–10 days). These times were separated by incrementally longer periods as the drill season progressed because refreezing in the relatively "short" ice shelf hole was not as rapid as initially expected.

3.2.15 ARA Hot-Water Drill System

The Askaryan Radio Array (ARA) is a large neutrino detector under construction near the Amundsen-Scott station at the South Pole (ARA Collaboration 2012). The ARA was designed to detect and measure high-energy neutrinos coming from space by observing the radio pulses they generate as they travel through the ice. Their radio pulses come from the so-called Askaryan Effect, in which a shower of particles in a dense material coherently emits radiofrequency energy. The detector was planned to consist of 37 stations with antennas deployed at depths of up to 200 m under the ice surface. Each station required six 150-mm-diameter dry boreholes (to prevent the freezing-in of the ARA equipment); thus, ARA-37 will include 222 holes and cover ~100 km^2.

In the 2010–2011 Antarctic field season, the ARA project borrowed the Rapid Air Movement (RAM) mechanical drill, with which ice chips are blown out of the hole by the exhaust air from the high-flow-rate turbine, to test methods for producing holes for the radio antennas. However, the drill could not go deeper than 63 m (Talalay 2016). Then, a few holes with a hot-water drill system designed by the Physical Sciences Lab, University of Wisconsin-Madison were drilled; however, equipment for pumping the holes dry

Fig. 3.59 Testing of the HWD system, McMurdo Ice Shelf, February 2006: **a** two single-sheave capstans; **b** hot-water reamer (*Photo* T. Falconer from Final comprehensive environmental evaluation (CEE) for ANDRILL 2006)

Fig. 3.60 Reaming of an access hole: **a** hot-water reamer tool; **b** view from the cellar of the drill rig looking down into the hole in the ice shelf and at the ring reamer in operational position around the riser (*Photo* D. Reid and T. Falconer from Antarctica New Zealand Pictorial Collection n.d.)

was not available, and the initial season's boreholes were drilled only to a depth of ~ 30 m, where the pooling of water prevented deeper drilling. A prototype station and a first-detector station were deployed.

The system was modified and required three sleds instead of one equipment sled, as used in 2010–2011 (J. Cherwinka, personal communication, 2014). The new system included a snow melter, a 6360-L water tank, a high-pressure pump (100 L/min) powered by a separate Honda engine, six boilers with a total heating power of 290 kW for heating water up to 87 °C, and a reel with a hose and a drill head borrowed from ANARE. After the boilers, the water flow was divided into two lines: the first line went back to the snow melter (55 L/min) and the second one (45 L/min) went to the drill hose and stem. The hose was wound on the reel and used in sequence for drilling or pumping out water.

In the 2011–2012 season, the system was tested at Amundsen-Scott station. Two first test holes were drilled near IceCube Lab to depths of 190 and 70 m. The drill head in the 190-m-deep hole got stuck because the hose reel leaked. The first instrumentation hole was drilled down to 210 m over 18 h and a submersible pump was lowered into the hole. However, after 13 h of pumping, the soft starter failed, resulting in the loss of flow up the hose. The pump, power cable, and hose were frozen in the hole. The hole was usable to 70 m. Then, using hand equipment, four holes were drilled and pumped down to 100 m, one hole per day, and one was drilled and pumped down to 60 m. The average

rate of penetration was ~ 20 m/h. Instrumentation was deployed in all holes to depths ranging from 37 to 87 m.

The hot-water drilling system was again upgraded. The drill head was integrated with a submersible pump (72 L/min, 5.6 kW, 480-V three-phase driven motor) allowing it to drill and pump water out simultaneously (Fig. 3.61). Hot water sprayed out of the nozzle and traveled some distance back up the hole to the pump, where it was pumped back to the surface. The hole diameter developed between the nozzle and the pump. The length of the drill head was 2.8 m and its weight was 82 kg. When drilling in firn, the conic-spray nozzle was installed at the head tip. When drilling in ice, a 10-m-long hose extension with a heavy drill stem and jet nozzle was attached to the drill head.

The water was almost fully recirculated (water lost in firn was ~ 2700 L/hole) and the melt tank was used only for start-up and as backup if needed. Three boilers were used to pump hot water (55 L/min, 88 °C) into the hole, one boiler was used as a spare and another one for preheating. The hose reel had a ~ 3.5-m-diameter drum without a level winder. A motorized sheave installed beside the drum controlled the speed during drilling; the drum operated in torque mode and was a slave to the sheave. To pump in and out, double hoses and a combo cable were fastened using custom bundles. The total weight of the ARA drill was ~ 15 t.

During the 2012–2013 season, 12 production holes were drilled to depths typically just over 200 m (Fig. 3.62).

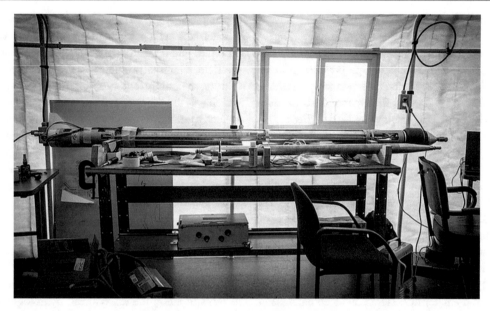

Fig. 3.61 Drill head (in the background) and drill stem (in the foreground) inside a workshop at the South Pole, January 2013 (Donenfeld 2013)

Fig. 3.62 Overview of the ARA hot-water drill, January 2013; the inset shows double hoses and a combo cable fastened with bundles (*Credit* J. Cherwinka)

Drilling was carried out by five people in one 12-h shift; the drilling of one hole took ∼7 h. The penetration rate in firn was 48 m/min, and drilling slowed down to 33 m/h after the pump was on (at ∼35 m). At a depth of ∼50 m, the firn nozzle was changed to the hose extension with the drill stem, and drilling was continued at penetration rates of 30–45 m/h. Fuel consumption was 360 L of AN8 fuel and 22 L of gasoline per hole.

3.2.16 JLU Shallow Hot-Water Drill

The shallow hot-water drill of Jilin University (JLU) was based on a commercially available high-pressure Kärcher HDS 6/14C washer that could deliver water at temperatures in the range of 80–155 °C and with a flow rate of 4–10 L/min at a pressure as high as 14 MPa (Talalay et al. 2018) (Fig. 3.63). In the field, the washer was fixed on a polyethylene Siglin cargo sled with a 200-L drum of diesel fuel. The rated fuel consumption of the washer was 3.5 kg/h. A mast with a sheave on top and a reel with a 100-m-long polyurethane rubber hose were fixed on a small aluminum sled. The high-pressure (18 MPa) hose reinforced with a steel net had an ID of 10 mm and an OD of 22 mm and weighed ~0.3 kg/m in air. The reel was driven by a 0.25-kW AC motor through a frequency inverter to smoothly adjust the lowering/hoisting speed in the range of 0–40 m/h. The 2-m-long, 60-mm-diameter drill stem contained changeable calibrated nozzle tips at the end. The drill stem was made from stainless steel with a lead hollow core and weighed 42 kg.

For easy operation, the system was equipped with three JHBM-50 load cells (3 × 500 = 1500 N) to determine the optimal distance between the nozzle and bottom of the borehole and a 5000-ppm encoder to measure the length of

Fig. 3.63 JLU shallow-depth hot-water drilling system (Talalay et al. 2018)

downhole hose (BC50S). A submersible pump, a return hose, a water tank with a heating unit for melting snow and ice if required, and a small 2-kW generator were also required in the field. The total weight of the system was approximately 290 kg, including the heaviest non-dismountable portion, which was the Kärcher washer, weighing 108 kg. Laboratory studies revealed that, under a hot-water flow rate of 10 L/min with a temperature of 60 °C, the 1.8-mm and 2-mm nozzles created boreholes 98–114 mm in diameter at a penetration rate of 34–37 m/h, whereas the 2.5-mm nozzle produced boreholes 146–156 mm in diameter at a penetration rate of approximately 25 m/h.

3.2.17 Summary

Some publications mentioned shallow-depth hot-water drills as tools to study the mechanical, hydrological, and thermal conditions at ice bases and to install instruments in glaciers (Boulton et al. 2001; Fischer et al. 2001; Huss et al. 2007; Murray et al. 1997, 2000; Murray and Porter 2001; Porter et al. 1997; Röösli et al. 2014; Ryser et al. 2014; and others) but, unfortunately, they did not provide any details about the drilling equipment used.

In general, shallow-depth hot-water drilling systems include four major components: (1) a source of heat (furnace and heat exchanger), (2) a pump, (3) flexible hose, and (4) a drill stem with a nozzle tip. Usually, the heater and the pump are installed on the surface of the glacier. On the other hand, the LGGE Electrochaude drill used a downhole electromagnetic pump and a nichrome spiral heating wire that provided near-bottom circulation of warm water. Such methods avoid heat and pressure losses in the delivery hose; however, the capacity of small-sized downhole equipment is much less than that of full-scaled equipment installed on the surface. Shallow-depth hot-water drills designed for special purposes (ANDRILL, ARA) were quite complex, bulky, and power-hungry systems and differed from other portable drills.

The penetration rates of shallow-depth hot-water drills (typically 40–70 m/h) are an order of magnitude higher than the penetration rates of electrically heated thermal drills and the flushing action of the water jets allows them to drill in glacier ice, which contains some sand and small pebbles. The required typical water consumption of these drills is 10–20 L/min. The best practice is to use local surface meltwater if available or to recirculate water back from drilled hole because melting snow is hard work and doubles the fuel consumption of the system. When drilling in firn, water leaks out from the hole and a melt tank is needed. Maintaining a constant and clean water supply for the drill is crucial because foreign particles in the water can wreak havoc on pumps and, particularly, check valves. If the water empties from the hole, which sometimes happens in englacial voids and when approaching the bottom of mountain glaciers, drilling becomes extremely difficult. The weight of the drill hose and stem can increase dramatically during such events because the buoyancy forces created by the water in the borehole are lost. Drillers should be mindful of the possibility of such occurrences at all times.

Operations are also difficult at temperatures below the freezing point of water. At temperatures below -10 °C, it is necessary to clean the circulation system by blowing or pumping antifreeze when drilling is interrupted, even for a short time. The real problems occur during unforeseen shut-downs due to electrical, fuel, or water supply problems because ice plugs quickly form, making it difficult or impossible to reestablish the flow when the original problem is fixed. In general, shallow-depth hot-water drilling systems are secure and reliable: the risk of breakdowns is very small and jams were seldom. The downhole drill itself cannot overheat or burn out. However, the equipment required is heavier than that for electrically heated thermal drills and maintenance is more specialized.

3.3 Intermediate Hot-Water Ice Drilling Systems

Intermediate hot-water ice drilling systems are more powerful and better controlled setups than shallow-drilling systems and were developed for drilling boreholes in cold ice to depths of up to 1500 m for studying marginal parts of the Antarctic and Greenland ice sheets, for accessing holes through ice shelves, and for the exploration of subglacial lakes (Table 3.2).

3.3.1 Browning Hot-Water Drill

The hot-water drill of J. Browning, Thayer School of Engineering at Dartmouth, USA, consisted of a single 184-kW boiler, a heat exchanger, a downhole pump, a booster pump, and reels of hose with a drill nozzle (Fig. 3.64). The 76-mm-diameter, 3-m-long drill head with the nozzle was attached to 460 m of reel-mounted 2″ hose. A cable winch carried the load. The hose was clamped to the cable every 30 m. The boiler consumed 284 L/h of fuel oil to heat 265 L/min of water from 2 to 98 °C (Browning et al. 1979; Koci 1984).

Browning's hot-water drill system replaced the flame jet of the Ross Ice Shelf Project (RISP) (see Sect. 5.1.2) and, in 1978–1979, it was used to drill three access holes through 420 m of ice at camp J-9. In the initial drilling tests, small-nozzle diameters yielding high water-jet velocities were used. However, the results were disappointing. Then,

the jet velocity was reduced, and this proved to be the key to successful drilling.

First, a return hole was drilled to a depth of 52 m with a large cavity at the bottom. A single axial water jet can make no more than a ragged, small-diameter hole in firn. A showerhead nozzle was used for the first 30 m of each hole. A 7.4-kW submersible pump suspended in the return hole pumped water to the boiler. The main hole passed through this cavity, allowing water to circulate from the drill to the pump. The penetration rate was ∼42 m/h. The average hole diameter was ∼0.9 m.

3.3.2 PICO Intermediate Hot-Water Drilling System

This system was built by the PICO, University of Nebraska and was especially designed for a hot-water drilling project at Crary Ice Rise, near camp J-9 in the southeastern part of Ross Ice Shelf (Boller and Sonderup 1988). The winch was wrapped with a maximum of 600 m of hose. The drill hose was standard Synflex 3000 hose with an ID of 24 mm, which was modified by wrapping it with electrical conductors, a Kevlar strength member, and an outer neoprene jacket (Koci 1989). The outer diameter was nearly 50 mm, requiring a minimum bending radius of 0.5 m. Because the hose was buoyant (0.33 kg/m), weights were added to the drill stem. The pump produced a pressure of 1.3 MPa at a flow rate of 85 L/min. This allowed for a drilling rate of ∼24 m/h, creating holes with a diameter ranging from 260 to 280 mm. Water was recirculated from a water well 40 m below the surface and pumped to the surface via a submersible pump.

Heat was provided by six heaters (two Alladin and four Hotsy heaters) connected in parallel. The inlet water temperature at the heaters was 20 °C and the outlet water temperature was 90 °C. By a 1″-diameter insulating hose, heat loss was reduced so that the water temperature at the nozzle fell off by only 2 °C per 100 m of water depth; thus, the water temperature at the drill was ∼86 °C. In addition, wires incorporated in the jacket allowed for making measurements while drilling (Hancock and Koci 1989). The drilling system contained an instrumentation package that could measure hole diameter, inclination, the depth of the drill, the water temperature inside and outside the drill, and the inlet water temperature. All measurements were displayed and recorded on a portable computer.

The system was quite large and bulky despite being able to be separated into smaller pieces; the single largest piece was the hose reel and the hose which, which together weighed 1.7 t. In addition, a 16-kW pump and a 10-kW generator were required, and both weighed 200 kg. The each

Table 3.2 Parameters of selected intermediate hot-water ice drilling systems

Institution or drill name	Years	Drilling site(s)	Flow rate/ (L min^{-1})	Max depth/m	Borehole diameter/mm	ROP in ice/ (m h^{-1})	References
Thayer School of Engineering at Dartmouth, USA	1978–1979	Ross Ice Shelf	Up to 265	420	900	42	Browning et al. (1979)
PICO, University of Nebraska, USA	1987–1988	Crary Ice Rise	85	480	260–280	24	Boller and Sonderup (1988)
ETH, Switzerland	1988–1995	Jakobshavns Isbræ (Glacier), West Greenland	60	1630	120–180	70–110	Iken et al. (1989)
AWI, Germany	1991–2018	Ekström and Ronne ice shelves	60	632	130–400	36–40	Nixdorf et al. (1994a, b, 1997)
Caltech, USA	1988–2001	Ice streams and interstream ridges of West Antarctica	80	1189	100–150	NA	Bentley et al. (2009)
BAS, United Kingdom	1990 up to now	Antarctic Ice Shelfs, Greenland Ice Sheet	90	~770	>300	72–96	Makinson and Anker (2014)
UAF, USA	1996 up to now	Alaskan glaciers, Pine Island Glacier and Nansen Ice Shelf, Antarctica	80–90	620	>250 mm	NA	D. Pomraning, personal communication
AMISOR	1999–2010	Amery Ice Shelf, Antarctica	45–80	620	Up to 500 mm	NA	Craven et al. (2002a, b, 2004
WISSARD	2013–2015	Subglacial Lake Whillans and Ross Ice Shelf, West Antarctica	Up to 274	801	>300 mm	30–60	Rack et al. (2014), Rack (2016)

Note NA—data are not available

Fig. 3.64 Browning hot-water drill, Site J-9, Ross Ice Shelf Project: **a** drill hose passing over the wheels on a crane skid; the boiler is in the Langdon shelter, and the hose reel and cable winch are to the left of the Jamesway (Browing et al. 1979); **b** boiler for the drill (Bentley and Koci 2007)

heaters weighed 125 kg and burned 10 L/h of diesel fuel. This translated to a consumption of 2 L/m of diesel and 8 L/h of gasoline.

During the 1987–1988 field season, two holes were drilled through Crary Ice Rise (83°S, 170°W) to install thermistor cables (Bindschadler et al. 1988). Set-up and testing of the drilling equipment at the top of the local ice dome took 11 days. The drilling of the first hole was stopped because the inclinometers indicated that the drill was angled. Drilling was resumed and bedrock was reached at 370 m after 18 h of drilling. The glacier bed was sensed by the simultaneous occurrence of three events: a decrease in

Fig. 3.65 Rock sample recovered from the bottom of the 480-m-deep hole at Crary Ice Rise, Antarctica (Bindschadler et al. 1988)

tension in the drilling hose, an increase in the water pressure of the drilling system, and an indicated increase in the tilt of the inclinometers on the drill stem (Koci and Bindschandler 1989).

Then, the camp was moved ∼15 km to the second drill site and, in this place, the ice/bedrock interface was reached at 480 m after 20 h of drilling. In the second hole, a similar set of events occurred at the glacier bed, but the drill remained at this level for ∼1 h before being raised to the surface. When the drill was returned to the surface, the drill stem was coated with a thick mud, and a rock and a clast were lodged in the caliper arms of the drill (Fig. 3.65). In total, ∼1000 cm^3 of sediment material were spread over the drill stem filling most of the ledges and holes.

3.3.3 ETH Intermediate-Depth Hot-Water Drilling System

In order to demonstrate the possibility of drilling through 1500 m of ice with a modified ETH shallow-depth system (see Sect. 3.2.2), Iken (1988) provided test results with two centrifugal pumps connected in parallel. A total of 11 boreholes were drilled on Findelengletscher in 1987 with different nozzles, flow rates, and hot-water temperatures to depths of 13.8–48.8 m. The maximal penetration rate of 250 m/h was achieved with a nozzle 2.5 mm in diameter, a flow rate of 22.5 L/min, and a temperature at the heater exit of 69 °C. The ETH intermediate-depth hot-water drilling system and its modifications were used many times for drilling access holes in Alpine glaciers and Greenland.

In 1988, near the centerline of Jakobshavns Isbræ, West Greenland at distance of 45 km from the calving front, four holes were drilled to depths of 1200–1330 m, each taking from 12 to 18 h to drill (Iken et al. 1989). At this site, the

surface ice velocity was very high (1.1 km/year), but the ice was not highly crevassed. Meltwater streams provided a convenient water supply for the drilling. A few shallow holes were also drilled.

Hose sections of 100–200 m in length were added as the drill proceeded down the holes: they were coiled up in the shape of the number "8" on a tarp. Three piston pumps in parallel provided a total flow rate of 60 L/min of water, which was heated to 58–76 °C using three or four diesel-oil heating units. The 6-m-long drill stem used, with an OD of 30 mm, consisted of three sections of double-walled lead-filled tubes screwed together. Nozzles of different sizes (typically 4.5–5.5 mm) could be attached at the end of the 300-mm-long conical drill tip. The penetration rate was controlled by a capstan-type motor winch operated via hand controls (Fig. 3.66). The drill was lowered at a speed slightly slower than that at which vibrations occurred on the drill stem (whenever the drill tip almost touched the bottom of the hole, the drill started to vibrate). The total weight of the equipment (with the water tank and generators) was ∼1000 kg plus 30 kg/100 m of hose. The heaviest piece was the motor winch, which was mounted on a sled (180 kg).

The original diameters of the deep holes drilled ranged from 120 mm to 180 mm depending on the penetration rate. Boreholes froze at a fast rate, e.g., a hole with a diameter of ∼180 mm froze shut in less than 10 h. Thermistors were installed in several holes and temperatures were recorded. Ice temperature decreased almost linearly from −18 °C at a depth of 400 m to −22 °C at a depth of 1000 m.

In July–August 1989, several additional boreholes were drilled at three sites (A, B, and C) in Jakobshavn Isbræ on a transverse profile across the ice stream (Iken et al. 1993). These holes were drilled with a hot-water discharge rate of 60–80 L/min. Six to eight oil heating units provided an

Fig. 3.66 Capstan during an operation at Jakobshavns Isbræ, West Greenland, summer of 1989 (*Photo* P. Gnos)

input-water temperature of 80–90 °C. With this equipment, the time required for drilling holes with an initial diameter of 150–200 mm to depths of approximately 1550 m was 20 h. At the two outermost sites, namely A and C, bedrock was reached at depths of 1540 and 1630 m, respectively. A small amount (a few cm^3) of bedrock material was retrieved from the bottom of one of the holes using small recessions in the drill tip. At the center line (site B), the deepest hole drilled was 1560 m deep, thus terminating 940 m above bedrock because of the limited capabilities of the drilling system.

In July 1995, a total of eight holes were drilled at site D, 4 km from the ice-stream center line in Jakobshavn Isbræ (Lüthi et al. 2002). This site was located 5 km upstream of the 1989 drilling sites. At a constant flow rate of 80 L/min, the glacier bed was reached at a depth of ∼830 m within 6–8 h. The water immediately drained from all the boreholes, and the borehole water level stabilized at the flotation level. In August 1995, three 900-m holes were drilled at site E near the ice-stream line, where the ice thickness was ∼2500 m, close to the 1989 drilling site B. Five boreholes at site D were permanently instrumented with tilt sensors, thermistors, and pressure cells, from which data was measured using a specially devised telemetry system. All operations had to be completed within 2 h after drilling because the boreholes rapidly refroze.

3.3.4 AWI Hot-Water Drilling System

Several German hot-water ice-drilling systems were used in Antarctica and Greenland. The first system was developed in the Institute of Geophysics, Münster and was tested in the Ekström and Ronne ice shelves in 1985–1986 (Engelhardt and Determann 1987a, b; Bässler and Miller 1989). This drill was an updated version of a type of drill used on Variegated Glacier in Alaska (see Sect. 3.2.4). Two holes through the 208-m-thick Ekström Ice Shelf were drilled just south of Georg-von-Neumayer Station (subsequently replaced by Neumayer Stations II and III). Because of bad weather conditions, reaming of the holes from 50 mm to 100 mm for the installation of a thermistor chain was not done. Drilling operations took four days. During second trials, another two closely adjacent holes with a diameter of ∼100 m were drilled down to 50 m. In the second one, water from a neighboring hole was pumped out and used for recirculating.

In between these tests, hot-water drilling equipment with a 550-m hose drum was moved to the central Ronne Ice Shelf and, at the beginning, two unsuccessful attempts to reach the ocean cavity were made. The first hole reached a depth of 380 m. The second hole had to be abandoned after 60 m of drilling because of the technical problems with the winch. The third hole was drilled without any complication

and reached the bottom at a depth of 460 ± 5 m. Drilling took 12 h. The ice was increasingly slushy below 430 m. The penetration rate near the surface was 120 m/h and it gradually decreased to 30 m/h at a depth of 430 m because of the cooling of the hot water in the increased length of submerged hose. However, in the near-bottom part of the hole, the rate suddenly increased to 60 m/h because ice-water mixtures require less heat to be penetrated via melting than solid ice. The water level in the borehole was initially at a depth of 35 m, slowly rising to 30 m during drilling. When the borehole pierced the ice shelf, the water level dropped to 53.8 m below the surface. The upper part of the hole was reamed to 100 mm (owing to time constraints, it was not possible to ream the entire hole).

In 1989–1990, Ronne Ice Shelf was penetrated again five times (plus one additional hole that failed to reach the ocean cavity) near the ice front west of Filchner station, where the ice thickness was 239 ± 2 m (Grosfeld and Hempel 1991). Water at a temperature of 80–100 °C was pumped into the hose under a pressure of 9 MPa. Holes were drilled with a 60-mm nozzle (one was enlarged with a 150-mm nozzle). The drilling of each hole took 7–8 h; this indicates an average penetration rate of 30–35 m/h. Water consumption was 6 m^3/hole. After penetration, the water level in the boreholes dropped to sea level at 35.35 m below the ice surface. The holes were used for temperature measurements in the ice and water and to measure basal melt rates. One ocean current meter was deployed, but it became stuck in the ice while being pulled back for maintenance.

In 1991–1992, Ronne Ice Shelf was penetrated twice through 420 m-thick ice using the AWI hot-water drill (Nixdorf et al. 1994a). The holes were strongly inclined; in each one, drilling required 580 m of hose to reach the water. Unfortunately, the inclinometer was stuck in one of the boreholes before oceanographic instruments were lowered into the ocean.

During the 1992–1993 season, three holes were drilled at newly rebuilt and renamed Neumayer station using a modified hot-water drill consisting of a ∼5-m^3 water tank, six completely identical high-pressure cleaners, a winch drum, and the control electronics for the electrical winch, comprising a display of hose tension, penetration rate, and depth of the nozzle (Fig. 3.67) (Nixdorf et al. 1994). The SYNFLEX hose used had an ID of 3/4″ and weighed 0.04 kg/m in water. The principal modification made to avoid the large hole inclination that occurred when drilling on Ronne Ice Shelf in the previous season was to add winch electronic circuitry that prevented the nozzle from touching the bottom of the hole. The feeding speed of the drill stem could be adjusted either via automatic control or manually.

The high-pressure cleaners used consisted essentially of a generator, a high-pressure pump, and a JP-8-fueled burner with a heating capacity of 125 kW. Each burner had a feed

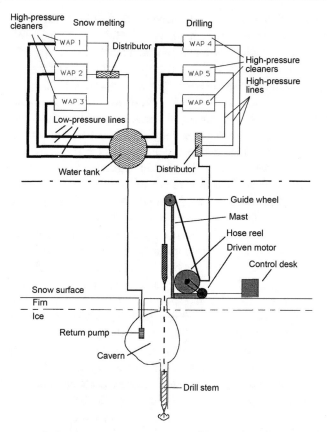

Fig. 3.67 Schematic of the AWI hot-water drill (Nixdorf et al. 1994a)

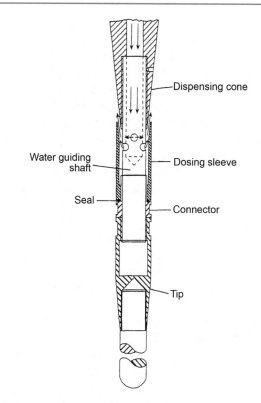

Fig. 3.68 Schematic of the AWI hot-water reamer (Nixdorf et al. 1997)

water hose and a low-pressure feed pump. The hot-water capacity of the system was, in total, 90 L/min. The water temperature at the outflow of each burner was ~ 95 °C, and the normal working pressure was ~ 8 MPa. The system used a combination of melted snow from a temporary water storage tank and recirculated meltwater. The ice shelf was ~ 240 m thick. Holes were at least 350 mm across and were inclined no more than a few degrees from the vertical. The actual drilling operations took 12 days. Fuel consumption totaled 14 t of diesel and 4 t of gasoline.

In 1994–1995, two other holes were drilled in Ronne Ice Shelf (Nixdorf et al. 1997). The system was again modified. At the inlet side of the hot-water heaters, washable filters were added. The water of the six individual pressurized hot-water lines was collected in a junction box. At this box, the driller controlled two valves: one for regulating the flow of water to the drill winch, and another one for controlling the flow to the meltwater tank in order to melt snow. The newly designed reamer allowed drillers to enlarge the diameter of the holes from 50 to 130 mm (Fig. 3.68). The water jet in the reamer was sprayed upwards, preventing the start of new holes. The first hole penetrated the ice shelf at a depth of 420 m. The drilling of this 50-mm-diameter hole took 10 h, and reaming it to 130 mm took 6.5 h. A total of 2775 L of diesel and 600 L of gasoline were required. The

second hole was drilled to a depth of 264 m (predrilling time: 5.5 h; reaming time: 5.5 h; diesel consumption: 1570 L; gasoline consumption: 300 L). In the first hole, there was a slush layer at the base. Instruments emplaced through the holes soon ceased transmitting data.

In 2002–2003, three access holes were drilled with the AWI hot-water drilling system through McMurdo Ice Shelf to the south of Ross Island in preparation for deep drilling as a part of the ANDRILL project (Barrett et al. 2005). The access holes were drilled first as narrow pilot holes 150 mm in diameter with a typical penetration rate of 30 m/h. They were then widened to at least 600 mm in diameter using a reamer at a slightly slower rate (~ 20 m/h). Additional reaming at a rate of 5 m/h was needed at the bottom of the hole as hot water was lost to the ocean beneath.

The drilling of the first hole to a depth of 70.5 m took 2.3 h and the subsequent reaming operations took 3.25 h. Upon completion, a specially designed reaming deflector attached to the reamer nozzle was used. Unfortunately, this deflector was lost at a depth of ~ 15 m and could not be recovered in subsequent attempts. Drillers decided to give up this hole, and thus another hole was drilled and reamed 2 m apart (so that the near-surface cavern with water could still be used) over a total time of 7.5 h. A caliper log showed that the diameter of the hole was larger than 560 mm from the top to a depth of 68.2 m. However, from there to the bottom of the ice shelf, the diameter narrowed almost linearly to

180 mm. The reaming attempts lasted ~4 h, but with little effect. A new reamer was built from two flanged steel salad bowls ~50 cm in diameter and bolted with a narrow gap around the flange to force hot water against the sides of the hole. The device worked well. The hole was then made available for measurements and sampling during the 10 following days with one in-between reaming operation.

At the next site, the hole made penetrated the ice shelf at a depth of 143 m. A pilot hole was drilled over 5.1 h, and reaming this hole took 7.3 h using the reamer nozzle and 3.9 h using the bottom reamer. The reaming process was hampered by falling temperatures, which caused some problems when starting the burners and froze the hose from the junction box to the melt water tank. A subsequent caliper log proved that the diameter of the hole was larger than 560 mm from the top to the bottom of the ice shelf. The hole was kept open during five days with one in-between reaming operation.

In 2005–2006, four holes were drilled north of Neumayer station near the ice front (ice thickness: 95–100 m) for the installation of oceanographic instruments below the ice (Bentley and Koci 2007). In July 2010, the AWI hot-water drill was used to drill two holes ~4 m apart at Russell Glacier in the land-terminating part of the West Greenland ice sheet (Smeets et al. 2012). Given the considerable size and weight of the drill, the repositioning of the complete drilling system was complicated, and thus it was decided to move the hose winch a few meters via helicopter. Initially, a narrow drill stem was used, after which a reamer was used to enlarge the diameter of the holes. The depth of the first hole was estimated to be 610 ± 1 m, and that of the second one was estimated to be 632 ± 1 m. After drilling, wireless probes were lowered into the holes.

Recently, in the season of 2017–2018, the AWI hot-water drill was used again for drilling two access boreholes through the 200-m-thick Ekström ice shelf in the vicinity of Neumayer Station (Fig. 3.69). The initial holes were drilled using a drill stem with a 4-mm-diameter nozzle, and then they were reamed three times to a diameter at least 400 mm. The rate of penetration with the initial drill stem was 36–40 m/h; the first reaming rate was in the range of 10–20 m/h, the second reaming rate was 4–10 m/h, and the final reaming rate was lower than 8 m/h. During drilling, the temperature of the hot water was ~90 °C, the inlet pressure was 2.5 MPa, and the flow rate was ~60 L/min.

3.3.5 Caltech Hot-Water Drilling System

Caltech team began modifying the USGS hot-water drilling system (see Sect. 3.2.4) to drill deeper boreholes and, during the summer of 1987, five intermediate-depth holes with depths ranging from 526 to 975 m were drilled down to the bed of the fast-moving Columbia Glacier, Alaska for basal-water pressure measurements (Meier et al. 1994). Water for drilling was pumped from nearby crevasses. After completion, each hole was reamed to a diameter of 75 mm. The drilling of the deepest hole took less than 17 h.

Humphrey et al. (1993) reported on an unplanned experiment in the subglacial shear zone of Columbia Glacier. During hot-water drilling, the drill became inadvertently stuck in the glacier bed and was dragged for five days through basal sediment. After its subsequent retrieval, an analysis of the bent drill stem yielded estimates of the sediment strength that were approximately one order of magnitude smaller than the applied shear stress.

Over the years, Caltech built an original hot-water ice drill. During the field seasons of 1988–2001, 119 holes, most of them deeper than 1000 m (maximum depth: 1189 m), were melted down to the glacier bed at 20 sites across the ice streams and interstream ridges of West Antarctica (Engelhardt 2004a). The boreholes were located on Whillans Ice Stream, Kamb Ice Stream, Bindschadler Ice Stream, Engelhardt Ridge, Raymond Ridge, and the Unicorn. The objective was to investigate and understand the mechanisms of ice-stream motion, so many measurements were made in the holes and especially at and in the glacier bed (Engelhardt 2004a, b; Engelhardt et al. 1990; Engelhardt and Kamb 1997, 1998; Kamb and Engelhardt 1991; Tulaczyk et al. 1998; and others). In the holes, where the glacier bed was soft, it was penetrated several meters and samples were recovered. At some locations, video imagery was recorded within the borehole, and ice cores were taken using a hot-water ice corer (see Sect. 3.5.3).

The first six boreholes were drilled in the field season of 1988–1989 on Whillans Ice Stream and were 230 m apart. Five of them reached the bottom at depths between 1030 m and 1037 m (Engelhardt et al. 1989). One was stopped at a depth of 950 m without reaching the bottom. The boreholes were ~100 mm in diameter. The total time for drilling and reaming each borehole to its completion averaged 55 h. Drilling was done in steps of ~400 m with intervening steps for reaming the boreholes to a diameter of 100 mm. A borehole of 100 mm in diameter in ice at −26 °C completely refroze in 10 h and was open for instrumental work for only 2–4 h. During drilling, the water level in the borehole stood at a depth of ~35 m below the surface. When the drill reached bottom, the water level in the boreholes dropped to ~105 ± 5 m below the surface.

One borehole was tested for preventing refreezing by introducing an antifreeze agent (ethanol), but this did not work. Because of the warming of the ice next to the borehole during drilling with hot water, the antifreeze is initially diluted by the melting ice from the borehole wall; then, as a cold wave arrives from farther outside the borehole, the diluted antifreeze solution crystallizes into frazil ice

Fig. 3.69 Overview of the AWI hot-water drill, Ekström Ice Shelf, Antarctica, January 2018 (*Credit* X. Fan)

platelets, which float up and form slush, blocking the passage of instruments.

After several modifications, the Caltech drill achieved the optimal configuration shown in Figs. 3.70 and 3.71. The water supply for drilling came from two sources. Parallel to the main borehole, a return hole was drilled, in which the overflow of the drilling water from the main borehole accumulated and from which water was pumped out (~90% of the required volume) and recycled. The return water system included a return water pump, hose, a spool, a motor drive, a sled, and a control box (Fig. 3.72). The remaining ~10% of water came from snow that was melted on the surface and replaced the volume that was lost when the borehole ice was melted to water. To maintain the required water volume balance, the return water was regulated so as not to drive down the water level in the borehole.

Typically, two reservoirs were used: a receiving or melting tank and a tank for filtration and temperature stabilization. Snow was periodically shoveled into the melting tank. Water was pumped from here to the clean-water tank (Fig. 3.73). Several levels of filtration and initial heating were applied in this tank. From here, water was pumped to high-pressure pumps. The level between tanks was regulated by the use of valves or by placing the second tank at a higher level and using a gravity return flow.

The integrated high-pressure pump system, comprising a pump, a motor, an accumulator to damp pulsations, freeze-protected pressure gauges, flow switches, and other safety items, was mounted a sled and was powered with

gasoline. Initially, drilling was done with a nozzle flow rate of 40 L/min. In subsequent years, this was increased to 80 L/min, delivered by four pumping systems. All four could be towed by one snowmobile.

There were eight boilers derived from the pressure-washer industry in the final heating bank of the Caltech system, which provided an output water temperature of 95 °C (Fig. 3.74). Each high-pressure pump supplied flow to two heaters plumbed in parallel. A central valve assembly allowed for the redirection of water as required (Fig. 3.75). Color coding prevented confusion during drilling operations. Red, blue, green, and yellow identified four parallel drilling sub-systems that were capable of being directed downhole or recirculated back to the reservoirs when in idle mode. A thermometer near the valve manifold measured the total output temperature of the surface system.

A capstan was used to pull the hose and took 80–90% of the tension force; thus, hoses suffered less than if they were wrapped under great tension. For portability and also availability reasons, hose lengths were limited to 300 m. There were four of these spools, each holding 300 m of high-pressure hose (Fig. 3.76). Each spool weighed ~250 kg. To adjust the tension against the capstan, magnetic particle clutches were used, and tension was modulated by turning a knob. Thermoplastic Synflex-Furon 3R80 hose with an ID of 19 mm was used. This hose had two braided layers of nylon, providing a working pressure of 15.7 MPa and sufficient tensile strength to drill at least 1200 m.

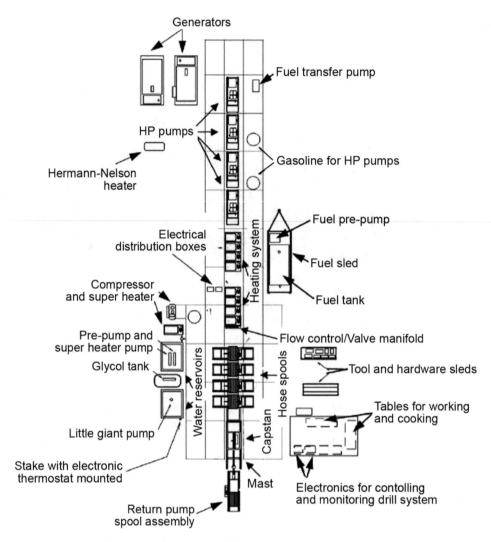

Fig. 3.70 Schematic of the Caltech drilling system (Bentley et al. 2009)

Fig. 3.71 Caltech hot-water drilling system, West Antarctica (*Photo* H. Engelhardt from Hubbard and Glasser (2005))

Fig. 3.72 Spool with return hose (Bentley et al. 2009)

Fig. 3.73 Clean-water tank with filtration system (Bentley et al. 2009)

Fig. 3.74 Boilers (Bentley et al. 2009)

Fig. 3.75 High-pressure distribution manifold (Bolsey n.d.)

Fig. 3.76 Spools with a 300-m-long high-pressure hose (Bolsey n.d.)

The 3.5-m-long drill stem consisted of brass machined sections (Fig. 3.77). The length and mass of the drill kept it vertical in the hole. The drilling system used two 20-kW generators. One generator supplied power while the other one stood by as backup.

3.3.6 Norwegian Hot-Water Drilling System

The drilling system of the Norwegian Polar Institute, Oslo included three Euroclean Delta Pluss burners, each consuming 5 L/h of jet fuel (Orheim et al. 1990). Hot water was pumped at a rate of 25 L/min at ~ 100 °C when all three heaters were coupled in series. A 450-m-long 1/2″ Hytrell steel-reinforced hose with a working pressure of 20 MPa was wound into a reel driven by a 0.75-kW three-phase 220-V AC motor that allowed for a variable tripping speed of the drill in the range of 0–10 m/min.

The drilling equipment had been used since 1985 in Norway and Svalbard, with surface temperatures down to -20 °C. In the 1989–1990 season, two holes were drilled in Fimbul Ice Shelf, bordering the coast of Queen Maud Land, Antarctica. A working platform and a windshield were built using wooden beams, plywood, and tarpaulin. A fiberglass hut was used as an office and drilling room.

Drilling down to 50 m was done using snow melted in a 3000-L plastic reservoir tank, and the water level was stabilized at 35 m. Then, the hole was enlarged and a return submersible Grundfos pump (25 L/min, 60-m water height, 0.75 kW, three-phase 220 V AC) was put down. Drilling was continued, alternating with reaming at ~ 50-m intervals. At a depth of 325 m, the pressure pump broke and was repaired over five days; however, after resumption, the hose got stuck. Drilling was continued in the same hole with a new hose, with the expectation that the first hose would be

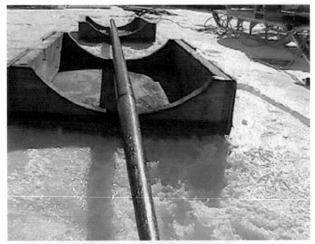

Fig. 3.77 Caltech downhole drill stem (Bentley et al. 2009)

melted out. This did not occur. The base of the ice shelf was penetrated with 397 m of hose in the hole after eight days of drilling. The hole diameter at the top was 0.8 m, decreasing downwards to ~50 mm at the base of the ice shelf. Further reaming of the hole was done over the next five days. After that, the second drill got stuck at 320 m (nearly the same depth of the first incident), and it was decided to abandon this hole.

Typical penetration rates were 30 m/h in the upper 200 m of the hole and 18 m/h in the lower 200 m. During drilling, the hose made a knot three times, probably when the drill stem was gripped in the hole while being lowered and the hose had passed it, so that the drill could thread itself through a loop on the hose. The total jet fuel consumption was 4 m^3. A cable with thermistors was lowered to a depth of 327 m below the surface. Two years after deployment, the minimum temperature was measured to be −27.9 °C, recorded by the deepest thermistor (Østerhus and Orheim 1992).

The loss of two hoses approximately 320 m in length required finding thinner ice for sub-ice instrumentation.

Therefore, a second site was chosen in Jutulgryta, an ice island between Fimbul Ice Shelf and slow-flowing grounded ice. Here, an access hole was drilled through 38-m-thick ice within a few hours. After penetration, sub-ice equipment was installed. It was found that the solid ice/water interface at a depth of 11.1 m was underlain by 27 m of slushy ice interspaced with thick water layers. In the two years after deployment, the data storage unit and other equipment on the surface and all instruments and sensors in and below Jutulgryta were recovered by melting a hole around the cable. The cable and instruments situated deeper than 10 m below the surface had been subjected to corrosion, but all instruments were in good condition.

In February 1993, another hole was drilled in Ronne Ice Shelf (Orheim et al. 1997). The hot-water drilling system used was essentially the same as the one used in 1990, but it was boosted with more heaters and equipped with 3/4″ hose. This was much heavier than the 1/2″ hose used three years before, which made it more difficult to detect whether the drill was hanging freely (a load sensor was not used). The estimated depth was 220–300 m. At a depth of 327 m (without ice-shelf penetration), it was decided to pull up the drill stem and half of the hose was recovered in a stick/slip fashion while hot water was kept running. However, at 161 m, the drill was irrevocably stuck. The incident most likely happened because of hole deviation. Then, the hole diameter was logged with a three-legged caliper down to a depth of 115 m. During logging, there was hose in the hole, and it is possible that some of the spikes in the record were caused by it. At all logged depths, the hole diameter was larger than 0.2 m.

3.3.7   BAS Intermediate-Depth Hot-Water Drilling Systems and Their Modifications

3.3.7.1   Initial BAS Intermediate-Depth Hot-Water Drilling System

To penetrate relatively cold ice (−26 °C) at Ronne Ice Shelf with a thickness of up to 600 m, an easily operable hot-water drilling system with minimal logistical support was developed in the BAS (Fig. 3.78) (Makinson 1993, 1994). The flexible coated-fabric water tank had a capacity of 12 m^3. It had the shape of a flat-topped cone with an inflatable annulus forming the upper lip of the tank, which rose as the tank was filled with water. It weighed only ~50 kg when empty.

A water recirculation system was used. The system consisted of two parallel holes interconnected by a cavity at a depth below both the percolation level and the sea level. A submersible multistage centrifugal pump, which was 1.5 m long and 90 mm in diameter, was suspended in the return hole. The pump capacity was 60 L/min from 80 m and it was electrically powered (240 V, 13.5 A) via a

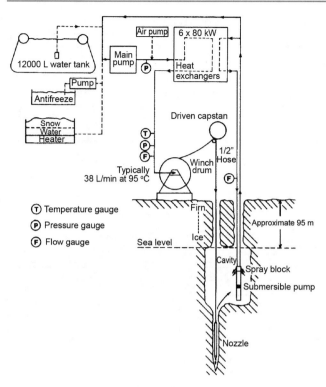

Fig. 3.78 Block diagram of the BAS drilling system used on Ronne Ice Shelf (Makinson 1993)

90-m-long cable with 6-mm² conductors. To assist in the recovery of the submersible pump and its umbilical, two snow mobiles were available to haul it from the hole. To avoid refreezing, a hot-water spray was introduced into the cavity and over the pump. This, together with the use of anti-icing paste (a form of grease) on the hoses and cables, reduced their chances of freezing to the sidewalls.

Four heat exchangers, each with burners rated at 80 kW, were used. A temperature gauge was able to cut out the system at high temperatures and low flows. The heat exchangers consisted of two concentric coils made from seamless tube rated for 14 MPa. Two units had 9.5-mm-bore tubing, whereas the other two had 13-mm-bore tubing with separate inlets and outlets, allowing the whole system to be configured for maximum efficiency.

The triplex plunger high-pressure main pump with ceramic plungers was rated at 38 L/min at 1000 rpm, with a maximum pressure of 15 MPa. The pump was self-priming with a pressure-relief valve and a pressure gauge and was powered by a hand-started Lister Petter AD2 diesel engine with a maximum power output of 11 kW via a V-belt speed reduction unit. Fuel consumption was 0.45 L/(kW·h) at 3000 rpm. The unit was air-cooled and weighed 80 kg.

The 5-kVA generator used was rated at 240 V, 21 A, and 50 Hz at 3000 rpm and weighed 30 kg. It was powered via a direct drive by a diesel engine identical to the one powering the high-pressure water pump, thereby reducing the number

of spares needed. It was used to provide power to the submersible pump and other power consumers.

The drilling system used 1/2″ Dunlop thermoplastic high-pressure hose with a fabric braid rated at 14 MPa. The operating range of the hose was from −40 to 93 °C with occasional uses at 125 °C. The hose couplings were swaged. Tensile tests showed that the hose could stand over 9 kN before failure, at which time the hose had elongated by 60%. The hose weighted 0.21 kg/m in air. The drilling winch had a drum capable of holding over 1000 m of hose and had a capstan wheel to raise and lower it. The drum and capstan were hydraulically powered and a series of gears provided any speed required during drilling.

The monitoring system included pressure gauges, flow indicators, temperature sensors, a water-depth sensor to measure the head of water above the submersible pump, and drilling depth and feeding speed indicators using a rotary pulse generator located on the capstan wheel axel.

A 30-mm single nozzle was used in ice with its head having a smooth parabolic profile. A 52-mm front-cone spray nozzle that operated at an angle of 30° was considered as ideal for drilling in firn and for enlarging holes (Fig. 3.79). An additional conical spray was added to the back of the nozzle in case the nozzle became trapped by water refreezing above it. It had a less powerful drilling capacity than the front cone spray but would still prove effective against hole closure, allowing for the recovery of the nozzle. The back conical spray operated continuously. To completely avoid the possibility of the nozzle cutting into the wall and forming a secondary hole, the hole was reamed using a 127-mm nozzle with its forward spray replaced by a "nose" section 0.25 m in length.

This system was used during 1990–1991 and 1991–1992 to penetrate ice that was 562 and 541 m thick, respectively, on Ronne Ice Shelf (Fig. 3.80). The holes were 200–250 mm in diameter and were drilled in 1–3 days. Repeated reaming kept the holes open for five days after drilling. Hole S1 (1990–1991; 77° 36′S, 65° 42′W), 300 km from the northern limit of the ice shelf, encountered 14 m of slush at the base, which were easily drilled through but quickly refilled the hole. Repeated drops of a heavy streamlined weight were needed to force the passing of oceanographic instruments through the slush.

No slush layer was found at Hole S2, ~100 km north of the previous drilling site (1991–1992, 76° 42′S, 64° 55′W). The improved drilling system created a 0.25-m-diameter access hole in two days at this site. However, because of problems with the submersible pump, the main hole had to be abandoned and a second hole was drilled. In order to drill a hole with an average diameter of 0.23 m, the penetration rate at Hole S2 was varied between 84 m/h near the surface and 42 m/h at the base of the ice shelf, performing drilling and then reaming.

Fig. 3.79 52- mm and 127-mm BAS nozzles with a cross section of the sprays (Makinson 1993)

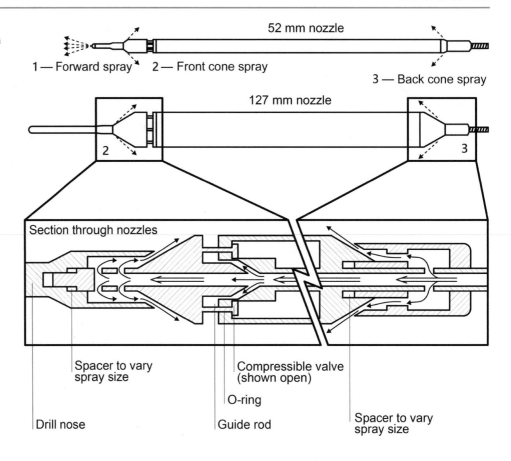

Fig. 3.80 BAS hot-water drill on Ronne Ice Shelf in the 1991–1992 season. In the foreground, the driven capstan using ½″ drilling hose can be seen; to its right is the 1″ hose of the submersible pump 80 m below with the winch drum, hydraulic controls, and four heat exchangers behind (Makinson 1993)

Over periods of seven to ten days, approximately 4 t of AVTUR fuel were burned at each site. The most unsatisfactory aspect of the drilling operations of 1990–1991 was the equipment sinking into the snow surface over the two-week operating period. This was usually a result of exhaust heat from equipment. In the subsequent season, the reorientation of exhausts and extensive use of insulating materials prevented the equipment and hoses from sinking into the snow. The time period for assembling the drill, which was separated into smaller components prior to starting work, was in the order of 10–14 days.

The caliper profiles obtained in 1990–1991 showed that the borehole had an irregular diameter profile, with narrowings reducing the average hole diameter by 50–100 mm. Initial closure rates of 9–11 mm/h were observed for a hole diameter of approximately 0.13 m in ice at −26 °C. In addition, a feature known as "necking", located at the base of the ice shelf, was revealed. It occurs when the drill reaches the base of the ice shelf and the hole connects with the sea below. A portion of the hot water from the nozzle dissipates into the sea, reducing the drilling action and consequently the hole diameter. Only with repeated reaming in this region can necking be reduced, and it is a feature that drillers should be aware of prior to the deployment of oceanographic equipment.

In 1995–1996, hot-water drilling at Ronne Ice Shelf was continued at site S3, ∼150 km southwest of site S1 (Nicholls and Makinson 1998). The total weight of all the drilling equipment was ∼5 t. The hose was changed to several 350-m lengths of ¾″ polyester-braid reinforced plastic hose. The 120-m-length closest to the nozzle was made of steel-braid reinforced rubber, providing weight to keep the drill hose vertical. The recirculation system was modified to use two return holes and two submersible pumps: one was used to feed the heat exchanger (the water was then returned to the cavity via both umbilicals); the second one lifted water from the cavity directly into the rubber water tank.

First, a 35-m-deep test hole was drilled. Then, a return hole was drilled and reamed down to 105 m, and a submersible pump was lowered on a plaited umbilical to a depth of ∼98 m. The umbilical consisted of a 1″ return hose, a ½″ hose supplying warm water to stop the pump from freezing, a cable for supplying power to the pump, and the signal cable for a pressure sensor that indicated the level of the head of water above the pump. Initially, there was a 64-m head of water. The second return hole, 0.6 m away from the first one, was drilled and reamed again to a depth of 105 m. Then, a connection was made between the holes via fan spraying and a second submersible pump was placed in position.

The test hole, previously drilled to 35 and 1.1 m away from the nearest return hole, was drilled and reamed to 105 m. A fan spray was lowered in that hole to the same depth as the submersible pumps in an attempt to connect it to the existing cavity. After 6 h, the spray had not connected the holes, and a fourth hole was drilled in the center of the 1.1-m gap to help make the connection. During this period, the engine on one of the high-pressure pumps broke and the drilling system was reconfigured for use with only one pump. The drilling of the main hole was continued. The

system configuration now consisted of five heat exchangers heating the water used to drill the main hole, and hot water was pumped down at 3.1–3.8 MPa and 85–90 °C with a flow rate of 38 L/min, or approximately half the original power and pressure.

When the hole penetrated the ice shelf at a depth of 825 m, the water level in the cavity rose from a depth of approximately 93 m to 84 m, and the load on the hose was reduced by up to 100 N for a period of 30–40 s. Approximately 10 h later, reaming of the hole was started, but the nozzle got stuck at 200 m. It was finally pulled free after a 30-min struggle. The smaller nozzle was then used to ream the hole, but one of the generators failed almost immediately. As both generators were needed to run the two submersible pumps, the system was reconfigured again for use with only one. Reaming of the hole was resumed on the next day, but difficulties were encountered. Some hole sections were drilled with ease, but others were very difficult to penetrate. With approximately 300 m left to drill, it was decided to recover the nozzle and the hole was abandoned.

An additional fuel stock was delivered to the site and the drilling of a new hole was started on the other side of the submersible pumps (the recirculation system to the cavity was maintained during the hiatus). Instead of drilling continuously to the base of the ice shelf, this hole was drilled and then reamed in stages, first to the depth of the cavity (105 m), then to 350 m, then to 600 m, and finally to the base at 825 m. The hole was reamed twice, allowing drillers to carry out oceanographic profiling and to deploy oceanographic mooring in the meanwhile. Near the base of the ice shelf, water was at −2.44 °C. Throughout the drilling of the second hole, the total amount of fuel burned in the heat exchangers and diesel engines was ∼10 m³.

In 1998–1999, two other holes (S4 and S5) were drilled through the southern Filchner-Ronne Ice Shelf with depths of 941 m and 763 m, respectively, using the same drill that was used for previous drilling campaigns (Fig. 3.81) (Nicholls et al. 2001). The first site, site S4 (80° 59.1′S, 51° 36.4′W), was between Berkner Island and Dufek coast, 12 km south of Berkner coast. Site S5 (S80° 20.0′S, 54° 46.4′) was 17 km west of the southwest coast of Berkner Island, between Berkner Island and Henry Ice Rise. In total, 6.5 m³ of aviation fuel were used at site S4 and 4.5 m³ were used at S5. The time from the start of the drilling operations to the start of oceanographic profiling was 3 days and 7 h at site S4 and 2 days and 7 h at site S5. Retrieving the CTD probe from the borehole was difficult at both sites, as it was clearly getting stuck at the bottom of the hole. At site S5, retrieval was not possible, and the wire had to be cut.

In 2002–2003, four more holes were drilled (Fox 1–4) in the northern Ronne Ice Shelf, ∼50 km inshore of the ice front (Fig. 3.82) (Nicholls et al. 2004). Their thicknesses were 300–400 m. During the drilling process at Fox 1, a

Fig. 3.81 BAS hot-water drill on southern Filchner–Ronne Ice Shelf in the 1998–1999 season (*Credit* K. Makinson)

Fig. 3.82 BAS hot-water drill on Ronne Ice Shelf in the 2002–2003 season (*Credit* K. Makinson)

hydraulic connection was made with the ocean cavity when the nozzle depth was estimated to be ~310 m. The remainder of the ice thickness, ~70 m, was made up of slushy ice. It was assumed that the hole had pieced into a slush-filled basal crevasse (the area near this site was heavily crevassed owing to the nearby boundary between the ice shelf and the coast). At the other three holes, weak basal melting was encountered.

3.3.7.2 Current BAS Intermediate-Depth Hot-Water Drilling System

The newest BAS ice-shelf hot-water drilling system had two operating modes, one for ice thicknesses up to 500 m and the other one for ice thicknesses up to 1000 m (Makinson

and Anker 2014). The basic 500-m drilling system weighed 3.2 t and, including comprehensive spares, windbreaks, and a Weatherhaven tent, weighed a total of 4.6 t. Appropriate additions, including three CAT pump units, added 1.8 t for upgrading to the 1000-m drilling system.

To reduce the number of diesel or petrol engines within the system and eliminate hydraulic systems, three-phase generators and electrical motors with variable-frequency drives were selected for all pumps. The primary electrical generators were 15-kVA, three-phase, super-silenced units. A single rotary multistage centrifugal pump (Caprari E6X25-4/24) with a 5.5-kW motor delivered 90 L/min (maximum pressure of 2.1 MPa) at 80 °C or 500 kW via 1″ Kutting thermoplastic drill hose (Fig. 3.83). The hose consisted of a polyester elastomer lining, a polyester braid, and a polyurethane outer jacket. The maximum working pressure of the hose was 6.9 MPa with a 4:1 safety factor for dynamic applications, with operating temperatures ranging from −40 °C to 100 °C and a weight of 0.52 kg/m in air. The hose was slightly negatively buoyant in water, which acted to counter the 0.02-kg/m buoyancy of the hot water in the hose.

A multistage centrifugal Caprari E4XP35/20 submersible pump with a 2.2-kW motor was used as a return pump. It was capable of delivering 90 L/min at 0.98 MPa and supplied water from depths of at least 60 m. The pump was connected to an umbilical consisting of H07RN-F three-phase power cable, 1¼″ thermoplastic return hose, and ¾″ thermoplastic hose to deliver hot water to the pump and cavity. A water level sensor and cable were also attached to the umbilical.

Two Exchange Engineering heat exchangers were used, each with an output of up to 250 kW, consisting of 100 m of ¾″ bore stainless-steel tube wound into a double coil and rated at 6.9 MPa. The oil burners used were Nu-Way MOL 350 units operating on 230 V and at a fuel pressure of

Fig. 3.83 Rotary multistage centrifugal pump with a 5.5-kW motor used as the main pump in the 500-m hot-water drill configuration and as the return pump in the 1000-m hot-water drill configuration (*Credit* K. Makinson)

860 kPa, each consuming 0.75 L/min of Jet-A1 fuel. Each unit was fitted with adjustable high-temperature and low-flow cut-off switches. A secondary burner with a 60-kW output provided heat to the borehole cavity and pump or to the water storage tank.

The hose drum was driven by a 0.37-kW motor via a 1:72 gearbox that provided constant tension to the hose between the winch drum and a capstan. The capstan, which had a circumference of 2.5 m, powered the drill hose up and down the hole and was driven by a 1.1-kW motor coupled to a 1:137 self-locking worm gearbox. The speed range of the winch was 0–54 m/h and could be controlled in increments less than 3 m/h.

The initial 500-m system with a heat output of ∼500 kW was built and tested in 2009 and was successfully used during the 2011–2012 austral summer season on the Larsen C and George VI ice shelves, which have a minimum temperature of −12 °C (Fig. 3.84). A total of five access holes at least 300 mm in diameter were drilled at three sites through almost 400 m of ice to provide access to the underlying ocean.

Various spray configurations were used during the different phases of drilling access holes to the ocean. Typically, 15° full-cone sprays were used initially to drill holes with the required hole diameter down to the depth of the cavity (∼60 m at <60 m/h) through porous firn and, in some areas, through thick ice lenses that had formed by the refreezing summer meltwater and were prevalent throughout the firn. To accelerate the lateral formation of the cavity, a 150° and/or a ring of horizontal sprays was used. Once in non-porous ice, a forward point 0° jet was used to provide good forward penetration. Typically, the pressure drop across the nozzle would be in the range of 0.5–1.0 MPa, yielding an exit velocity of 30–40 m/s.

To ensure that the hole was sufficiently large before deploying the permanent sub-ice-shelf instrument mooring, a bidirectional reamer was attached to the drill nozzle (Fig. 3.85).

During the reaming process, no forward-pointing sprays were used, removing the possibility of accidentally drilling into the sidewall of the hole and creating a secondary one. A spring-loaded reamer valve was activated when contact was made with a narrowing in the hole while travelling either down or up. The reamer consisted of a series of aluminum plates containing holes and channels that could be stacked together to provide a range of diameters from 150 to 350 mm in steps of 50 mm. When activated, hot water flowed through the network of holes and channels within the reamer. Water sprayed laterally at locations in the stack where spacers were placed, such as at the largest diameter of the reamer. Both springs were adjustable over a wide range of activation loads and were set prior to deployment. Typically, a force of ∼200 N was sufficient to fully activate the valves.

The hot-water drilling system was upgraded in 2013 with an additional drill hose and another generator, increasing the capacity of the pumping and heating modules from 500 to 1000 m. The thermal power of the system was enhanced by 50% by increasing the flow to 120 L/min and the temperature to 90 °C. Three positive displacement CAT1531 plunger pumps, each with a 5.5-kW motor and capable of delivering 40 L/min at 6.9 MPa, replaced the 5.5-kW centrifugal pump, which was redeployed to recover water from deeper (at least 102 m deep) borehole cavities and was capable of delivering 120 L/min at 1.7 MPa. Pulsation dampers and pressure relief valves were fitted to each of the high-pressure pump units.

This 1000-m drilling system was used on the southern Ronne Ice Shelf during the 2014–2015 field season at site S5 with a minimal ice temperature of −27 °C (Fig. 3.86), where an access hole had been built up in the 1998–1999 season. However, at that time, the retrieval of the CTD probe was unsuccessful. First, return and main holes were drilled at a distance of ∼0.6 m down to a depth of 85 m and then an interconnection was established using a spray nozzle (K. Makinson, personal communication, 2015). Drilling of the

Fig. 3.84 Current version of the BAS 500-m hot-water drilling system at Larsen C Ice Shelf, 2011–2012; on the left is the driven capstan using 1″ drilling hose, and on the right are two heat exchangers and a 10-m³ flexible coated fabric water tank (*Credit* K. Makinson)

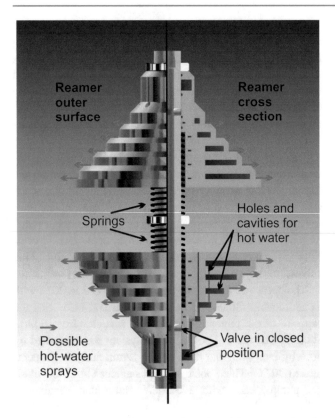

Fig. 3.85 One example of spray configuration of the reamer when the valve is in the off position (Makinson and Anker 2014)

Fig. 3.86 Driven capstan and bottom hole assembly with the reamer at site S5, southern Ronne Ice Shelf, in the 2014–2015 season (Hot water drilling n.d.)

main hole was continued and the ice shelf was penetrated at a depth of $\sim$770 m. Hot water with a temperature of 88 °C was pumped down at a rate of 110 L/min, which corresponded to a delivered heat power of 675 kW. The nozzle temperature dropped to 63 °C at the final depth. The rate of penetration was kept in the range of 72–96 m/h, and a total of $\sim$10 h were spent to drill to the ice base. A special drill brush 1 m in diameter was used to enlarge the hole at the base of the ice shelf to prevent the snagging of equipment during recovery from the hole. The estimated hole diameter after reaming was larger than 300 mm.

3.3.7.3 BAS System in the Norwegian Polar Institute

In the season of 2009–2010, the BAS hot-water drilling equipment was borrowed by the Norwegian Polar Institute and three holes were drilled at three sites in Fimbul Ice Shelf (Hattermann et al. 2012) (Fig. 3.87). A large and flexible rubber water tank (12 m^3) was placed in a shallow snow pit. The system included three CAT pumps powered by diesel engines; two were used in parallel and one was a spare. The maximum water flow rate from each pump was $\sim$38 L/min. Water was heated to $\sim$85 °C using seven heaters. One heater was used in the circuit of the submersible pump, sending warm water down the hole to prevent freezing around the pump and hose. The hose was driven by a

capstan with embedded sensors for drill speed, depth, and hose tension.

The first hole penetrated the ice shelf at 230 m; drilling and reaming took 13.5 h. The 60-m-deep return hole was drilled 0.6 m apart from the main hole. The hole was reamed out twice to 250 mm so that oceanographic instruments could pass through. The second hole was drilled 50 km away down to a depth of 395 m over $\sim$40 h (Fig. 3.88). The permanent instrumentation that would monitor the ocean circulation and temperatures for years was lowered down. The third access hole was drilled $\sim$120 km from the second site; the ice there was $\sim$250 m thick.

3.3.7.4 Hot-Water Drilling at Store Glacier, Greenland

As a part of the Subglacial Access and Fast Ice Research Experiment (SAFIRE), which was a joint project between the University of Cambridge and Aberystwyth University in Wales, a modified BAS hot-water dill was used to drill in Western Greenland at a site located 30 km from the calving front of the fast-flowing, marine-terminating Store Glacier (70° 31′N, 49° 55′W, 982 m asl) (Doyle et al. 2018). In late July and early August 2014, four adjacent boreholes with

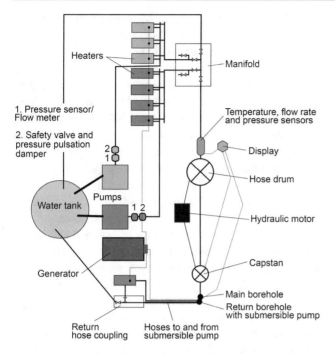

Fig. 3.87 Hot-water drill setup at Fimbul Ice Shelf (Nøst et al. 2009)

Fig. 3.88 Heaters running on the second site at Fimbul Ice Shelf in the season of 2009–2010 (*Photo* P. G. Gabrielsen from Nøst 2009)

depths of 603–616 m were drilled to the glacier bed within a 10-m² area. The site was located at a high-ablation area with a melting rate of ∼1–2 m/year and surface movement speeds of 600 m/year. The glacier surface was heavily crevassed and bumpy at a scale of meters to tens of meters (Fig. 3.89). An additional three boreholes deeper than 600 m were drilled to the glacier bed in July 2016 at a site located 50 m to the northeast of the 2014 drill site. These boreholes made a strong hydrological connection with the glacier bed during drilling, draining rapidly to ∼80 m below the ice surface.

Three pressure-heater units (Kärcher HDS 1000 DE) delivered a total of 45 L/min of water at 70–80 °C and

Fig. 3.89 Hot-water drilling at Store Glacier, Greenland, July 2014 (*Credit* K. Makinson)

11 MPa to a 2.1-m-long drill stem through a 1000-m-long hose with an ID of ¾″. To detect the glacier bed and measure the depth of the drilling, the length and weight of the spooled-out hose was recorded using a rotary encoder and a load cell located on a sheave wheel on the drilling rig at 2-s intervals. The drill's progress was governed by a mechanical winch.

The ice temperature profile exhibited a steep curve characteristic of fast ice flows with a minimum of −21.25 °C at a depth of 302 m, almost exactly midway between the surface and the bed. Owing to low englacial temperatures and fast refreezing rates, the installation of a thermistor string in one borehole failed. To overcome this problem, subsequent boreholes were drilled at a slower rate with a wider-angled, solid-cone water jet. In 2014, holes were drilled at an average rate of 72 m/h, allowing for 600-m-deep boreholes with an initial estimated diameter of ∼150 mm to be completed within 8.5 h. After drilling, it took ∼1.25 h to recover the drill from the glacier bed. In 2016, boreholes were drilled at slower average rates of 30–60 m/h, achieving slightly larger borehole diameters (200 mm).

Extended drilling efforts were also made to recover sediment cores from one of the holes drilled in 2016. This hole connected and drained at a depth of 611.5 m, below which drilling progressed intermittently at a slower (averaging 24 m/h) and more variable rate, including transient periods of partial unloading. At a depth of 657 m, the drill's progress ceased completely, which was interpreted as indicating contact with bedrock or consolidated sediments. The drill was then recovered to the surface and a sediment corer was lowered to the bed, but no sediment was retrieved. A subsequent attempt to obtain a sediment core resulted in the corer becoming irretrievably lodged in the borehole.

3.3.7.5 Hot-Water Drilling on Petermann Ice Shelf, Greenland

During the summer of 2015, the BAS system was used by an international team for access drilling at Petermann Ice Shelf, Northwestern Greenland, which is the second largest floating ice shelf in the Northern Hemisphere (Fig. 3.90) (Münchow et al. 2016). Water at ∼90 °C was pumped at a rate of 80 L/min, giving a total heating power of ∼0.5 MW. Access holes that were 300 mm wide were drilled at three sites. At two of these sites, holes were drilled through ∼100 m of ice within the central channel 16 and 26 km seaward of the grounding zone. The third site was located within 3 km of the grounding zone, again within the central channel, where the ice shelf was 365 m thick.

3.3.7.6 BAS System Used in the FISP Project

In 2014, the AWI initiated the Filchner Ice Shelf Project (FISP) in collaboration with the BAS for monitoring the present-day hydrographic conditions underneath Filchner Ice Shelf and determining the basal melting rates and the ice shelf's thermal regime. To install measuring devices below the ice over the course of three summer expeditions, 10 holes in ice that was up to 900 m thick were drilled. During the 2014–2015, field season three ∼300-mm-diameter access boreholes were drilled through ∼750-m-thick ice within a few kilometers from each other (Clarke et al. 2016). Approximately 6 m³ of aviation fuel and 800 L of petrol were used. Four other holes with depths of 853–891 m were successfully drilled at four locations and were equipped with instruments on both sides of the southern Filchner Ice Shelf in the austral summer season of 2015–2016. In the 2016–2017 season, hot-water drilling and the deployment of instruments were continued at three locations on the northern Filchner Ice Shelf, roughly 60 km behind the calving front.

Fig. 3.90 Hot-water drilling on Petermann Ice Shelf, Greenland, summer of 2015 (Nicholls 2015)

3.3.7.7 Hot-Water Drilling System of Victoria University of Wellington

To observe the ice/ocean interface directly at the base of Ross Ice Shelf, measure the ocean's properties, and take sediment sample from the sea floor, a drilling team from the Antarctic Research Center, Victoria University of Wellington, New Zealand redesigned the BAS hot-water drilling system and built a new modular drill (Silverwood 2018). This system was able to deliver up to 180 L/min of hot water at a temperature just below the boiling point (Fig. 3.91). This water was melted from nearby snow, which was shoveled into portable flexible coated fabric tanks and heated by a series of boilers powered by three generators.

To create holes with a diameter of not less than a predetermined size (250 mm), a reamer was fixed just above the drill stem (Fig. 3.92). The reamer was a gauge tool so that if it could pass down the hole, it was guaranteed to have a minimum diameter. In these cases, the operator was trying to drill as fast as possible while still holding back the full weight of the tool and hose. The reamer could spray water outwards and, if needed, upwards, enlarging the narrowing parts of borehole if refreezing happened.

The first drilling site, HWD-2, was established in the season of 2017–2018 in the middle of Ross Ice Shelf, ∼350 km away from Scott Base. Two holes were drilled; however, during the melting of the first hole (HWD-2a), the drill stem became jammed partway down. Nonetheless, the drilling team managed to withdraw the nozzle and begin again. The second hole, HWD-2b, pieced into the ocean cavity at a depth of 367.5 m and was kept open for 14 days for scientific observations by periodically reaming to remove the ice that formed on its walls. Surprisingly, a video camera recorded a thin layer of ice crystals at the base of the ice sheet, meaning that sea water was actually freezing onto the base of the ice instead of melting it, as it had been expected. The next drilling site, HWD-1, will be located near the ice shelf's grounding line, ∼1000 km from Scott Base, where hot-water drilling is planned for the 2019–2020 season.

3.3.8 Hot-Water Drilling System of the University of Alaska-Fairbanks

The University of Alaska–Fairbanks (UAF) system was designed for the subglacial exploration of the Black Rapids glacier, a surge-type glacier in east-central Alaska. It was deployed at the glacier in 1996, 1997, and 2002. The system used hose in the same as the Caltech drill, i.e., Synflex (Furon) thermoplastic hose with an ID of 19 mm (Bentley et al. 2009). The working flow rate of hot water was 60 L/min. Water was recovered from the borehole for reuse using a submersible pump.

Fig. 3.91 In preparation for hot-water drilling at site HWD-2, Ross Ice Shelf, December 2017 (Silverwood 2018)

Fig. 3.92 Bronze drill stem attached to the reamer at site HWD-2, Ross Ice Shelf, December 2017 (Silverwood 2018)

In 1996, seven holes were drilled through the ice to obtain accurate depth measurements and to prepare for the subglacial wireline drilling carried out in the spring of 1997, when four holes 100–150 mm in diameter were drilled at three locations to a maximum depth of 620 m (Truffer et al. 1999). In April and May 2002, the hot-water drill was used to make two 0.25-m-diameter holes down to the ice–till interface at depths of 497 and 498.9 m for subsequent till probing using a cable hammer (Harrison et al. 2004).

During December 2012 and January 2013, the UAF team drilled three 200-mm-diameter holes (PIG A–C) through the 400–500-m-thick Pine Island Glacier, one of the largest and fastest shrinking glaciers of the West Antarctic Ice Sheet (Stanton et al. 2013). The drilling team tested a hot-water

Fig. 3.93 Reamer is ready to be lowered into the borehole, Pine Island Glacier, January 2013 (*Photo* D. Pomraning; Campbell 2013)

drill in Windless Bight, near McMurdo Station, during the 2010–2011 Antarctic field season. This simple hot-water drill consisted of five high-output diesel water heaters feeding a ¾″ hydraulic hose connected to a long brass drill stem. The 500 m of hose were coiled into the shape of an "8" by hand near the borehole, making it a labor-intensive operation (Fig. 3.93). A well pump recirculated the melt-water back to the water pump and heaters. The average penetration rate was ∼70 m/h.

In the 2015–2016 and 2016–2017 field seasons, the UAF hot-water drilling system was used at Nansen Ice Shelf in collaboration with the Korea Polar Research Institute (KOPRI) to drill several access holes at least 250 mm in diameter through ∼400-m-thick ice at a site 10 km from the ice edge (Fig. 3.94). Again, the winch was not used, and the ¾″ drill hose was laid out in the shape of an "8" on the ground and went right over the reel (D. Pomraning, personal communication, 2018). The 100-m lengths of hose were connected via high-pressure couplings. Hot water at a temperature of 70–80 °C was pumped down with a high-pressure pump powered by a 30-kW diesel engine at a

rate of 80–90 L/min and a maximal pressure of 5.5 MPa. Five heat exchangers, each with burners rated at 88 kW, were used; each burner consumed ∼9.5 L/h of diesel fuel. Water was recirculated using a submersible pump with a 3.7-kW electric motor rated at 95 L/min from a depth of 60 m. The electric power for the submersible pump and other units was provided by 12.5-kW diesel generator. The initial 150-mm-diameter hole was enlarged to 250–300 mm by performing two passes with a reamer.

Recently, the UAF hot-water drilling system was prepared to be used in the five-year program that began in October 2018 with the deployment of fuel and supplies and the establishment of field camps on Thwaites Glacier, West Antarctica.

3.3.9 AMISOR Hot-Water Drilling System

An Australian modular hot-water drilling system (Fig. 3.95) was designed within the AMISOR (Amery Ice Shelf Ocean Research) Project with the capability of drilling boreholes at least 250 mm in diameter to depths of ∼1000 m in cold ice (Craven et al. 2002a, b, 2004). The AMISOR hot-water drilling system was powered by a Hatz 12-kVA three-phase diesel generator with an identical Hatz motor driving a CAT 1051 high-pressure surface pump for delivering water at rates of up to 45 L/min. Up to six 80-kW Kärcher Volcano industrial heaters were plumbed in parallel via inlet and outlet manifolds to heat water to ∼80 °C. The system used 1″ ID 580N-16-SL thermoplastic hose in 183-m lengths to deliver hot water to a 90-kg stainless-steel drill stem via a motor-driven hose drum winch and an electronically con-trolled capstan drive. The hose and capstan winches were individually driven by 4.7-kW Baldor servo motors and controllers driving Brevini 50:1 reduction gearbox. This system allowed for very fine control (better than 1 cm/s at full torque) of the drilling feed rates and for hauling rates greater than 1 m/s. A sensor block provided continuous readouts of water temperature, pressure, and flow rate to the operator. A load sensor was fitted to the hose pulley.

A set of spraying systems nozzles covering a range of outlet diameters delivered either a solid stream or a full-conic spray to the borehole depending on whether dril-ling was underway in a water-filled ice cavity or in the upper air-filled porous firn section of the hole. The water supply came from melted snow stored in a Structure-Flex 12-m³ portable water tank that was free-standing on insulated boards set out on the snow surface. A subsurface water-recovery system was employed, which used a Grundfos SP5A-38 4-kW submersible pump in an adjacent shallow hole with a hydraulic connection near sea-level depth with the main borehole. A small amount of hot water was fed back to the auxiliary borehole to prevent it from freezing around the submersible pump.

Fig. 3.94 Hot-water drilling system with hose in the foreground, water basins, pump, and heaters, Nansen Ice Shelf, January 2017 (Truffer 2017)

The drilling project took 11 years between 1999–2000 and 2009–2010 at eight drilling sites. During the 1999–2000 season, the hot-water drilling system was tested at a site 5 km west of Sansom Island, Sandefjord Bay, at the corner of Amery Ice Shelf. A combination of heavy katabatic ground drift and snow-bearing winds from the incursion of synoptic systems across the nearby coast resulted in the repeated burial of the plant and equipment (generator and pumps) required to operate the drill. This forced a redesign of the system layout involving the use of Weatherhaven protective tents to house all the major components of the system, effectively weatherproofing the drill (Fig. 3.96). These initial tests proved the basic integrity of the system with a shallow borehole drilled to a depth of 47 m, which

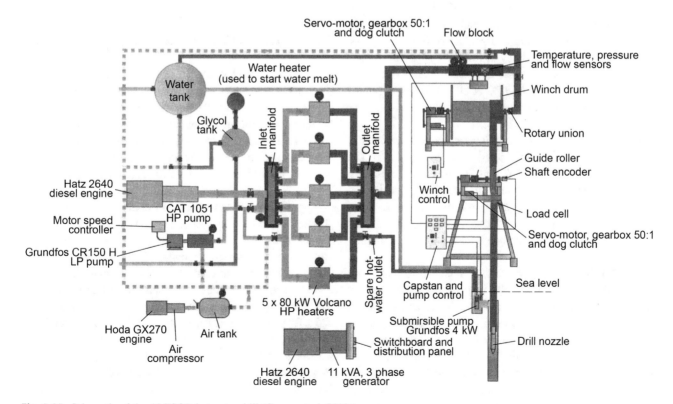

Fig. 3.95 Schematic of the AMISOR hot-water drill (Craven et al. 2002a)

Fig. 3.96 Inside the AMISOR
hot-water drilling tent (Treverrow
and Donoghue 2010)

Fig. 3.96 Inside the AMISOR hot-water drilling tent (Treverrow and Donoghue 2010)

was beyond the porous firn cut-off limit, where meltwater began to pool in the base of the hole.

In 2000–2001, a first access borehole 300–350 mm in diameter and 373 m deep was drilled at site AM02 (69° 42.8′S, 72° 38.4′E; 57 m asl), ~50 km west of Sansom Island. A total of 14 h were required from the commencement of drilling the main borehole past the well depth (at 45 m) until shelf breakthrough, including reaming to a depth of 250 m. With repeated reaming, the borehole was enlarged to nearly 500 mm across and was maintained open for one week for sampling and measurements. Thermistor measurements within the body of the ice shelf at site AM02 were typical of those at basal melt sites, with a minimum internal temperature of −20.3 °C at 80 m increasing monotonically to the pressure-melting point at the base, reaching −2.12 °C.

In 2001–2002, a 479-m-deep borehole was melted at site AM01 (69° 26.5′S, 71° 25.0′E; 65 m asl), 50 km west-northwest of site AM02 and 90 km from the calving front. A hydraulic connection to the ocean was made at a depth of 376 m, approximately 100 m above the base, indicating a highly porous or "honeycomb" structure in the lowest layer of ice. The main borehole was again maintained open for one week.

In 2003–2004, a new 480-m-deep hole was drilled at the site AM01. In 2005–2006, after increasing the flow rate of the drill from 45 to 80 L/min, holes AM03 (70.56°S, 72.39° E) and AM04 (69.90°S, 70.29°E) were drilled, the former through 722 m of ice with no marine ice and an apparently melting base and the latter through 603 m of ice with 203 m of marine ice and a hydraulic connection to the ocean at approximately 525 m. There was a minor basal "necking" effect at hole AM03 owing to insufficient heat supply at this

high depth, but the passage of instruments in and out was not a problem. At hole AM04, the ice from a depth of 550 m to the base at 603 m was very irregular, and thus instruments bumped against the walls as they were lowered and raised. In the 2009–2010 season, the two last AMISOR boreholes were drilled at sites AM05, 20 km east from Jetty Peninsula, and AM06 (71° 21′E, 70° 15′S), short of 63 km to the east of AM05. At these sites, the ice was ~600 m thick.

3.3.10 WISSARD Hot-Water Drill

A clean, 1000-kW-class hot-water drilling system (Fig. 3.97) was developed by the Science Management Office, University of Nebraska–Lincoln, for use in the Whillans Ice Stream Subglacial Access Research Drilling (WISSARD) project to gain access to Subglacial Lake Whillans and to the grounding zone of Ross Ice Shelf beneath ~800 m of ice in West Antarctica (Rack et al. 2014).

The WISSARD hot-water drill comprised the following primary modules (Fig. 3.98): (1) a melt tank; (2) a water supply tank; (3) a water filtration and decontamination unit; (4) two heater–pump units; (5) a hose reel unit, and (6) a command-and-control module. There were also several auxiliary modules that were transferred from the IceCube project (see Sect. 3.4.2): (1) two containerized 225-kW generators; (2) a power distribution module, and (3) a day fuel tank. Additional modules included: (1) a storage traverse unit with installed shelving; (2) a workshop module; and (3) two 40″-long ski-mounted work decks connected by another deck to where a knuckleboom crane was mounted.

Fig. 3.97 Aerial view of the WISSARD camp showing the hot-water drilling system (line of blue and red containers), work decks, and science labs (center, left), as well as a tent city and a cluster of traverse units in the 2014–2015 season (*Credit* M. Albert)

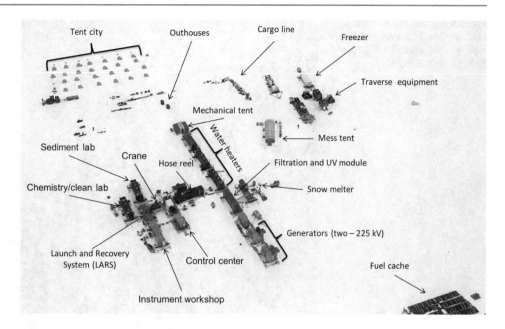

Fig. 3.98 Schematic diagram of the WISSARD hot-water drilling system: (MT) melt tank; (WST) water supply tank; (WFU) water filtration and decontamination unit; (HPU-1, HPU-2) two heater-pump units; (HRU) hose reel unit; (C&C) command-and-control module; (MWM) MECC (mobile expandable contaier configuration) workshop module; (GEN-1, GEN-2) two containerized generators; (PDM) power distribution module; (DFT) day fuel tank; (P) pumps (Rack 2016)

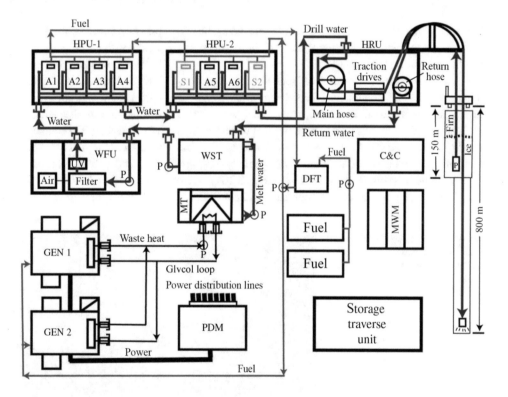

The initial water supply was provided by melting snow in the insulated 1.9-m³ melt tank (MT) using waste-heat thermal input from hot glycol (60–70 °C) pumped through a series of heat-exchange plates installed in the tank in the recirculating loop back to the two generators (Blythe et al. 2014). The melt tank was loaded with snow using a CAT297C multi-terrain loader or via shoveling (Fig. 3.99). The meltwater was fed into the water supply tank (WST), which was a 14-m³, insulated, oval stainless-steel

refurbished water tank. The tank had a main hatch and two observation hatches on one end. During drilling, the main portion of the water was sent to the water supply tank from the borehole using a Grundfos submersible pump (maximum flow rate: 302 L/min).

Water from the water supply tank was pumped to the water filtration and decontamination unit (WFU), where it was filtered through two water-filter modules (2.0 and 0.2 μm), which were connected via hoses to two (185 and

Fig. 3.99 Lead driller D. Duling at the melt tank; the water supply tank is in the background (*Photo* F. Rack from Reed 2015)

245 nm) germicidal UV light sources, before being heated and pressurized and sent to the main drill hose and down into the borehole (Fig. 3.100).

Three-phase 480-V power was routed from two containerized 225-kW generators (GEN-1 and GEN-2) to the power distribution module (PDM), where lines were routed to all of the other modules. Fuel from the fuel bladders filled the day fuel tank (DFT) and was distributed to the fuel manifolds in the heater–pump units. Two specially modified 40″ containers (HPU-1 and HPU-2) (Fig. 3.101) were designed to each accommodate four large Alkota 12257K

pressure washers, marked as A1–A6 in Fig. 3.98. Two spare washers (S1 and S2) were planned to be installed in the HPU-2 container, but they were not purchased owing to budget limitations. Each washer provided a flow rate of 45.6 L/min, yielding a maximum of 274 L/min from the six washers at a peak operating pressure of 13.7 MPa through a 147.2-m-long 304 stainless-steel coil with an ID of 19.1 mm using a General TSF2421 pump with a stainless-steel manifold and valves. Normal operating pressures were typically much lower ($\sim$10 MPa). Water flowing to the main hose reel was typically at a temperature of 78 °C, whereas the

Fig. 3.100 View inside the water filtration and decontamination unit (Rack et al. 2014)

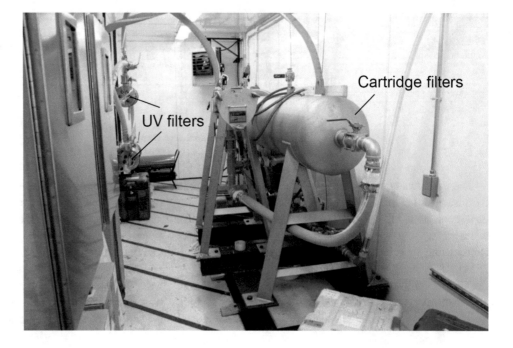

Fig. 3.101 Heater-pump unit HPU-1 (*Credit* F. Rack)

water being drawn back out of the borehole by the return water pump was just above the freezing point.

The 40′ shipping container of the hose reel unit (HRU) was modified with double doors at each end of the container on either side, in addition to the double doors at either end of the container (Fig. 3.102). The HRU incorporated the main hose reel with bottom-feed level winding, dual-traction drives, the return hose reel, and the bulk of the wiring and panels of the control system (Fig. 3.103). The main hose reel's motion was controlled via programmable logic controllers and an encoder buffer board connected to a

series of variable-frequency drives that were integrated with the motors for the dual-traction drive system (Burnett et al. 2014). A rail system was installed on the roof of the container to fix two IceCube crescents and guide the main drill hose into the borehole and the return hose out of the borehole. The rails could be extended up to 9.14 m beyond the end of the container.

Two custom-designed traction drives controlled the passage of the main drill hose through two pairs of tapered urethane-coated belts forming the upper and lower closed loops, which were pushed together using the upper lever arm

Fig. 3.102 F. Rack on a work deck in front of the hose reel unit (*Credit* G. Schilling from Antarctic album 2013)

Fig. 3.103 Layout of the hose reel unit, showing the extension rail and the crescent dolly in position over the borehole (Rack 2016)

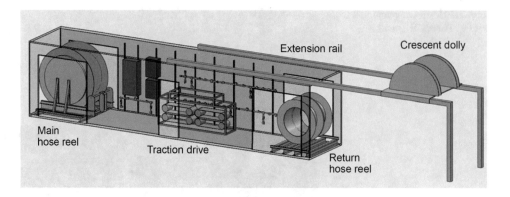

and a spring to grip the drill hose tightly between them during deployment, drilling operations, and recovery. The drill hose exited the traction drive and passed through a set of rollers and the hose-cleaning system before being fed onto one of the two IceCube crescents. The return water hose, which was connected to a submersible pump, came out of the borehole and over another crescent to the return hose reel.

The 1000-m-long main hose had an ID of 31.8 mm. The inner liner material was Rilsan PA11 BESNO P40 TLO polyamide resin that was surrounded by a braided construct consisting of aramid and highly tensile synthetic fiber to limit the amount of hose stretch to less than 5%. The outer cover consisted of abrasion-resistant and fungus-resistant polyether-urethane material. The maximum bending radius of the hose was 0.305 m.

The drill hose passed from the crescent down to the work deck and through the germicidal UV collar mounted below into the borehole (Fig. 3.104). The UV collar used 12 mercury amalgam UV lamps encased in quartz tubes that were arranged vertically in a circular pattern around the collar. There was reflective coating on the inside surface of the collar, which served to boost the intensity of the UV light towards the center of the collar. The safety of the personnel working around this UV system was of primary concern. An interlock shut-off switch was integrated into the lids of the collar to turn off the lamps when the lids were open, and a manual shut-off switch was used to turn off the lamps when the hose was stationary for any significant amount of time to avoid damage to the hoses or cables from prolonged exposure to UV light.

The traverse from McMurdo Station to the Subglacial Lake Whillans camp (84.240°S, 153.694°W), which had a distance of 1014 km, took 14 days with 12 tractors towing 26 sleds (Fig. 3.105) (Rack 2016). The drilling operations began on January 22, 2013 to create a shallow hole (approximately 30 m deep), which was used for the assembly of scientific tools away from the location of the main borehole. Following this, a 120-m-deep return hole was melted through the firn and into the ice and a submersible pump was deployed in it. Then, the main borehole

was drilled to the same depth, parallel to the return hole but separated by a distance less than 1 m, and water was circulated into the borehole for several hours until a connection between the two holes was indicated by changes in the load cell monitoring the weight of the drill hose and the pressure sensor monitoring the level of water in the borehole. The submersible pump was then turned on and water was recirculated back to the surface to confirm the connection.

Afterwards, the main borehole was deepened at rates from 30 m/h to more than 60 m/h to ~700 m, which was ~100 m less than the estimated total thickness of the ice over the lake, and borehole visual observations were carried out using a video camera recorder and a downhole optical caliper tool, which confirmed that the borehole was well formed, vertical, and with the desired diameter. Then, a wireline water-sampling system was lowered into the borehole to measure the contamination of the drill water. The main hose and drill head were redeployed into the borehole to melt the final 100 m of ice and, on January 27, 2013 the hot-water drill made a connection to the base of the ice sheet at a depth of ~801 m, which was indicated by changes in the measured values of the load cell and the water-pressure sensor in the return borehole (water rose by 28 m in less than 1 min) (Tulaczyk et al. 2014). Approximately an hour before the anticipated breakthrough, the water level in the borehole was pumped down to ~110 m below the ice surface to make sure that borehole's water pressure was lower than the lake's water pressure.

Following the initial 30 h of the scientific operations, it was determined that the borehole required reaming, and thus the drill was realigned and the borehole was enlarged over the course of 20 h before returning to deploying scientific tools until the ending of operations on February 1, 2013. The subglacial water layer was 2.2 m thick beneath the borehole. The geothermal heat flux below Subglacial Lake Whillans, which was determined from the thermal gradient and the thermal conductivity of sediment under the lake, was 285 ± 80 mW/m^2, which is significantly higher than the continental and regional averages (Fisher et al. 2015).

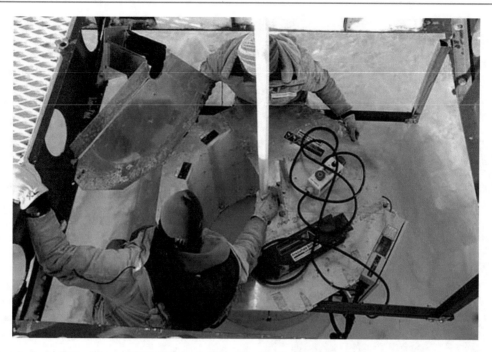

Fig. 3.104 Rillers deploy an UV collar to sterilize the hose that goes down the borehole after three overlapping doors on the top of the collar are closed (Rack 2016)

Fig. 3.105 Aerial image of the traverse from McMurdo Station to the Subglacial Lake Whillans camp, January 2013 (Rack et al. 2014)

In December 2014, the University of Nebraska–Lincoln drill team had spent the day to drill a 0.38-m diameter hole that extended vertically through the 14.6 m of ice thickness present at the Site E at McMurdo Ice Shelf, 30 km away from McMurdo Station (Spears et al. 2016). The primary aim of this project was to deploy the Icefin under-ice unmanned underwater vehicle that was designed specifically for deep-field operations and can be deployed and recovered vertically through a small-diameter hole drilled in the ice shelf.

In January 2015, a drilling team from the University of Nebraska–Lincoln provided clean access to a depth of ~740 m near the grounding zones of Whillans Ice Plain and Ross Ice Shelf, 115 km west from the previous site. Cameras sent down the drilled hole revealed an unsuspected population of fish and invertebrates living beneath the ice sheet, the farthest south location in which fish have ever been found. The surprising discovery of fish in waters that are extremely cold (−2 °C) and perpetually dark poses new questions about the ability of life to thrive in extreme environments.

After general maintenance on the drill components (high-pressure pumps, filtration system, and shipping containers), the WISSARD hot-water drill was used in the Subglacial Antarctic Lakes Scientific Access (SALSA) project in 2018–2019. The primary goal of the project was to penetrate into Mercer Subglacial Lake, one of the largest and most dynamic subglacial lakes on Whillans Ice Plain, and to deploy deep SCINI remotely-operated vehicle. Lake Mercer is located about 40 km to the southeast from Subglacial Lake Whillans and is twice the size of Lake Whillans, and five times deeper—but was probably connected to the ocean at the same time as Lake Whillans (Fox 2018, 2019).

Fig. 3.106 SALSA team members D. Harwood and T. Campbell take mud samples from the drill head recovered from the borehole after breakthrough, December 2018 (*Photo* K. Kasic from Metcalfe 2019)

Prior drilling, the drill head was decontaminated by spraying of 3% hydrogen peroxide. UV lamps on the collar were turned on to decontaminate the drilling hose. Since starting the borehole, most of the focus has been on keeping the drill balanced and heading toward the lake at 15 m/h penetration rate and 136 L/min flow rate (SALSA Chief Scientist Reports 2019). It took 76 h of drilling to reach the roof of the lake on 26 December, 2018 when the water rose in the borehole on ~14 m indicating breakthrough. Upward reaming started immediately after breakthrough at the rate of 60 m/h below 600 m and 48 m/h above 600 m (Fig. 3.106). Based on subsequent measurements, the ice thickness was determined at 1086 m and the lake depth was ~13–14 m.

3.3.11 Intermediate-Depth Hot-Water Drilling System of the University of Wyoming

In 2010, a new intermediate-depth hot-water drilling system was built in the University of Wyoming based on the design

of their shallow-depth hot-water drilling system (see Sect. 3.2.12); the new drill yielded greater efficiency in regards to drilling time, weight, and personnel required for operation (Harrington et al. 2015; Harper et al. 2017; Meierbachtol et al. 2013). Hot-water was pumped into the hose under a pressure of 6.9 MPa and jetted from an ~80-kg drill stem (Fig. 3.107). Surface meltwater, which was temporarily housed in a 7.7-m^3 tank, was used for drilling. The weight of the drill stem and the hose hanging from the drill tower was measured with a digital load cell embedded in the tower's drive wheel. Depth marks on the drilling hose and a sensor cable string measured hole depth with an uncertainty less than 1% (5 m).

Between 2010 and 2015, 32 boreholes down to the bed of the ice sheet were drilled at six sites, with 2 to 10 boreholes drilled at each site, on Isunnguata Sermia, which is a land-terminating outlet glacier of the western Greenland Ice Sheet. The depth of the holes ranged from just under 100 m at locations less than 1 km from the ice sheet margin up to 830 m at a location approximately 46 km inland from the terminus of Isunnguata Sermia. Three sites were located in a region of draining supraglacial lakes. At one site, ~4 km from the margin, a hole was drilled within the area with the deepest bedrock, but it unfortunately did not penetrate the full ice thickness and reached a depth of 700 m.

The load increased steadily throughout the drilling operations as additional hose was lowered into the borehole. The load was nearly completely relieved from the tower when the drill stem encountered the bottom of the hole. During the drilling process, boreholes were filled with water up to the surface as drill water was continuously added to the hole, giving the base of the borehole a relatively high head (110%) with respect to the ice overburden pressure.

When the advancing borehole intersected bedrock, one of the three following series of events occurred: (1) a sudden draining of the borehole and the equalization of the borehole pressure with that of the subglacial drainage system; (2) delayed draining, either fast or slow, and the eventual equilibration of the hole pressure with bedrock pressure; or (3) no draining, with the hole maintaining a static and high head indefinitely. The nozzles at the end of the drill stem all exhibited substantial damage due to blunt-force impact with hard surfaces at the glacier bed (Fig. 3.108). The holes remained open after drilling for a limited time (temperature distribution depended on site location and was as low as approximately −14 °C); they refroze in just 2–3 h. The diameter of the holes was estimated to be ~150 mm.

Uncommon thermal tracing was carried out at a site located ~34 km east of the terrestrial terminus of Isunnguata Sermia (Meierbachtol et al. 2016). Two temperature sensors were embedded in the drill stem to track the heat transfer between the drilling water and the borehole during the advance of drilling and extraction operations

Fig. 3.107 Starting a new hole at Isunnguata Sermia, western Greenland Ice Sheet, summer of 2012 (*Photo* J. Harper from Davis 2013)

Fig. 3.108 Drill nozzles showing the damage resulting from near-perpendicular impact with the glacier bed; for comparison, the leftmost nozzle was used to drill more than 1 km in several different boreholes but was never in place when intersecting the glacier bed (Harper et al. 2017)

(Fig. 3.109). One sensor was located on the inside wall of the drill stem to measure the temperature of the drilling water prior to exiting the nozzle, and another was pressed against the outside wall of the stem to measure the borehole water temperature. Temperatures were logged at 1-min intervals by a data logger located in the cavity of the hollow drill stem. Sensor resolution was 0.065 °C. The drill water temperatures decreased from approximately 65 to 47 °C over the course of drilling the 675-m-deep borehole because of the increasing conductive losses as the drill hose was payed into the water-filled hole.

3.3.12 Summary

Intermediate-depth hot-water drilling is considered to be the most suitable way to access the seawater cavities beneath ice shelves and subglacial lakes in the margin of the Antarctic and Greenland ice sheets. This method is quick, allowing for drilling holes in a matter of days. Very deep holes with rather large diameters can be drilled using light and simple equipment that can be operated by a small crew. The alternative is to use mechanical drilling systems, which require multiple seasons to achieve similar boreholes and the use of

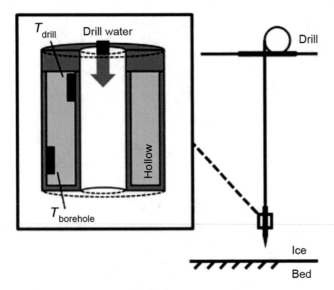

Fig. 3.109 Instrumented-drill-stem schematic indicating the layout of the water temperature sensors (Meierbachtol et al. 2016)

drilling fluids that are potentially harmful to the environment. Hot-water drilling offers more options for controlling contamination than mechanical systems. Although ultra-fine filtration at high flow volumes poses challenges, it can be done with existing technology.

However, hot-water holes are rather irregular. It is difficult, therefore, to use them for certain measurements (such as borehole closure or inclinometry). Another important issue is estimating the refreezing rate for the safe retrieval of the drill stem and other instruments. The hole cannot be left to refreeze until its diameter becomes less than the size of the tools and instruments being lowered through it. Most studies have aimed to model the water refreezing rate of boreholes drilled through glaciers or ice shelves (Yen and Tien 1976; Humphrey and Echelmeyer 1990; Li 1993; Hughes et al. 2013; Greenler et al. 2014).

To date, in only a few studies have the refreezing rates in ice boreholes filled with water been directly measured. A case in point was the caliper profiles taken shortly after a hole was drilled using a hot-water drilling system through the 562-m-thick Ronne Ice Shelf, Antarctica (Makinson 1993). These measurements indicated initial refreezing rates of 9–11 mm/h for a hole diameter of approximately 130 mm in ice at −26 °C. After the hole had been maintained open for two days and the host ice was warmed, the refreezing rate decreased to 4–5 mm/h. These rates are slightly lower than the water refreezing estimates obtained via modeling. For example, Humphrey and Echelmeyer (1990) predicted that boreholes would totally refreeze within 4–23 h at an ice temperature of −25 °C and an initial diameter of 100–240 mm, that is, at rates of 25–10.5 mm/h.

In operations where there is a risk of losing the drill in the hole because of refreezing above it, a device known as a back drill can be incorporated into the top of the drill stem. When during the retraction of the drill the back drill encounters a constriction in the hole, a spring-loaded sliding mechanism redirects the jets of water upwards. Caltech and the BAS used a double conical device to ream holes to the minimal safety diameter on the way down and up.

Where relatively deep boreholes are required, more than one hose length can be drilled in stages. Additional hose sections may be added between the drill and the hose entering the borehole as required, as opposed to attaching the full anticipated hose length to the drill from the outset. During such procedures, the drill stem need not be removed from the borehole, but it should be raised a few meters up from the bottom of the borehole.

The traditional way to pull the drill hose to and out of the borehole is to use capstan drives. These drives allow the hose to be wound onto the hose reel without experiencing significant tension from the weight of the hose (in single-grooved capstan drives, the ratio between the pulling and holding forces is 1.8–2.5 depending on the friction between the hose and the wheel), which minimizes the probability of hose breakups and reduces the size of the motor required for the operation of hose reels using a very low torque. A new method to pull the hose was suggested by the WISSARD drill team (Rack et al. 2014). They replaced the capstan with a traction drive that, unlike capstans, also allows the hose connections to pass through the belts without difficulty, providing flexibility in drilling operations when multiple connections are required.

The drill tip needs to be designed to allow for the nozzle to be readily replaced. The degradation of the nozzle can result in a non-uniform or non-centered water jet, which in turn results in the degradation of the penetration rate and in borehole deviation. The verticality of the hot-water hole depends largely on the care of the driller to keep the drill properly suspended because the stream of water melts the ice directly beneath the nozzle and the probe follows through this hole. In general, the longer and heavier the drill stem is, the straighter the resulting borehole is, but the more difficult it becomes to manage on the ice surface.

The hose limits the flow capacity of the system. Given that the pressure in the hose will rise roughly with the square of the flow, one hits a hard limit for a given hose size very quickly. Increasing hose size would require more heaters. In addition, it would also imply a substantial increase in hose spool sizes, the associated drives, and also a thorough rework of the capstan and sheaves. To a large extent, this would result in an entirely new drill system.

Water supplies can be accomplished via different methods. The easiest way is to use meltwater from surface ponds if available. In cold environments, water has to be melted from snow to initiate the drilling process. Typically, the drill will have to penetrate firn, where the hot water applied drains through the pores and cannot be recovered for recirculation. This firn layer may span many tens of meters. While drilling, however, drillers are not as concerned with the absolute depth of the firn as with the pooling level, which is the point at which there is enough density to establish a water level. Above that level, the firn acts as a sponge, drawing away water. This level may be anywhere up to 30–40 m or possibly more depending on the location and properties of the firn.

In cases in which drillers have a hole large enough to accommodate a return-water pump, the implementation of water recycling is probably advisable. In cases in which boreholes are to be small, setting up a recycling system via a return hole at a very short distance away from the main hole is likely the best alternative. However, the process of drilling a return hole through firn and down below the pooling level requires fuel and effort. If the ultimate objective is a hole with a depth of 200–300 m, it is probably best to melt water on the surface and dispense it with a subsurface pump. When contemplating depths deeper than 300–400 m, however, pumping will seem very attractive. However, in these cases, a melt tank is also necessary to a far lesser extent to compensate for the density differences between the ice and the water replacing it.

When transporting or otherwise shutting down a hot-water drill, it is essential to purge water from the system and to fill all hoses, burners, gauges, valves, and pumps with antifreeze. The most commonly used fluids for preventing freezing are: (1) ethylene glycol (because it is poisonous, it raises environmental concerns); (2) propylene glycol, which is often classified as a food-grade liquid and is becoming more commonly used (it tends to become very viscous at colder temperatures; however, it was nonetheless used by IceCube at South Pole); and (3) ethanol, which is very thin and easy to pump around, but its flammability is an issue.

3.4 Deep Hot-Water Ice-Drilling Systems

Deep hot-water ice-drilling systems are aimed to create holes deeper than 1500 m through the Antarctic and Greenland ice sheets and ice shelves for subglacial exploration and other scientific experiments (e.g., the installation of neutrino detectors, as done in the IceCube Project; Benson et al. 2014). To date, the deepest successful use of a hot-water drill reached a depth of ∼2500 m in the US IceCube project, melting hole with a diameter larger than 0.6 m at the South Pole (Table 3.3). The drillings of these holes were stopped 300 m above the base of the ice sheet.

3.4.1 AMANDA Project, South Pole

The motivation behind the Antarctic Muon and Neutrino Detector Array (AMANDA) and the IceCube Neutrino Observatory was to realize the idea of building a large-scale neutrino telescope to observe high-energy neutrinos from astrophysical point sources. With this aim in view, strings of widely spaced photomultiplier tubes (PMTs) have to be placed into deep hot-water holes drilled in clear ice. High-energy neutrinos going through Earth will occasionally interact with ice and create a muon; such muons will emit Cherenkov light when passing through the array, so they can therefore be tracked by measuring the arrival times of these Cherenkov photons at the PMTs.

A shallow array, AMANDA-A, was deployed at a depth of 800–1000 m in 1993–1994 in the exploratory phase of the project (Andres et al. 2000). Studies of the optical properties of the ice carried out with AMANDA-A showed that the high concentration of residual air bubbles at these depths leads to the strong scattering of light, making accurate track reconstructions impossible. Therefore, in the season of

Table 3.3 Parameters of deep hot-water ice drilling systems

Institution or drill name	Years	Drilling site(s)	Flow rate/ (L min^{-1})	Max drilled depth/m	Borehole diameter/m	ROP in ice/(m h^{-1})	References
AMANDA	1993–2004	South Pole	Up to 350	2400	0.45–0.6	14.6	Koci et al. (1997), Koci (2002)
IceCube	2004–2011	South Pole	760	2500	>0.6	∼90	Benson et al. (2014)
BAS, RABID project	2004–2005	Rutford Ice Stream, West Antarctica	135	2025	NA	72–120	Smith (2005)
BAS	2012–2013	Subglacial Lake Ellsworth, Antarctica	180–226	∼300	>0.36	NA	Siegert et al. (2014)
BAS, BEAMISH project	2013–2019	Rutford Ice Stream, Antarctica	160	2200	>0.3	NA	Smith and Fothergill (2016)
Chinese hot-water deep drilling system	2010–up to now	Amery Ice Shelf, Antarctica	Up to 260	In progress	>0.4	∼38	P. Talalay, unpublished

Note NA—data are not available

1995–1996, a deeper array consisting of 80 PMTs arranged on four strings (AMANDA-B4) was deployed at depths ranging from 1545 to 1978 m, where the concentration of bubbles was predicted to be negligible according to the extrapolation of the AMANDA-A results. The detector was upgraded in 1996–1997 with 216 additional PMTs on six strings. This detector comprising four plus six strings was named AMANDA-B10. AMANDA-B10 was upgraded in the season of 1997–1998 with three strings instrumented between 1150 and 2350 m, starting the third and final stage of the project, the AMANDA-II stage. In 2000, nine additional strings were added, increasing the PMT count to 677 at 19 strings and forming a circle 200 m in diameter. In 2005, after nine years of operation, AMANDA officially became a part of its successor project, the IceCube Neutrino Observatory (see the next section).

During the initial stage, the old Crary Ice Rise hot-water drilling system (see Sect. 3.3.2) was resurrected by the PICO, University of Alaska–Fairbanks, to prove the concept of drilling holes with a diameter larger than 0.4 m (the first AMANDA modules with PMTs had an effective diameter of 0.3 m) to depths of 1 km without requiring a large capital investment (Koci 1994). First, a sub-ice reservoir was established slightly deeper than the pooling depth ($\sim$40 m at the South Pole) and the submersible Grundfos 40S75-25 pump was set in it. The pump delivered water at a rate of up to 220 L/min from the hole to the tank, where it was pre-heated and more water was added to compensate for phase changes. At this stage, the water was also degassed, which helped to keep the ice clear as it refroze. The diameter of the holes was considered large enough to accommodate an instrument-supporting cable, a 1¼″ 76-m-long return hose with winded round four-wire power cable, and the main hose.

Twin triplex pumps powered by Lister 4 CD diesel engines took water from the tank and pumped it through the heaters and 1200 m of hose. The operating pressure was $\sim$5 MPa at a flow rate of 200 L/min. Water was heated to 88 °C (which is the boiling point at the South Pole at that altitude) using six Whitco Model 75 oil-fired heaters, each rated at 150 kW at sea level (because of the actual altitude of $\sim$2835 m asl, the heaters were de-rated to $\sim$100 kW each). Preheat was supplied by up to twelve 80-kW (50-kW at that altitude) oil-fired heaters, bringing the total heat input power to over 1 MW. Hot water was supplied to the drill through thermoplastic hose with an ID of 30 mm. Tension and drilling rate were controlled using Kevlar-reinforced electromechanical cable, which also transmitted information from the instrument package; it consisted of three-axis flux-gate magnetometers, inclinometers, and rate sensors. The position of the drill relative to the hole wall was monitored using ultrasonic devices (Zagorodnov et al. 1992). These devices also provided information on hole diameter at two locations different from the hole caliper.

During the 1991–1992 season, two $\sim$800-m-deep test holes were drilled at the South Pole (Fig. 3.110). Unfortunately, the use of old equipment and hoses that had been subjected to many seasons of ultraviolet radiation caused many problems. Heat input was limited to 0.5 MW, which was half of its estimated value and matched the heat losses in the hole. The penetration rate decreased with depth until all the available heat was used in reaming operations. The drilling of the first hole lasted 16 days and required 45 m³ of fuel. The drilling of the second hole was faster (seven days) but ceased at the same depth because the total heat input once again matched the hole losses.

The heat input of the new AMANDA drill was twice that of its predecessor (Fig. 3.111) (Koci et al. 1997; Koci 2002; Makovicka et al. 1998). The major differences between this system and the previous one were the use of thick-walled 1½″ diameter hose, new and larger hose reels and winches, and the capability of monitoring over 30 parameters during drilling (circulating water pressure, temperature and flow in the major points of the water loop, depth, electric cable tension, drill lowering/raising speed, bottom temperature inside and outside the drill stem, hole diameter, etc.). The

Fig. 3.110 Drill stem of the initial AMANDA hot-water drilling system in the 1991–1992 season (*Credit* A. Karle)

Fig. 3.111 Schematic of the AMANDA hot-water drilling system: (T), (P), (F) stand for temperature, pressure, and water flow transducers, respectively (Koci et al. 1997)

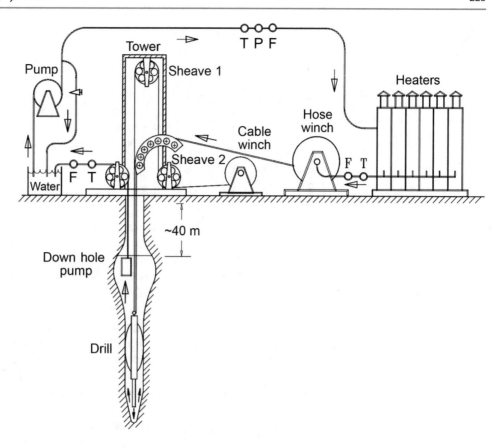

drill stem consisted of two sections: (1) weight to provide pendulum steering and vertical stabilization in the hole, and (2) instrumentation (Fig. 3.112). The single nozzle at the bottom of the drill emitted a hot-water jet at the front of the drill.

As before, a smaller preheating system controlled the temperature of the inlet water flowing to the main system for regulating the final water temperature. The preheating stage was centered around a large 32-m³ main supply-holding tank, and several smaller resupply-holding tanks. The total supply of water available for drilling contained in these holding tanks could be up to 62.5 m³. This water was maintained at a temperature of ~ 34 °C using eight oil-fired water heaters. Water from the main tank was pumped to the main heating plant using three high-pressure triplex pumps, each pumping up to 150 L/min (the maximal flow rate was as high as 350 L/min at a pressure of 7 MPa). The main heating plant included 12 industrial-grade, high-pressure, oil-fired water heaters manifolded together, which heated the water to its final drilling temperature of ~ 85 °C. This hot water was pumped into high-pressure hose, which was used in sections cut to custom lengths if needed (typically 150 m). Approximately 2400 m of hose was spooled onto three winches and used sequentially as the hole was drilled deeper. The hose was run over a roller guide crescent at the tower structure and was terminated by the drill stem. Owing to heat

losses over this amount of hose, the final drilling temperature was ~ 75 °C.

During the 1993–1994 field season, four deep holes with depths of 1012–1050 m were drilled (Fig. 3.113). The drilling procedure included the following phases: (1) penetration to the depth of the water-impermeable firn layer and downhole pump installation; (2) drilling to a depth of 250 m and the first reaming on the way up and down; (3) drilling to a depth of 550 m and the second reaming, again on the way up and down; (4) drilling to a depth of 720 m and the third two-fold reaming; and (5) drilling to the final depth of 1050 m and the final reaming while raising the drill from the hole. A drilling rate of 14.6 m/h was experimentally found as appropriate for drilling plumb holes. The reaming speed was 231 m/h. The total drilling time (including the final reaming, which took approximately 6 h) of the holes varied from 77 to 96 h. The fuel consumption was reduced to 17 m³ per hole.

In the 1995–1996 season, four holes were drilled to depths ranging from 1900 to 2200 m. At the same time, new and slightly larger (0.33 m) PMTs modules were designed, thus requiring holes with larger diameters (>0.45 m). Drilling time varied from 90 to 110 h per hole, with a fuel consumption of 32–38 m³ per hole.

In the following season, six holes were drilled to a depth of 1950 m, further reducing drilling times to 75 h and fuel

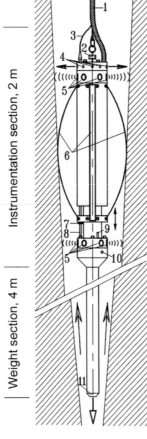

Fig. 3.112 Schematic diagram of the AMANDA hot-water drill stem: (1) hose; (2) load cell; (3) data cable; (4) upper nozzles; (5) sonar transducers; (6) caliper spring; (7) magnet; (8) linear displacement transducer: (9) return-water flow temperature transducer; (10) pressure transducer; (11) nozzle (Koci et al. 1997)

Instrumentation section, 2 m

Weight section, 4 m

In 1999–2000, to decrease hydraulic resistance, the ID of the lower 700-m portion of the main hose was increased to 2″, whereas the rest of the hose still had an ID of 1½″. The number of oil-fired Whitco Model 75 heaters was also increased from 12 to 16, increasing the maximal heating power input to 2.5 MW. The nine last holes were drilled to depths of ∼2400 m, taking an average of 100 h per hole, and the AMANDA-II project was completed.

3.4.2   IceCube Project, South Pole

Similar to its predecessor (the AMANDA project), the Ice-Cube Neutrino Observatory (or simply IceCube) project was aimed at deploying strings with PMTs spaced at depths between 1450 and 2450 m in holes melted in the ice using hot-water drills, transforming 1 m^3 of ice deep below the South Pole into a highly sensitive optical instrument. The design philosophy for the IceCube Enhanced Hot Water Drill (EHWD) was based on the AMANDA drilling experiences with several critical differences: (1) it had double the thermal capacity (increased from 2.4 to 4.7 MW), which was required to drill large-diameter (>0.6 m) holes faster; (2) it had a continuous drill hose on a single hose reel, eliminating the need to add/remove hose segments throughout the drilling/reaming operations; (3) it had two drilling structures, allowing for a streamlined drilling–deployment flow.

The two separate EHWD structures were (Fig. 3.114): (1) the seasonal equipment site (SES), which provided electricity and a stable supply of hot pressurized water, and (2) the tower operations site (TOS), where the hole was drilled. The two sites were linked by long cables and

consumption to ∼25 m^3 per hole. In 1997–1998, four other holes were drilled to a depth of 2400 m. In addition, hole diameter was enlarged to ∼0.6 m to allow for the extra time needed to deploy 42 POMs between depths of 1400 and 2400 m.

Fig. 3.113 Aerial view of the AMANDA drilling site, South Pole, in the 1993–1994 season (*Photo* R. Morse from AMANDA 2000)

Fig. 3.114 Schematic of the Enhanced Hot Water Drill (EHWD) (Benson et al. 2014)

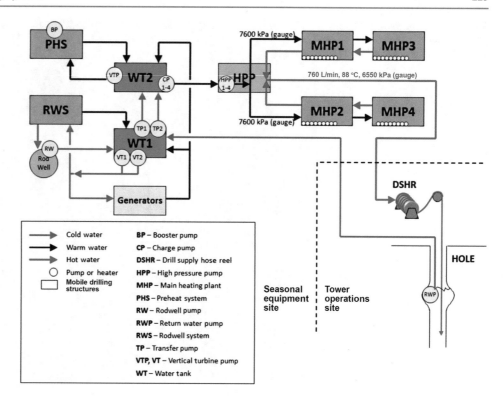

insulated hoses. The SES consisted of generators, water tanks, the pump and heating buildings, a central control building, mechanical and electrical shops, spare parts storage units, and a Rodwell water supply system. All the equipment was packaged into shipping containers.

The TOS included the drill tower and the attached operations building, the hose and cable reels, and the downhole drill assembly (Figs. 3.115 and 3.116). There were two towers and one set of drill reels. After drilling, the drill reels were moved to the location of the next hole, where

Fig. 3.115 Schematic of the tower operations site (TOS) showing surface and downhole equipment (Benson et al. 2014)

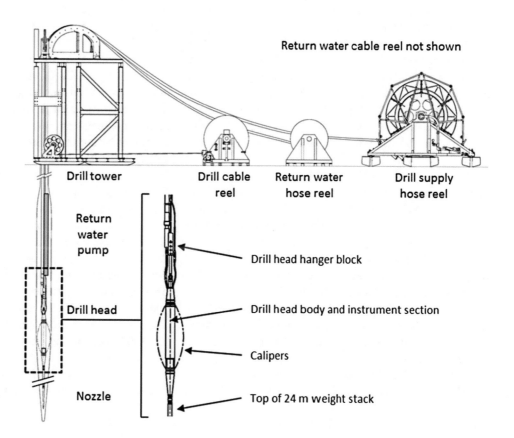

Fig. 3.116 Aerial view of the tower operations site (TOS) (Halzen and Klein 2008)

the second tower had already been staged. The first tower stayed at its existing location to support the deployment of instrumentation. Once deployment was finished, the first tower could be moved to the next location while drilling at the second tower was underway. This leapfrog sequence of the tower structures reduced hole turnover time and allowed for nearly continuous drilling operations.

Owing to the massive size and complexity of the SES, it remained stationary throughout each drilling season. At the end of the drilling season, the SES was decommissioned and repositioned within a virgin sector of the IceCube array that was staged for the following drilling season. The distance between the SES and TOS had a practical limit estimated to be ~450 m and determined by the pressure and voltage drops through the SES–TOS link.

The water tanks (WT1 and WT2) used shipping containers as their body frame and were sealed on the inside with welded stainless-steel paneling. The walls were filled with foam insulation. The capacity of each water tank was 38 m³. On both ends, a screened pump bay provided the first level of system filtration. WT1 was accompanied by a snow ramp that allowed for large equipment (as big as a Caterpillar 953) to dump snow directly into the tank.

Water from WT1 and WT2 was pumped using four Myers D65-16 high-pressure positive-displacement triplex piston pumps (HPP), each capable of pumping 246 L/min, to the main heating plants. Nominally, all four pumps ran at 190 L/min (760 L/min total) and a gauged pressure of 7.6 MPa. The pumps were driven by 45-kW Baldor Super-E

induction motors and were controlled by Unico vector motor drives. Water was supplied to each pump from its own submersible charge pump in WT2, maintaining the intake gauge pressure of the large pumps above 276 kPa to prevent cavitation. The pumps were equipped with pressure relief valves that vented into a heated hose leading back to WT2; however, the primary freeze protection scheme for this critical line was a throttled leak flow (7–20 L/min) bleeding from the charged intake side of each high-pressure pump.

The four main heating plants (MHP1–4) included thirty-five Whitco hot-water heaters plumbed in parallel. Each heater was capable of a power output of 125 kW at a maximum temperature of 88 °C and was powered by a fuel-fired burner unit consuming ~13 L/h. To improve the efficiency of these standard water heaters, a molded ceramic combustion liner was installed to increase temperature and improve flame quality by insulating and reflecting radiation back into the combustion region. A secondary stainless-steel condensing heat exchanger was also added. These enhancements raised the thermal efficiency from 78 to 93% and saved the project ~379 m³ of fuel. Downstream of the MHP buildings, the heated water was routed back to the HPP to combine the flows in a simple manifold, which measured the pressure, flow, and temperature and directed the flow via insulated surface hose to aramid-reinforced EPDM-rubber drilling hose with an ID of 64 mm wound onto a hose reel.

The drilling hose itself could only reliably carry a load of 7 kN when pressurized and bent over the TOS crescent.

Therefore, most of the load was carried by the 25-mm-diameter Vectran-reinforced Hytrel-jacketed power and communication cable, which had a working load of 45 kN. A tape periodically wrapped around the hose and cable during drilling (and removed during reaming) transferred the load of the hose to the cable.

The giant hose reel (DSHR) had a size of 10.28 m 5.79 m × 4.59 m and a mass of 43.5 t (the reel's components were the largest pieces ever transported to the South Pole using a C-130 Hercules transport aircraft) (Fig. 3.117). The DSHR held the full-length hose (>2500 m), which was composed of twenty two 122-m-long sections. The reel drum motors were driven by Unico vector drives with custom logic that used feedback from payout encoders and load cells to move the cable at the commanded drilling speed while maintaining hose tension at a constant ∼7 kN.

The hose was routed from the reel up to the top of the tower and over a crescent into the hole. The tower was an aluminum welded and bolted structure with a multi-wall transparent polycarbonate shelter with skis under the deck. It was towed as a unit from hole to hole. The height of the tower was limited by air-traffic requirements for drilling in close proximity to the South Pole skiway, so a half-circle crescent with a belt of shaped plastic blocks supported the hose over a 1.4-m radius. The cable was routed through 0.9-m circular sheaves located on the tower base and upper platform. Cold water from the top of the hole at a depth of 60–80 m was pumped by a 3.7-m-long 37-kW Grundfos submersible pump and returned to the SES through another surface hose. The return hose and power cable went over another crescent at the top of the tower.

The IceCube downhole drill assembly had instrumentation in the drill head to measure hole diameter, supply-water temperature, supply-water pressure, hole-water column pressure, hole-water temperature, tilt relative to the vertical, horizontal magnetic field (for measuring the rotational angle), and the load exerted by the drill head on the hose (Fig. 3.118). The diameter of the hole was measured by eight caliper springs symmetrically located around the drill body. A 24-m-long heavy stainless-steel weight stack hung beneath the drill head produced a pendulum effect to keep the drill vertical. It was also needed to allow time for heat transfer from the hot water into the ice to open the hole enough for the 0.29-m-diameter instrumented drill body to pass. The weight stack was assembled from a tapered front piece that housed the nozzle, a single 3-m section with an OD of 102 mm, and five 3-m sections with an OD of 127 mm (Fig. 3.119). A single 32-mm-diameter nozzle was used, and heat transfer limited drilling to ∼90 m/h. Out of water, the drill head by itself and with the weight stack weighed 227 and 795 kg, respectively, and the assembled weight when submerged was 522 kg.

Additional secondary subsystems maintained the water storage in the reservoirs: a preheating system (PHS) controlled the temperature and level of WT2, which was the stable reservoir, while a Rodwell system (RWS) was responsible for providing make-up water to the system and controlling the temperature and level of WT1, which was the

Fig. 3.117 In the foreground: hose reel (DSHR); in the background: tower (at the center) and blue IceCube Lab (at the left), where computers collect, archive, and sort the massive quantities of data produced by the telescope, December 2008 (*Photo* J. Haugen from IceCube (2008))

Fig. 3.118 Weight stack attached to the instrumentation section of the EHWD downhole assembly, December 2005 (*Photo* A. Karle from Antarctic Photo Library n.d.)

less stable reservoir. The RWS provided make-up water from a Rodriguez well (see Sect. 3.2.9) and was responsible for maintaining this well over the course of each drilling

season. An independent firn drill was built to allow for predrilling firn holes in parallel with deep drilling operations (see Sect. 1.1.4).

During the 2004–2005 to 2010–2011 austral summers, the EHWD was used to drill eighty-six ∼2500-m-deep vertical holes (the horizontal deviation of the borehole axis from top to bottom was less than ∼2 m) and 5160 digital optical modules were frozen into the ice. IceCube logistics has delivered ∼4.3 thousand tons to the South Pole, half of which was fuel. Typically, there were approximately 30 drillers, with a significant fraction of the drill team returning for most of the seven seasons. In addition, there were approximately 20 other IceCube personnel at the South Pole, including scientists, engineers, IT personnel, graduate students, and postdoctoral researches.

The production drilling sequence was to drill, ream, and move to the next location. Independent firn drilling stayed ahead of deep drilling by no fewer than a couple of holes, and often the Rodwell and the first few holes of the season had already been firn-drilled the prior season. Hole production rate was 48 h per hole on average, and the quickest cycle time was 32 h. Figure 3.120 summarizes the performance of the EHWD over the course of the construction of IceCube. The trends clearly illustrate early difficulties followed by continual improvements derived from equipment reliability and crew experience. The lessons learnt allowed for the peak production shown in 2009–2010 (20 holes).

As a result of its safety culture, the IceCube project had only four lost-time drilling-related safety incidents in ∼52 on-ice person-years. Some minor technical problems were related to motor demagnetizing, drive programming, and air in the fuel system (J. Cherwinka, personal communication,

Fig. 3.119 EHWD downhole assembly retrieving from the hole (*Photo* T. Gustafsson from Antarctic Photo Library n.d.)

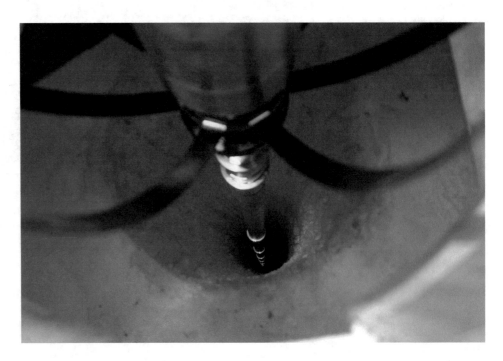

Fig. 3.120 Seasonal performance of the EHWD (Benson et al. 2014)

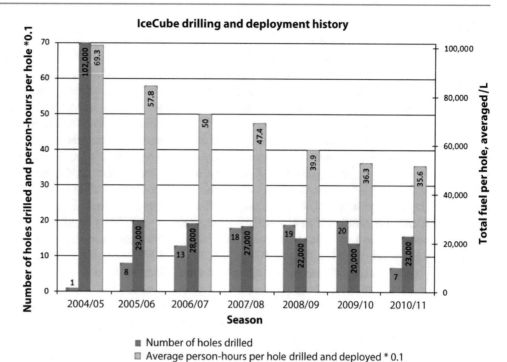

2014). One of the major problems was that hose bending under tension and pressure often caused fractures near the fittings (Fig. 3.121), even though the hose fittings had aluminum strain-relief assemblies to help carry the bending load as the fittings passed over the crescent. Recently, the EHWD was stored on Berms at the South Pole and is waiting for its next mission. However, some components were transported to McMurdo and were reused in the WISSARD project (see Sect. 3.3.10).

Fig. 3.121 Hose fracture near a fitting (*Credit* J. Cherwinka)

3.4.3 RABID Project, Rutford Ice Stream, Antarctica

Rutford Ice Stream is a fast-flowing glacier that drains part of the West Antarctic Ice Sheet into Ronne Ice Shelf. It is ~200 km long, 20–25 km wide, 2–3 km thick, and flows at approximately 300–400 m/year (Smith 2005). At the beginning of the 2000s, the integrated RABID project (the name of the project, "RABID", had no real meaning; it came from the operating name of the big fuel depot prior to the drilling operations for that project; A. Smith, personal communication, 2018), which was a collaborative initiative between the BAS and Swansea University, UK, was established with the aim to study the dynamics of Rutford Ice Stream and the basal conditions and the climatological and glacial history of the West Antarctic Ice Sheet. The primary objective of the project was to install downhole instrumentation through deep access holes and measure basal sliding, bed and ice-column deformation, basal water pressure, and ice temperature.

Two holes were proposed at two locations, where different conditions had previously been identified beneath the ice. At each location, the first hole would be relatively narrow and drilled quickly and an instrument string would be deployed in it. The first hole would also indicate the exact ice thickness, which would be particularly useful when drilling the second hole. The second hole would be wider and would be used to retrieve basal sediment, followed by the deployment of a tethered stake.

The drilling system used, modified from an intermediate-depth-scale one (see Sect. 3.3.7), was able to deliver 135 L/min of hot water at 85–95 °C, providing a heat output approaching 1 MW. The ID of the hose was increased to 1.25″. A large-diameter (1.5 m) capstan driven by an electric motor was used to carry the hose up and down the hole, and two storage drums driven hydraulically and each capable of holding up to 450 m of hose were used (Fig. 3.122). The hose was in contact with ∼¾ of the circumference of the capstan wheel, or approximately 3.5 m. This contact area generated sufficient friction to hold the hose with minimal back tension from the hose storage drums. As drilling progressed, additional 200-m lengths of hose were preheated and added to the empty hose drums. Prussik loops were fixed at each hose coupling as backup in case of coupling failure. After the main capstan wheel, the hose passed over three small rollers held in a frame below the winch before the hose entered the hole.

Other equipment included 16 oil-fired heat exchangers, four triplex ceramic plunger water pumps, four diesel generators (10 kVA each) plus small petrol generators, two 12-m³ flexible coated fabric tanks, three submersible water pumps (66 L/min each), a 50-kg brass drilling nozzle, a reamer unit made from stainless steel and aluminum, two small downhole data loggers, and the associated hoses, fittings, manifolds, etc. (Fig. 3.123). The maximum operating pressure of the system was 5.5 MPa.

In the season of 2004–2005, hot-water drilling equipment was installed near the middle of the ice stream, approximately 35 km upstream from the grounding line (78° 08.4′S, 83° 55.2′W). The drilling of the first hole was started on January 8, 2005. First, a cavity with a depth of ∼60 m was created and three submersible pumps were lowered into it to recover and recirculate the drilling water. Then, the drilling of the main hole was started. The capstan wheel was erected over a drilling pit that was ∼2 m deep and used for attaching the drilling nozzle or borehole instruments and could be accessed by ladder.

At a depth of a few hundred meters above the glacier bed, it was found that the tension in the drill hose was no longer increasing as expected, and it was decided to return the drill to the surface to see if this was due to sediment entrained within the ice. With the drill at the surface, there was no evidence of any sediment, and the drill was returned to the bottom of the hole. Unfortunately, in the following hours, the weather suddenly deteriorated very rapidly. The wind picked up from calm to over 12 m/s and snow began falling heavily. A recurrence of the unexpected behavior of the hose tension finally led to the decision to abandon the hole. The final depth reached in this hole was ∼1900 m. The drill was recovered, followed by the submersible pumps, and everything was packed away.

The likely explanation for the reduction in tension is borehole narrowing as the water jet temperature dropped to 50 °C when approaching a depth of 2000 m and the reamer located 50 m above the drill nozzle slowed down the penetration. The purpose of the reamer was to ensure that the hole drilled in the ice was of at least a certain minimum diameter. Any constriction in the hole less than this minimum caused the reamer to operate; some hot water was diverted to the restriction, thereby widening it to the correct value. While this happened, there were likely to be reductions in both the drilling rate and the hose tension.

The drilling of the second hole began on January 16, 2005. A new pit was made ∼10 m away from the first one and the capstan, hose drums, and controls were moved there. As greater depths were reached, the drill monitors again showed hose tensions lower than expected at times. In each case, the sequence of raising the drill and then lowering it again appeared to return operations to normal. The only major incident during this period happened when the drill was at a depth of 1980 m. The hose coupling at approximately 40 m below the surface failed. The prussiks held the hose and prevented it from being lost down the hole. The drill was raised, the coupling was fixed, and drilling re-commenced. The rate of penetration gradually decreased from 120 m/h at a depth of 700 m to 72 m/h at a depth of 2000 m.

When the drill reached a depth of 2025 m (∼136 m above the glacier bed), the hose tension was again lower than expected and the drill was raised for a short distance. During raising, as the hose coupling passed over the top of

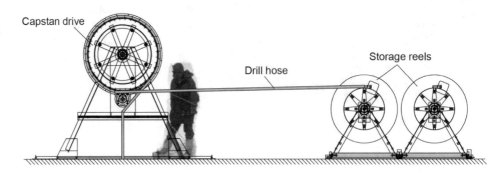

Fig. 3.122 Schematic of the hot-water drill hoisting equipment used at Rutford Ice Stream, Antarctica, 2004–2005 (Makinson 2003)

Fig. 3.123 RABID drill setup at Rutford Ice Stream, Antarctica, 2004–2005: **a** layout of burners, pumps, generators, and water tanks; **b** hose drums and capstan; **c** drilling hose feeding off the drum and onto the capstan wheel; **d** drill in operation during poor weather (Smith 2005)

the capstan wheel (the coupling that had just been replaced), it failed. Once again, the backup prussik held the hose as planned. It was decided to attach another prussik to the hose in the drilling pit. As this prussik was being attached, the first prussik failed and 1900 m of hose were lost down the hole, along with the associated drilling nozzle, reamer, and data loggers (Fig. 3.124). Close inspection of the failed prussik showed that abrasion had weakened it. The loss of the hose was accepted as final and the rest of the drilling system was decommissioned and switched off.

The hose's operating specifications were 93 °C and 2000 psi (13.8 MPa). A hose coupling had been tested with a static load of 10–13 kN at approximately 25 °C but with no water in the hose. At the time of the hose failure, the drill

was operating at ∼75 °C, 5.5 MPa, and a load of 3.5 kN. Hose failure was most likely the result of the softening of the outer sheath at 90 °C, which was up to 2 mm thick in some parts of the circumference. Moreover, some delamination of the sheath from the rest of the hose had occurred next to the couplings; both these effects caused slippage from the coupling.

A borehole video camera was lowered down the hole to locate the top of the hose. This showed that it was resting at a depth of 589 m below the surface. Then, a 300-m-long thermistor string was lowered into the upper part of the hole (Barrett et al. 2009). Once this was done, two days were spent retrieving ice cores from selected depths in a third shallow hole using a hot-water ice-coring drill based on the

(a)   **(b)**

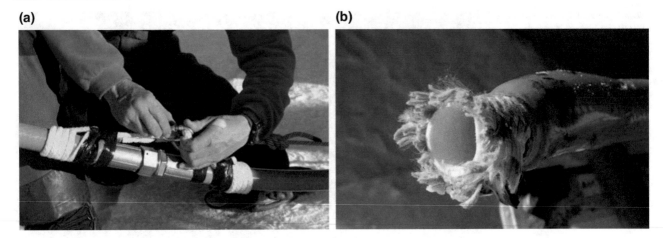

Fig. 3.124 Hose coupling of the RABID drill hose: **a** example of prussik loops used as a backup in case of coupling failure; **b** hose end after coupling failure (Smith 2005)

Caltech design (see Sect. 3.5.3). The fuel consumption of the whole system during drilling was ∼200 L/h.

3.4.4 Clean Access to Subglacial Lake Ellsworth, Antarctica

As a result of intensive remote surveying and modelling, the UK scientific community chose Subglacial Lake Ellsworth, West Antarctica as the best candidate for exploration (Siegert et al. 2012). It is 14.7 km long and up to 3.1 km wide, with a maximum water depth of 156 m. The aim of the project was to deploy purpose-built sterile probes and samplers through a borehole (>0.36 m in diameter), which would collect samples throughout the lake's water column and a sub-lake sediment core up to 3 m long. The drilling site (78° 58.0′S, 90° 34.5′W) was selected in the central part of the lake to maximize the probability of recovering a long undisturbed sedimentary record from the lake floor. At the drilling site, almost 2000 m asl, the ice sheet was 3155 m thick and the underlying lake was 146 m deep. To access the lake, a specially designed hot-water drill with clean deployment methodologies was developed to enter the lake cleanly and safely.

In November 2011, some of the heavy items (boiler, hoses, and winch containers) were transported to a designated location in Lake Ellsworth and were stored for overwintering (the hot-water drilling equipment weighed ∼45 t). In November of the following year, the lake probes and scientific equipment were delivered to the site along with engineers from the BAS and the National Oceanography Centre, who assembled the equipment in the order shown in

Figs. 3.125 and 3.126. The majority of the drilling and sampling equipment were housed in lightweight 20′ shipping containers, making transport and deployment relatively straightforward.

Most of the facilities were standard commercial equipment with minor modifications to meet environmental conditions. Power was provided by four standard generator sets running on AVTUR fuel: 1 × 20-kVA generator (GEN1) to run the camp and 3 × 100-kVA generators (GEN2–4), which had a combined output capable of sustaining the drilling operations. The drilling strategy included drilling of two closely spaced holes (∼2 m apart) up to ∼300 m deep and interconnected by a cavity to enable drilling water recirculation (Fig. 3.127); a hydrological head 284 m below the ice surface was required to equilibrate the borehole and lake water pressures to ensure minimal exchange between the two upon the moment of access.

The hot-water drilling equipment was centered on a single 1.5-MW oil-fired boiler (Fig. 3.128). The boiler burnt 250 L/h of AVTUR fuel and required an auxiliary 108-kW electric heater. Even though the boiler had been tested in the UK, several painful incidents happened. The boiler controller failed almost immediately upon starting the boiler. Soon afterwards, the spare control board also failed. Various valves cracked, and the pressurization system did not work properly owing to faulty gauges. In addition, the air/fuel mix was difficult to balance, resulting in smoke and reduced heat output. It appeared that the boiler was not properly conditioned to face the cold environment and high altitude. A third board was ordered from the suppliers and it took 15 days for it to arrive at the site. After the boiler was repaired, drilling was resumed.

Fig. 3.125 Schematic layout of the field camp (Siegert et al. 2014)

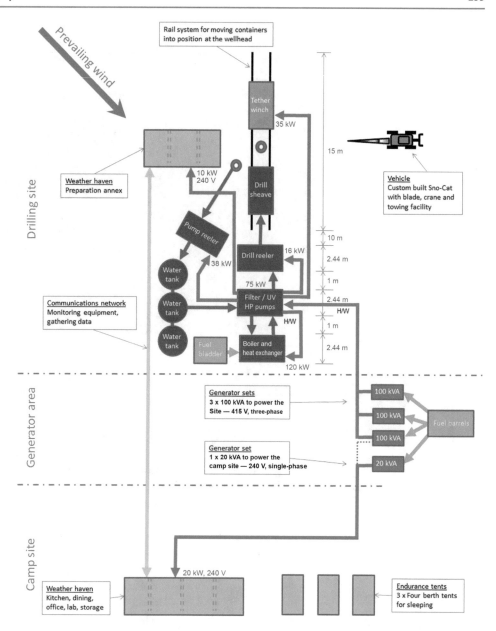

The initial water was created by melting snow into three 30-m³ interconnected flexible storage tanks sized 5.5 m 1.5 m. The central tank contained a heating coil fed from the primary circuit of the boiler. The water tanks were satisfactory, but filling them manually was hard work (Fig. 3.129) and the uneven surface meant that they tilted to one side, obstructing some of the valves. During idle time, it was necessary to keep 60-m³ of water warm, which costed fuel.

First, a 340-m umbilical hose (Fig. 3.130) was used to create a 300-m-deep hole, at which point a reservoir of water was created. Hot water was pumped at a rate of 180–226 L/min by a bank of four (three on duty and one in stand-by) Cat high-pressure triplex plunger pumps connected in parallel; any of the three could be used to reach the maximal required pressure of 13.8 MPa (Anonymous author 2012). A flush pump was used to cool the pumps using an ethanol-based coolant to keep the drill water sterile. Pumps were electrically driven with an inverter control.

A return hot-water drill nozzle was incorporated along with a stainless-steel Caprari 33-stage 6″ submersible pump. The drill head was added below the pump so that it could drill its own hole and cavity (Fig. 3.131). The pump was driven by a 22-kW AC motor with a VSD controller. Water recovery was rated up to ~235 L/min from a depth of 300 m. To allow for faster drilling to the cavity's level, an additional lance was attached to the head's bottom. The umbilical system drilled 60 m, but the lance was bent in the process (Fig. 3.132). After changing it with the spare one, this happened once again; and the lance was then removed.

Fig. 3.126 Aerial view of the Lake Ellsworth field camp, December 2012 (*Credit* P. Bucktrout from Siegert et al. 2014)

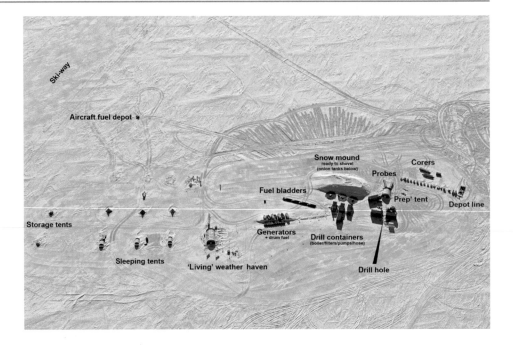

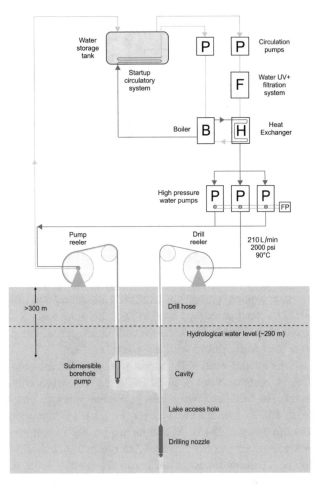

Fig. 3.127 Schematic diagram showing the setup of the Lake Ellsworth hot-water drilling system: heaters (H); pumps (P); filters (F); Boiler (B) (Siegert et al. 2014)

The reasons of this problem were lack of good winch-speed control at slower drilling speeds (hydraulic drive limitations) and damage to the winch load cells, which made it difficult to operate the drill. Once below 70 m, which was the water-impermeable depth, the pump on the umbilical was activated to send water back to the surface and avoid further loss of water to the ice sheet.

After the return hole was established, the main hole was started. An auger was used to drill down to 4–5 m and then the drill stem attached to the main hose was set down. A single-piece 1¼″, 3400-m-long polyurethane-jacket hose, which was self-supporting in air-filled holes (60-kN integral strength member), was wound onto the hydraulically controlled reel. The reel had local and remote controls with the ability to adjust the feeding speed in the range of 0–2 m/min when drilling and 0–12 m/min when hauling. It was estimated that temperature of the water would decrease from 90 °C at the surface to 55 °C at 2200 m and to 40 °C at the lake. The brass drill stem had a single forward-facing nozzle with an adjustable annulus curtain spray. The total weight of the nozzle assembly was ∼200 kg.

Cleanliness was assured via the following series of measures. First, all the water used to drill the holes was passed through a four-stage filtration system (20, 5, 1, and 0.1 μm) to remove particles (each stage housed seven polypropylene filters) and was radiated with strong ultraviolet (UV) light (254 nm, 200 W) to kill any remaining microorganisms. Secondly, all the hoses were wiped with ethanol prior to their descent into the ice sheet. Thirdly, the main hole was air-locked, and the upper air-filled component was radiated with UV light through a specially designed

Fig. 3.128 Containerized oil-fired boiler (*Credit* D. Blake)

Fig. 3.129 Addition of snow into the water tank via shoveling (*Credit* D. Blake)

probe. Fourthly, all the probes and corers were built in ultra-sterile conditions and transported to the ice sheet in robust bags. Fifthly, a deployment system was configured using a series of airlocks created from glove hatches and seals and it was cleaned with hydrogen peroxide vapor to ensure that the probes could be placed within the clean hole and extracted from it without exposure to the atmosphere.

The main and return holes had to be linked to the sub-surface cavity before further drilling into the lake. This was necessary to recirculate water to the surface and to regulate the water level of the main borehole prior to lake entry. At a depth of 300 m, two cavities were created in the return hole with an estimated volume more than 100 m^3 after $\sim$40 h of reaming and in the main hole with estimated volume at least 70 m^3 after $\sim$20 h of reaming. Unexpectedly, the linking of the two subsurface cavities failed and the umbilical hose became frozen in the ice, making it impossible for the borehole pump to recover water from the subsurface reservoir.

After a few days of attempting to solve all the problems, drilling was abandoned on December 25, 2012. The cause of the failure to link the main and return boreholes was most likely that one or both holes were not vertical as a result of the problems encountered with the winch controls and drill sensors (Schiermeier 2014). Even if the link between two holes would have been established higher, the stocked fuel was not enough to continue to the lake surface. Consequently, the umbilical hose was cut near the borehole mouth, and the full deep-drilling capabilities of this 3.4-km hot-water drilling system and sampling equipment remain untested.

3.4.5 BEAMISH Project, Rutford Ice Stream, Antarctica

The project for measuring basal conditions on Rutford Ice Stream, called the BEd Access, Monitoring and Ice Sheet History (BEAMISH) (2013–2019), is led by the BAS and has almost the same goals and approaches as the RABID project (see Sect. 3.4.3). The overall goal of BEAMISH is to find out how Rutford Ice Stream changed in the past and how it is flowing today (Smith and Fothergill 2016; K. Makinson, personal communication, 2016).

Readily available industrial equipment was used to build the drilling system (Fig. 3.133): six generators, four boilers, four CAT pumps, two submersible pumps, hoses with reels, and spares. Local snow is initially melted and the water is stored in a number of large surface storage tanks ($\sim$10 + 10 + 30 m^3). Water is then pumped at a rate of $\sim$160 L/min and a maximal pressure of 6.7 MPa and heated to a maximal temperature of 90 °C via four water heaters with a total heating capacity of $\sim$1 MW. Secondary heaters are used to provide heat to the submersible borehole pumps. Water is then pumped through a single-length 2300-m 1¼″ drill hose with a Hytrel outer cover to a nozzle. The reel has mechanical level wind and is driven by a three-phase AC motor controlled by an inverter in either torque mode or drive mode. While drilling, the capstan drive feeds the hose at rates in the range between 45 and 120 m/h (Fig. 3.134).

The drill water from the nozzle and the subsequent meltwater returns towards the surface via two return boreholes. A water recirculation cavity was created below

Fig. 3.130 Umbilical hose feeding into the return hole (*Photo* P. Bucktrout from Kaufman 2012)

Fig. 3.131 Hot-water drill nozzle incorporated with a submersible pump (*Credit* D. Blake)

the local hydrological level (200–275 m below the ice surface, between the ice floatation pressure or the sea level pressure). Two submersible borehole pumps (~ 200 L/min from a depth of 275 m) were installed in the cavity and returned water to the surface storage tanks. These tanks were maintained at several degrees above the freezing point using the heat recovered from the water heater exhausts.

The electrical generators used were standard off-the-shelf three-phase 15-kVA units housed in acoustic cabinets and fueled by unleaded petrol. There were seven generator units; each unit used up to 7 L/h, with a fuel consumption of ~ 1.8 m^3 per hole. In total, ~ 7.5 m^3 of AVTUR fuel were used per hole including the heating of the drill water.

The work on Rutford Ice Stream in the 2018–2019 season had mixed results in terms of instrumentation deployments though there were still many successes and a total of three access holes ~ 0.3-m-wide were successfully drilled to the bed (the desire was to drill a total of four holes at two sites) (Fig. 3.135) (K. Makinson, personal communication, 2019). Two first boreholes were drilled in the vicinity of the main

study location (78° 08.383′S, 83° 54.901′W), and the second drilling site was located 2 km apart. The first hole was finished at the depth of 2152 m below the surface on 8 January, 2019 after 63 h of continuous round-the-clock drilling operation. Sediments were collected both by the drill head collectors and sediment corer.

3.4.6   Chinese Hot-Water Deep Drilling Project at Amery Ice Shelf

The hot-water deep-drilling project at Amery Ice Shelf employs a multidisciplinary approach to the study of the region. It includes drilling access holes using a hot-water drilling system, ice sampling of the most interesting intervals using a hot-water corer, oceanographic observations, sampling of sediments beneath the ice shelf, and real-time video camera recordings of the boreholes and ocean cavities. Boreholes with a minimal diameter of 0.4 m should be kept open for three or four days for making observations. It has been planned to drill 10–12 holes from the edge of the ice shelf to the grounding zone up to a depth of 2100 m at

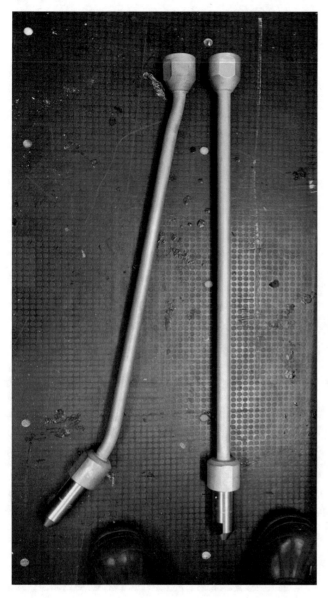

Fig. 3.132 On the left: drill lance with a nozzle that was bent while drilling a return hole via the umbilical hose; on the right: unwrapped drill lance for comparison (*Credit* D. Blake)

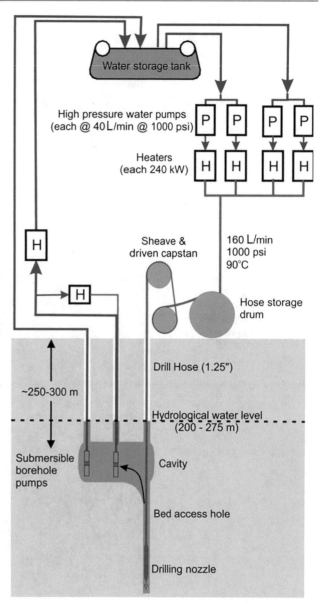

Fig. 3.133 Schematic of the BEAMISH hot-water drilling system (Smith and Fothergill 2016)

intervals of nearly 50 km. All equipment will be transported from site to site using tracked vehicles.

The general concept of their hot-water drilling system requires a supplementary borehole to be drilled to enable water to be returned to the surface. This hot-water drilling system is composed of three standard air-cooled generators, four oil-fired boilers (three in use and one as a spare) with an individual heat power of 0.7 MW, four high-pressure Cat 3521C pumps with a rated flow of 87 L/min each, fuel tanks, a water tank, reels with hoses, winches with electrome-chanical cables, high-pressure pumps, and a central control building (Fig. 3.136). All the equipment is installed in 20′ standard shipping containers.

In order to avoid water loss in the upper snow–firn layer, an independent firn electrothermal drill was built for pre-drilling two holes ~1 m apart down to a depth of nearly 50 m (see Sect. 1.2.32). The first hole will be then drilled using the main hot-water drilling system down to a depth of 150–200 m (slightly deeper than the hydrological water sea level at the point of the drill site), where a reservoir of water would be created. The maximum depth of the water supply return hole is assigned to be 300 m. Upon creation of the subsurface reservoir, a second (main) hole will continue to be drilled through the ice shelf. The main-hole drill nozzle is equipped with sensors to measure borehole diameter, pressure, and temperature.

Fig. 3.134 BEAMISH hoisting system (*Credit* K. Makinson)

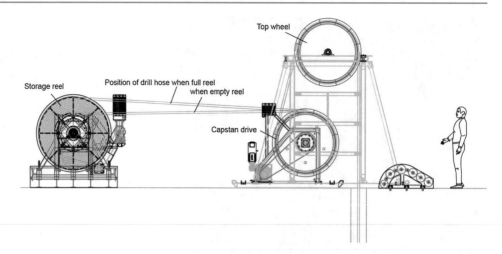

Fig. 3.135 BEAMISH hot-water drilling setup on Rutford Ice Stream in West Antarctica, January 2019 (Scientists drill to record depths in West Antarctica 2019)

The main reel has an electromechanical drive system and holds a 2200-m-long single-length hose (Fig. 3.137). The maximum hose lifting/lowering rate at full winding on the drum is 30 m/min, and drilling rate can be controlled in the range of 0–120 m/h. The main hose (ID/OD: 38 mm/60.5 mm) with a thermoelastoplast outer jacket has incorporated 8 mm × 1 mm electrical signal lines for communication with the downhole drill nozzle. The general concept of this hot-water drilling system was tested at an experimental site of Jilin University in Changchun. Four 9-m-deep holes were drilled in an ice well at an average penetration rate of ∼38 m/h. The diameter of the holes measured after drilling varied in the range of 0.40–0.46 m. Drilling at Amery Ice Shelf is planned as soon as full financial and logistical support is obtained for the project.

3.4.7   Summary

Deep hot-water drills, with their inherent speed, offer the only feasible alternative for rapid accessing to subglacial environments through the thick ice of the Antarctic and Greenland by using and recycling multiple times the water melted from the surrounding ice sheet. To ensure clean access to subglacial lakes, the drilling water can be ultra-filtered, UV-treated, and pasteurized before being used to melt the access hole. Thus, it can contain significantly less microbial and particulate content than the surrounding ice.

Nevertheless, deep hot-water drilling and clean accessing the base of deep ice sheets involves considerable technical and logistical challenges, especially when water is present at the bed. Unlike electromechanical drills, for which certain

Fig. 3.136 Schematic of the Chinese deep hot-water drilling system

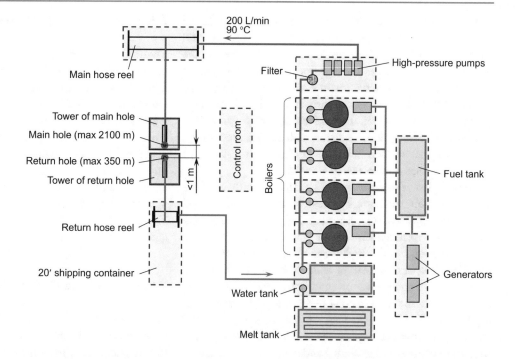

Fig. 3.137 Main reel of the Chinese deep hot-water drilling system

methods to recover stuck drills are known (Talalay et al. 2015), the recovery of the hose used in hot-water drilling system is not possible. Even if a system capable of attaching to the failed end of the hose down the hole is available, the drill nozzle at the bottom of the hole will have frozen in the ice very quickly (presumably in <10 min), after which it would never be freed.

Precise reeler control and an accurate level wind on and off the reels are essential for efficient deep hot-water drilling (Makinson et al. 2016). Reelers can be powered either hydraulically or electrically and, although hydraulic systems can be used to prevent problems with electrical noise, some implementations have suffered from problems with large temperature fluctuations adversely affecting fluid viscosity, causing low torque at slow speeds and poor positional accuracy. However, these features can be overcome with an appropriate design. Currently, most deep hot-water drilling systems use electrical variable-frequency drives. To increase the precision of the winching process, a capstan or linear traction drive could be installed together with an accurate level wind mechanism.

Reliable drill instrumentation, communication, and monitoring systems are essential for safe and successful deep hot-water drilling. Key parameters critical to the drilling process, such as water temperature, flow rate, drill speed, depth, load, and hole water level must be reliably measured and have built-in redundancy, including simple mechanical gauges as well as digital sensors with local readouts and adequate spares. Downhole instrumentation just behind the drill nozzle is sometimes used on hot-water drilling systems to measure hole water temperature and pressure above the nozzle, drill water temperature arriving at the nozzle, hole inclination and diameter. In such cases, drillers have full information, which, in conjunction with surface monitoring, adds confidence to their ability of drilling a vertical hole of the required diameter as fast as possible.

Various issues encountered in the field require solutions to prevent their reoccurrence. Besides the ones mentioned above, these involve: (1) drill hose coupling failure, which requires the development of an improved hose/coupling system or using single-length thermoplastic hose; (2) using a multi-boiler system instead of a single boiler; (3) mainland testing of all equipment in conditions that most closely resembling those found in polar regions; and (4) the training of drilling teams and drawing up recommendations and drilling protocols that explain the correct conduct and procedures to be followed in normal and abnormal (to the extent possible) situations.

The need for deep drilling and the recovery of bottom samples requires a high heat input and an insulated hose to allow for drilling beyond 1500 m. Both are necessary to prevent freeze-up within the hole and to preserve heat within the hose for drilling. The maximum thermal power that can be delivered to the drill nozzle depends on the ID and length of the hose (Makinson et al. 2016). In overcoming the frictional losses in the hose that increase with the square of flow velocity, the practical pressure limit of 15 MPa for standard commercial equipment is soon reached. For example, for the same thermal power system, increasing the ID of the hose from 32 mm to 38 mm would decrease the operating pressure, pumping power, and electrical power generation requirements by ∼60%. However, the larger size and increased weight of the continuous drill hose, the reeler, and the supporting infrastructure would be increased by at least 50%, making it incompatible with available logistics. Thus, an optimal balance must be determined.

3.5   Coring Ice Drills with an Intermediate Heat-Transfer Agent

These drills could be used for recovering of short ice cores from the surface of glaciers. Another application of coring hot-water ice drills is the sampling of holes with a desirable depth that were drilled with full-diameter nozzles. In such cases, the drill stem has to be pulled up to the surface and changed to the coring device. Optionally, hot water can be supplied by the same hydraulic hose that is used for the hot-water drilling of the borehole.

3.5.1   Browning Hot-Water Coring Drill

Two different techniques for coring using hot water were investigated together using the hot-water drill of Thayer School of Engineering at Dartmouth, USA (see Sect. 3.3.1). During the 1978–1979 season, several coring technologies were tried (Browning et al. 1979). The simplest technique, which was found unsatisfactory, used an annular "cookie cutter" drill to free a cylinder of ice at the bottom of the hole; this cylinder would float to the surface for retrieval. However, several ice cores were easily obtained with a 460-mm-diameter annular drill attached to the bottom of a 2.4-m-long cylindrical container. The recovered cores had a diameter of ∼300 mm.

3.5.2   IGSFE Ice-Coring Device

A shallow-depth ice-coring device was designed and tested in the IGSFE, Irkutsk, USSR at the beginning of the 1980s to take short cores from the surface of glaciers (Fig. 3.138) (Kravchenko 1984). A hollow ring was fixed to the drill stem of the portable hot-water drill (see Sect. 3.1.3). The following sizes of the cutting ring ensured the optimal heat transfer from hot water to ice: $D_R = 28.9d^{1.43}$; $d' = 0.08d$; $D \geq 3d$ (designations are shown in Fig. 3.138). Drilling was performed under the own weight of the cutting ring and stem, and thus the driller had to control only the verticality of the stem. It was necessary to cut the core every 1 m by turning the stem and then removing it from the hole. Drilling was continued to a depth equal to the available length of the stem.

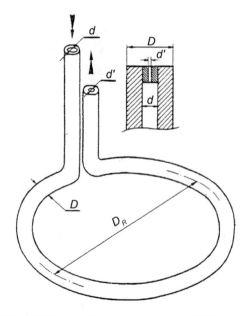

Fig. 3.138 IGSFE ice-cutting ring (Kravchenko 1984)

3.5.3 Caltech Hot-Water Ice-Coring Drill and Its Modifications

In the early 1990s, Caltech designed a hot-water ice-coring drill to use it in combination with a full-scaled hot-water drilling system in operations in which complete ice cores were not required (Engelhardt et al. 2000). The drill consisted of a cutting drill head, four thermally insulated 6-mm stainless-steel tubes along the outside of a Lexan core barrel, and a back-drilling head (Fig. 3.139). The cutting drill head had 40 jets, each 1 mm in diameter, on an annulus with a diameter of 108 mm (Fig. 3.140). The axes of the jets were tilted outward at angle of 12° from the vertical axis of the drill. Two core catchers were built into the cutting head.

The backdrilling head could redirect the flow of hot water from a downward to an upward direction in case the borehole got blocked by refreezing, and it consisted of two parts that could slide against each other. During normal drilling operations, the outer part was held in the upward position by a spring (Fig. 3.141a). If during pull-up, however, the outer part was held by an obstruction in the borehole, the inner part that was attached to the high-pressure hose was pulled up against the spring until it came to a definite stop (Fig. 3.141b). In this position, the water passage to the drilling head was blocked, but it was redirected through a pattern of 36 jets against the obstructing ice above the corer.

Hot water was supplied by the same high-pressure drilling hose with an ID of 19 mm that was used for the hot-water drilling of the borehole. Two corers were built for retrieving ice cores 70 mm and 94 mm in diameter. The core-barrels were 2 and 4 m long. The hot-water ice-coring drill was tested for the first time in the 1992–1993 Antarctic drilling season. Although the heaters and pumps could supply up to 65 L/min of water at 95 °C at the surface, the corer worked better at a lower flow rate of 40 L/min.

In the 1993–1994 season, the hot-water ice-coring drill was successfully used to retrieve 18 ice cores up to 2 m long and ~70 mm in diameter. The ice cores were drilled at depths of 300 and 950 m in the heavily crevassed chaotic shear zone between Whillans Ice Stream and the Unicorn (83° 37′S, 138° 03′W). The average coring rate was 48 m/h.

During the 1997–1998 season, a series of hot-water-drilled ice cores, vertically spaced 100 m apart, were drilled from the top to the bottom of the ice sheet at three separate sites on Siple Dome (Gow and Engelhardt 2000). The deepest core was retrieved from within 15 m of the ice/rock interface, located 1004 m beneath the surface of the dome, and contained some rock fragments.

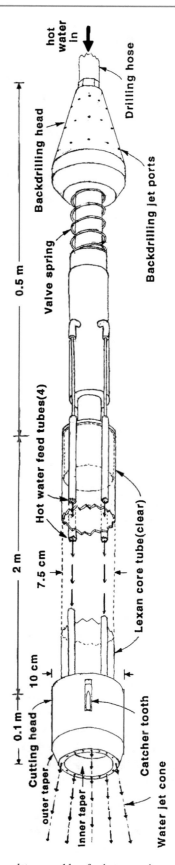

Fig. 3.139 Complete assembly of a hot-water ice corer (Engelhardt et al. 2000)

Fig. 3.140 Cutting drill head: **a** schematic drawing (Engelhardt et al. 2000); **b** during testing (*Credit* K. Makinson)

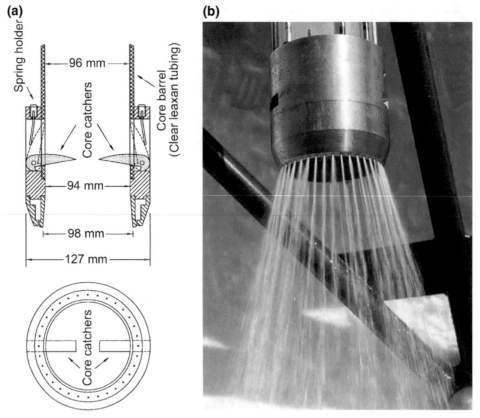

To prevent the melting of the core during its transit to the surface (∼ 1 h was needed to raise the core from a depth of 1000 m), the water in the hole was cooled to ∼ 5 °C prior to coring. The Siple Dome cores were 90–100 mm in diameter and 3–4 m long.

During the 2000–2001 season, several cores were retrieved from boreholes at Kamb Ice Stream, including a short section of clear basal ice with a single layer of dispersed sediment clots (Fig. 3.142). However, attempts to recover ice with higher sediment content covering the subsequent portion of the 10-m thick basal ice layer were unsuccessful (Vogel 2009). After detailed inspection of the ice corer, it was discovered that the core catcher had jammed.

Over the last two decades, a significant number of cores have been drilled using different versions of the Caltech ice-coring drill in Iceland (Gaidos et al. 2004) and at several Antarctic sites, including Ronne Ice Shelf (Nicholls et al. 2012), Rutford Ice Stream (Fig. 3.143) (Smith 2005), Amery Ice Shelf (Craven et al. 2005, 2009; Treverrow et al. 2010), and Ross Ice Shelf (Fig. 3.144) (Rack 2016). Although some differences exist in the coring drill head and the core barrel among these modified versions, the drills' design has remained largely faithful to the prototype design. The WISSARD hot-water drilling system had a separate reel with an additional 1000-m-long continuous hose with an ID of 19.1 mm to be ultimately used with a hot-water ice corer (Rack et al. 2014).

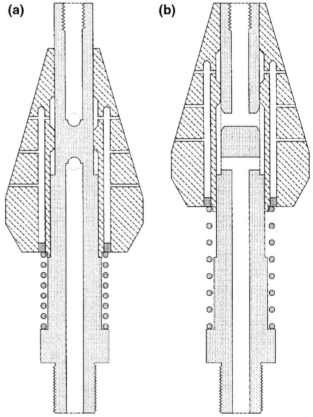

Fig. 3.141 Design of the backdrilling head in two plane views perpendicular to each other: **a** coring position with the spring fully extended; **b** backdrilling position with the spring fully compressed (Engelhardt et al. 2000)

Fig. 3.142 Ice core retrieved from the top of the basal ice layer at Kamb Ice Stream showing highly dispersed, thin, banded debris layers (Vogel 2009)

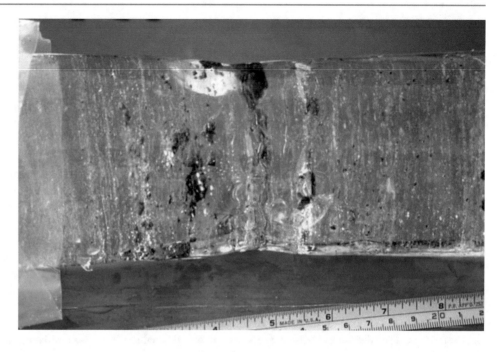

Fig. 3.143 Hot-water ice-coring drill at Rutford Ice Stream, Antarctica, in the 2004–2005 season (*Photo* A. Taylor, British Antarctic Survey)

Fig. 3.144 Hot-water ice-coring drill at the site near the grounding zone of Ross Ice Shelf, January 2015 (*Credit* F. Rack)

3.5.4   SPSMI Hydrothermal Ice-Core Drill

The hydrothermal method looks similar to hot-water drilling methods, but hydrophobic drilling fluid is used instead of the water as the circulation medium. Kudryashov and Menshikov (1994) from the St. Petersburg State Mining Institute (SPSMI) proposed a hydrothermal cable-suspended thermal drill in which a high-power thermal energy source is placed in a thermally isolated long boiler above a core barrel. The liquid (for example, kerosene-based drilling fluid) that fills the borehole is heated in the boiler and is used a source of heat to melt the hole down. Thermally isolated pipes fixed on the surface of the core barrel carry this fluid to a specially shaped ring-shaped drill head (Fig. 3.145), which forms the hole while leaving an ice core in the center. The hydrodynamical action of the jet promotes the efficient melting of ice.

The hot fluid is supplied to the bottom of the hole under the action of the decompression created by a pump located in the water tank. A mixture of melted water and post-working fluid is lifted to the water tank, where the working fluid and water are separated by the difference of densities. After separation, the pure fluid returns to the boiler. Lab tests were carried out with a drill head having the following sizes: $d_1 = 80$ mm, $d_2 = 76$ mm, $D_1 = 110$ mm, $D_2 = 114$ mm, and $h = 15$ mm (designations are shown in Fig. 3.145). These tests showed that penetration rate strongly depends on the flow rate of the hot fluid. At a flow rate of 2.54 L/min and an initial liquid temperature of 97 °C, a penetration rate of 2.55 m/h was achieved. In this case, diameter of the borehole was 118–120 mm, and the diameter of the core was 68 mm.

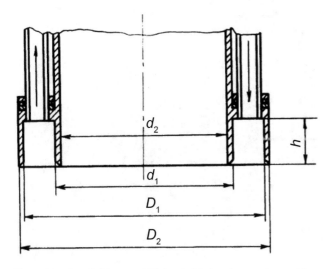

Fig. 3.145 Hot-fluid core drill head (Kudryashov and Menshikov 1994)

3.5.5   Summary

Using hot-water ice corers is effective for recovering ice samples from targeted depths in combination with a hot-water drilling system. They are light (weighing only 30 kg when empty), fast, and sufficiently portable to be used in remote locations and on traverses. In conjunction with a hot-water drill that can provide deep access holes quickly, they can provide a series of ice cores for site-selection studies and for the determination of all the ice properties when complete cores are not essential. Because borehole drilling and ice coring is done with hot water only, contamination with the other fluids used for mechanical drilling cannot occur and ice-cutting chips are not produced, thus simplifying the core-handling procedures.

The major drawback of hot-water ice coring is the varying quality of the recovered cores. For example, the nominal core diameter of the Australian corer was 100 mm, but the diameter of the recovered cores at Amery Ice Shelf was sometimes reduced to ~60 mm in some sections of the cores in which there had been a mismatch between the vertical velocity of the coring attachment and the water temperature of the drilling system (Treverrow et al. 2010). At that time, cores varied from well-formed cylindrical samples over 1 m long to samples 0.2 m in length with tapered ends. Smith (2005) pointed out that once a section of core has been taken, the next few meters of ice below it will have been too disturbed to be retrievable. This next ~1 m needs to be drilled away before deeper ice can be successfully retrieved.

Although hot-water coring systems had demonstrated some advantages over their mechanical counterparts, they are not considered as a real alternative for continuous coring from the point of view of core quality in the first place, but also regarding logistics, cost, efficiency, and power consumption.

With hydrothermal ice corers, one of the main problems of hot-water drilling systems, which is the refreezing of the water in the borehole, can be avoid. Such drilling programs should include transportation of the drilling fluid to the site (for drilling to depths of 3000–4000 m, this could amount to 50–60 m^3 depending on the diameter of the hole and fluid losses during drilling), which could compensate the weight of the heavy generators and fuel required to melt snow and heat the water before dumping it into the hole.

References

Ágústsdóttir AM, Brantley SL (1994) Volatile fluxes integrated over four centuries at Grímsvötn volcano, Iceland. J Geophys Res 99:9505–9522

AMANDA (2000) University of Wisconsin-Madison News. Retrieved 8 Oct 2018 from https://news.wisc.edu/newsphotos/amanda.html

Andres E et al (2000) The AMANDA neutrino telescope: Principle of operation and first results. Astroparticle Phys 13:1–20

Anonymous author (2012) Hot water drilling in Antarctic ice. World Pumps 2012(6):32–35

Antarctic Album (2013) Drilling into Subglacial Lake Whillans. Posted Live Science staff on 27 Jan 2013. Retrieved 8 Oct 2018 from https://www.livescience.com/26623-antarctica-lake-whillans-drilling.html#

Antarctic Glaciology (n.d.) Antarctic Research Center, Victoria University of Wellington. Retrieved 20 Feb 2018 from https://www.victoria.ac.nz/antarctic/outreach/image-gallery/antarctic-glaciology

Antarctic Photo Library (n.d.) US National Science Foundation. Retrieved 8 Oct 2018 from https://photolibrary.usap.gov/#26-2 and https://photolibrary.usap.gov/#5-1

Antarctica New Zealand Pictorial Collection (n.d.) Retrieved 25 Feb 2018 from http://antarctica.recollect.co.nz/nodes/view/46440 and http://antarctica.recollect.co.nz/nodes/view/16069

ARA Collaboration (2012) Design and initial performance of the Askaryan Radio Array prototype EeV neutrino detector at the South Pole. Astropart Phys 35(7):457–477

Ashmore DW, Hubbard B, Luckman A et al (2017) Ice and firn heterogeneity within Larsen C Ice Shelf from borehole optical televiewing. J Geophys Res Earth Surf 122:1139–1153

Astaf'ev VN, Surkov GA, Truskov PA (1997) Torosy i stamukhi Okhotskogo morya [Ice ridges and stamukhas of Sea of Okhotsk]. St.-Petersburg, Progress-Pogoda (in Russian)

Barrett BE, Nicholls KW, Murray T et al (2009) Rapid recent warming on Rutford Ice Stream, West Antarctica, from borehole thermometry. Geophys Res Lett 36:L02708

Barrett PJ, Carter L, Damiani D et al (2005) Oceanography and sedimentation beneath the Mcmurdo Ice Shelf in Windless Bight, Antarctica. Antarctic Data Series No. 25, Antarctic Research Centre, Victoria University of Wellington, New Zealand

Bässler K-H, Miller H (1989) Evaluation of hot water drills//Ice Core Drilling. In: Rado C, Beaudoing D (eds) Proceedings of the third international workshop on ice drilling technology, Grenoble, France, 10–14 Oct 1988. Laboratoire de Glaciologie et Geophysique de l'Environnement, Grenoble, pp 116–122

Benson T, Cherwinka J, Duvernois M et al (2014) IceCube enhanced hot water drill functional description. Ann Glaciol 55(68):105–114

Bentley CR, Koci BR (2007) Drilling to the beds of the Greenland and Antarctic ice sheets: a review. Ann Glaciol 47:1–9

Bentley CR, Koci BR, Augustin LJ-M et al. (2009) Ice drilling and coring. In: Bar-Cohen Y, Zacny K (eds) Drilling in extreme environments. penetration and sampling on earth and other planets. Wiley-VCH Verlag GmbH & Co., KGaA, Weinheim, pp 221–308

Beverly C (1982) ADOM (Air Deployed Oceanographic Mooring). ADM (Advanced Development Model) thermal ice drill. Test results. Marine Systems Engineering Laboratory, University of New Hampshire, Sept 1982. Project NR-294–063

Beverly C (1983) Autonomous computer controlled ice drill performance tests. In: Proceedings of IEEE OCEANS'83 Conference, 29 Aug–1 Sept 1983, San Francisco, USA, pp 1057–1059

Beverly CN, Westneat AS (1982a) Autonomous computer controlled ice drill. Marine Systems Engineering Laboratory, University of New Hampshire, Durham, USA, Sept 1982. Report on Project NR-294-063

Beverly C, Westneat A (1982b) Autonomous computer controlled ice drill. In: Proceedings of IEEE OCEANS'82 Conference, 20–22 Sept 1982, Washington, DC, USA, pp 1261–1264

Bindschadler RA, Koci B, Iken A (1988) Drilling on crary ice rise, Antarctica. Antarct J US 23(5):60–62

Björnsson H (1991) Skýrsla um starfsemi Jöklarannsóknafélags Íslands 1990 [Report on the activities of the Icelandic Glaciological Society 1990]. Jökull 41:105–108

Blake EW, Clarke GKC, Gérin MC (1992) Tools for examining subglacial bed deformation. J Glaciol 38(130):388–396

Blatter H (1987) On the thermal regime of an Arctic valley glacier: a study of White Glacier, Axel Heiberg Island, N.W.T., Canada. J Glaciol 33(114):200–211

Blythe DS, Duling DV, Gibson DE (2014) Developing a hot-water drill system for the WISSARD project: 2. In situ water production. Ann Glaciol 55(68):298–302

Boller WL, Sonderup JM (1988). Hot-water drilling on the Siple Coast. Antarct J US 23(5):62–63

Bolsey RJ (n.d.) Hot water ice drilling. Retrieved on 12 Mar 2018 from http://www.deepicedesign.com/hot-water-ice-drilling/

Boulton GS, Dobbie KE, Zatsepin S (2001) Sediment deformation beneath glaciers and its coupling to the subglacial hydraulic system. Quatern Int 86:3–28

Browning JA, Bigl RA, Sommerville DA (1979) Hot-water drilling and coring at site J-9, Ross Ice Shelf. Antarct J US 14(5):60–61

Burnett J, Rack FR, Blythe D et al (2014) Developing a hot-water drill system for the WISSARD project: 3. Instrumentation and control systems. Ann Glaciol 55(68):303–310

Campbell D (2013) Warming ocean thawing Antarctic glacier, researchers say. UAF News and information. Posted on 18 Sept 2013. Retrieved 18 Sept 2018 from https://news.uaf.edu/warming-ocean-thawing-antarctic-glacier-researchers-say/

Chandler D, Hubbard B, Hubbard A et al (2008) Optimising ice flow law parameters using borehole deformation measurements and numerical modelling. Geoph Res Lett 35:L12502

Choi CQ (2010) Glaciers may have soggier bottoms than thought. Live science, Planet Earth. Posted on 29 Sept 2010. Retrieved 6 Mar 2018 from https://www.livescience.com/10139-glaciers-soggier-bottoms-thought.html

Clarke GKC, Blake EW (1991) Geometric and thermal evolution of a surge-type glacier in its quiescent state—Trapridge Glacier, Yukon Territory, Canada, 1969–89. J Glaciol 37(125):158–169

Clarke GKC, Collins SG, Thompson DE (1984) Flow, thermal structure and subglacial conditions of a surge-type glacier. Can J Earth Sci 21:232–240

Clarke R, Corr H, Nicholls K (2016) Initial environmental evaluation science projects on the Filchner-Ronne Ice Shelf, Antarctica. British Antarctic Survey, Cambridge, UK and Alfred Wegener Institute, Bremerhaven, Germany

Climate change scientists from Aber are heading to the Himalayas to study the world's highest glacier (2017) Aber Times. Posted on 19 Apr 2017. Retrieved 18 Feb 2018 from http://www.abertimes.co.uk/climate-change-scientists-aber-heading-himalayas-study-worlds-highest-glacier/

Copland L, Harbor J, Sharp M (1997) Borehole video observation of englacial and basal ice conditions in a temperate valley glacier. Ann Glaciol 24:277–282

Craven M, Elcheikh A, Brand R et al (2002a) Hot water drilling on the Amery Ice Shelf—the AMISOR project. Mem Natl Inst Polar Res Spec Issue 56:217–225

Craven M, Elcheikh A, Brand R et al (2002b) Hot water drilling on the Amery Ice Shelf, East Antarctica. In: Smedsrud LH (ed) Proceedings of 16th Forum for Research into Ice Shelf Processes (FRISP), 25–26 June 2002, Bergen Geophysical Institute, University of Bergen, Norway, Report No. 14 (Report Series R27)

Craven M, Allison I, Brand R et al (2004) Initial borehole results from the Amery Ice Shelf hot-water drilling project. Ann Glaciol 39:531–539

Craven M, Carsey F, Behar A et al (2005) Borehole imagery of meteoric and marine ice layers in the Amery Ice Shelf, East Antarctica. J Glaciol 51(172):75–84

Craven M, Allison I, Fricker HA et al (2009) Properties of a marine ice layer under the Amery Ice Shelf, East Antarctica. J Glaciol 55(192):717–728

Davis M (2013) The Greenland Ice Sheet and surface meltwater. Posted on 22 Feb 2013. Retrieved 29 Sept 2018 from http://www.polarfield.com/blog/greenland-meltwater

Doble MJ, Forrest AL, Wadhams P et al (2009) Through-ice AUV deployment: operational and technical experience from two seasons of Arctic fieldwork. Cold Reg Sci Tech 56:90–97

Donenfeld J (2013) Building the Askaryan Radio Array at the South Pole. Posted on 31 Jan 2013, Retrieved 18 Feb 2018 from https://www.jeffreydonenfeld.com/blog/2013/01/building-the-askaryan-radio-array-at-the-south-pole-2/

Down the Hole (n.d.) Rodwell adventure videos. Retrieved 31 May 2018 from http://www.southpolestation.com/trivia/rodwell/rodwell.html

Doyle SH, Hubbard B, Christoffersen P et al (2018) Physical conditions of fast glacier flow: 1. Measurements from boreholes drilled to the bed of Store Glacier in West Greenland. J Geophys Res Earth Surf 123:324–348

Engelhardt H (2004a) Thermal regime and dynamics of the West Antarctic ice sheet. Ann Glaciol 39:85–92

Engelhardt H (2004b) Ice temperature and high geothermal flux at Siple Dome, West Antarctica, from bore hole measurements. J Glaciol 50 (169):251–256

Engelhardt H, Determann J (1987a) Borehole evidence for a thick layer of basal ice in the central Ronne Ice Shelf. Nature 327(6120):318–319

Engelhardt H, Determann J (1987b) Heisswasserbohrungen und geophysikalische Untersuchungen auf dem Filchner und Ekström Schelfeis [Hot water drilling and geophysical investigations on the Filchner and Ekström shelves]. In: Fütterer D (ed) Die Expedition ANTARKTIS-IV mit FS "Polarstern" 1985/86: Bericht von den Fahrtabschnitten ANT-IV/3-4 [The expedition ANTARKTIS-IV of RV "Polarstern" 1985/86, Report of Legs ANT-IV/3-4]. Bremerhaven, Alfred Wegener Institute for Polar and Marine Research, vol 33, pp 126–130

Engelhardt H, Kamb B (1997) Basal hydraulic system of a West Antarctic ice stream: constraints from borehole observations. J Glaciol 43(144):207–230

Engelhardt H, Kamb B (1998) Basal sliding of Ice Stream B, West Antarctica. J Glaciol 44(147):223–230

Engelhardt H, Fahnestock M, Humphrey N et al (1989) Borehole drilling to the bed of ice stream B, Antarctica. Antarct J US 24 (5):83–84

Engelhardt H, Humphrey N, Kamb B et al (1990) Physical conditions at the base of a fast moving Antarctic ice stream. Science 248:57–59

Engelhardt H, Kamb B, Bolsey R (2000) A hot-water ice-coring drill. J Glaciol 46(153):341–345

Eyles N, Rogerson R (1977) Artificially induced thermokarst in active glacier ice: An example from northwest British Columbia, Canada. J Glaciol 18:437–444

Falconer T, Pyne A, Olney M et al (2007) Operations overview for the ANDRILL McMurdo ice shelf project, Antarctica. Terra Antarct 14 (3):131–140

Field Update: Dec. 12, 2010 (2010). ANDRILL, Antarctic Geological Drilling. Retrieved 24 Feb 2018 from http://www.andrill.org/science/ch/news/12-12-10

Final comprehensive environmental evaluation (CEE) for ANDRILL (2006) McMurdo Sound Portfolio. In: Huston M, Gilbert N, Newman J (eds). International Antarctic Centre, Christchurch, New Zealand

Fischer UH, Porter PH, Schuler T et al (2001) Hydraulic and mechanical properties of glacial sediments beneath Unteraargletscher, Switzerland: Implications for glacier basal motion. Hydrol Process 15:3525–3540

Fisher AT, Mankoff KD, Tulaczyk SM et al (2015) High geothermal heat flux measured below the West Antarctic Ice Sheet. Sci Adv 1: e1500093

Fisher D, Jones S (1971) The possible future behaviour of Berendon Glacier, Canada—a further study. J Glaciol 10(58):85–92

Fountain AG (1994) Borehole water-level variations and implications for the subglacial hydraulics of South Cascade Glacier, Washington State, USA. J Glaciol 40(135):293–304

Fountain AG, Schlichting R, Jansson P et al (2005) Observations of englacial flow passages—a fracture dominated system. Ann Glaciol 40:25–30

Fox D (2018) Life below the ice. Nature 564:180–182

Fox D (2019) Tiny animal carcasses found in buried Antarctic lake. Nature 565:405–406

Francois RE (1977) Arctic underwater operational systems. Arctic Systems. In: Amaria PJ, Bruneau AA, Lapp PA (eds) Proceedings of conference, 18–22 Aug 1975, Memorial University of Newfoundland, St. John's, Newfoundland, Canada. Plenum Press, New York, pp 81–102

Francois RE, Harrison JG (1975) A thermal drill for making large holes in sea ice. In: Proceedings of IEEE OCEAN'75 Conference, 22–25 Sept 1975, San Diego, USA. The Institute of Electrical and Electronics Engineers, Inc., New York, pp 303–310

Fudge TJ, Humphrey NF, Joel T et al (2008) Diurnal fluctuations in borehole water levels: configuration of the drainage system beneath Bench Glacier, Alaska, USA. J Glaciol 54(185):297–306

Gaidos E, Lanoil B, Thorsteinsson Th et al (2004) A viable microbial community in a subglacial volcanic crater lake, Iceland. Astrobiology 4(3):327–344

Gillet F (1975) Steam, hot-water and electrical thermal drills for temperate glaciers. J Glaciol 14(70):171–179

Govoni JW, Tucker WB III (1989) An update on portable hot-water sea ice drilling. Cold Reg Sci Tech 16:175–178

Gow AJ, Engelhardt H (2000) Preliminary analysis of ice cores from Siple Dome, West Antarctica. In: Proceedings of international symposium on physics of ice core records, 14–17 Sept 1998, Shikotsukohan, Hokkaido, Japan. Hokkaido University Press, Sapporo, pp 63–82

Greenler L, Benson T, Cherwinka J et al (2014) Modeling hole size, lifetime and fuel consumption in hot-water ice drilling. Ann Glaciol 55(68):115–123

Grosfeld K, Hempel L (1991) Untersuchungen des Filchner-Ronne-Schelfeises mit Hilfe von Heiss-wasserbohrungen [Investigation of the Filchner-Ronne ice Shelf with the help of hot water holes]. In: Miller H, Oerter H (eds) Die Expedition ANTARKTIS-VIII mit FS "Polarstern" 1989/90: Bericht vom Fahrtabschnitt ANT-VIII/5 [The expedition ANTARKTIS-VIII of RV "POLARSTERN" 1989/90: Report of Leg ANT-VI 11/5]. Bremerhaven, Alfred Wegener Institute for Polar and Marine Research, vol 86, pp 67–69

Gusmeroli A, Murray T, Jansson P et al (2009) Geophysical techniques to measure the hydrothermal structure of Storglaciären, spring and summer 2009. Tarfala Research Station, Annual Report 2008/2009

Gusmeroli A, Murray T, Jansson P et al (2010) Vertical distribution of water within the polythermal glacier Storglaciären, Sweden. J Geophys Res 115:F04002

Haeberli W, Fisch W (1984) Electrical resistivity soundings of glacier beds: a test study on Grubengletscher, Wallis, Swiss Alps. J Glaciol 30(106):373–376

Haehnel RB, Knuth MA (2011) Potable water supply feasibility study for Summit Station, Greenland. USA CRREL Report ERDC/CRREL TR-11-4

Halzen F, Klein SR (2008) Astronomy and astrophysics with neutrinos. Phys Today 61(5):29–35

Hancock WH, Koci B (1989) Ice drilling instrumentation. In: Ice core drilling. In: Rado C, Beaudoing D (eds) Proceedings of the third international workshop on ice drilling technology, Grenoble,

France, 10–14 Oct 1988. Laboratoire de Glaciologie et Geophysique de l'Environnement, Grenoble, pp 38–50

Hansen DP (1987) Thermal hole opener. Cold Reg Sci Tech 14:51–56

Hanson B, Hooke RLB, Grace Jr EM (1998) Short-term velocity and water-pressure variations down-glacier from a riegel, Storglaciären, Sweden. J Glaciol 44(147):359–367

Hantz D, Lliboutry L (1983) Water ways, ice permeability at depth, and water pressures at Glacier D'Argentiere, French Alps. J Glaciol 29 (102):227–239

Harper JT, Humphrey NF (1995) Borehole video analysis of a temperate glacier's englacial and subglacial structure: implications for glacier flow models. Geology 23(10):901–904

Harper JT, Humphrey NF, Pfeffer WT (1998) Three-dimensional deformation measured in an Alaskan glacier. Science 281 (5381):1340–1342

Harper JT, Humphrey NF, Pfeffer WF et al (2001) Spatial variability in the flow of a valley glacier: deformation of a large array of boreholes. J Geophys Res 106(B5):8547–8562

Harper JT, Humphrey NF, Pfeffer WT et al (2005) Evolution of subglacial water pressure along a glacier's length. Ann Glaciol 40:31–36

Harper JT, Bradford JH, Humphrey NF et al (2010) Vertical extension of the subglacial drainage system into basal crevasses. Nature 467:579–582

Harper JT, Humphrey NF, Meierbachtol TW et al (2017) Borehole measurements indicate hard bed conditions, Kangerlussuaq sector, Western Greenland Ice Sheet. J Geophys Res Earth Surf 122:1605–1618

Harrington JA, Humphrey NF, Harper JT (2015) Temperature distribution and thermal anomalies along a flowline of the Greenland ice sheet. Ann Glaciol 56(70):98–104

Harrison WD, Truffer M, Echelmeyer KA et al (2004) Probing the till beneath Black Rapids Glacier, Alaska, USA. J Glaciol 50 (171):608–614

Hattermann T, Nøst OA, Lilly JM et al (2012) Two years of oceanic observations below the Fimbul Ice Shelf, Antarctica. Geophys Res Lett 39:L12605

Hodge SM (1979) Direct measurement of basal water pressures: progress and problems. J Glaciol 23(89):309–319

Hooke RL, Pohjola VA (1994) Hydrology of a segment of a glacier situated overdeepening, Storglaciären, Sweden. J Glaciol 40 (134):140–148

Hooke RL, Holmlund P, Iverson N (1987) Extrusion flow demonstrated by bore-hole deformation measurements over a Riegel, Storglaciären, Sweden. J Glaciol 33(113):72–78

Hooke RL, Miller SB, Kohler J (1988) Character of the englacial and subglacial drainage system in the upper part of the ablation area of Storglaciären, Sweden. J Glaciol 34(117):228–231

Hooke RL, Pohjola VA, Jansson P et al (1992) Intra-seasonal changes in deformation profiles revealed by borehole studies, Storglaciären, Sweden. J Glaciol 38(130):348–358

Hot Water Drilling (n.d.) British Antarctic Service. Retrieved 16 Sept 2018 from https://www.bas.ac.uk/polar-operations/sites-and-facilities/facility/hot-water-drilling/#about

Hot Water Drilling at Langhovde Glacier, East Antarctica (2012) Retrieved 26 Feb 2018 from http://www.ice.lowtem.hokudai.ac.jp/~sugishin/research/hokudai2/langhovde/langhovde.html

How P, Benn DI, Hulton NRJ et al (2017) Rapidly changing subglacial hydrological pathways at a tidewater glacier revealed through simultaneous observations of water pressure, supraglacial lakes, meltwater plumes and surface velocities. Cryosphere 11:2691–2710

Hubbard B, Glasser N (2005) Field techniques in glaciology and glacial geomorphology. Wiley, England

Hubbard B, Sharp MJ, Willis IC et al (1995) Borehole water-level variations and the structure of the subglacial hydrological system of Haut Glacier d'Arolla, Valais, Switzerland. J Glaciol 41(139):572–583

Hubbard B, Binley A, Slater L et al (1998) Inter-borehole electrical resistivity imaging of englacial drainage. J Glaciol 44(147):429–434

Hubbard B, Tison J-L, Pattyn F et al (2012) Optical-televiewer-based identification and characterization of material facies associated with an Antarctic ice-shelf rift. Ann Glaciol 53(60):137–146

Hubbard B, Luckman A, Ashmore DW et al (2016) Massive subsurface ice formed by refreezing of ice-shelf melt ponds. Nat Commun 7:11897

Hughes KG, Langhorne PJ, Williams MJM (2013) Estimates of the refreezing rate in an ice-shelf borehole. J Glaciol 59(217):938–948

Humphrey N, Echelmeyer K (1990) Hot-water drilling and bore-hole closure in cold ice. J Glaciol 36(124):287–298

Humphrey N, Kamb B, Fahnestock M et al (1993) Characteristics of the bed of the lower Columbia Glacier, Alaska. J Geophys Res 98:837–846

Huss M, Bauder A, Werder M et al (2007) Glacier-dammed lake outburst events of Gornersee, Switzerland. J Glaciol 53(181):189–200

IceCube 2008 (2008) University of Wisconsin-Madison News. Retrieved 8 Oct 2018 from https://news.wisc.edu/newsphotos/iceCube2008.html

Iken A (1988) Adaptation of the hot-water-drilling method for drilling to great depth. Eidg. Tech. Hochschule, Zürich. Versuchsanst. Wasserbau. Hydrol Glaziol Mitt 94:211–229

Iken A, Bindschadler RA (1986) Combined measurements of subglacial water pressure and surface velocity of Findelengletscher, Switzerland: conclusions about drainage system and sliding mechanism. J Glaciol 32(110):101–119

Iken A, Truffer M (1997) The relationship between subglacial water pressure and velocity of Findelengletscher, Switzerland, during its advance and retreat. J Glaciol 43(144):328–338

Iken A, Röthlisberger H, Hutter K (1977) Deep drilling with a hot water jet. Z Gletscherkd Glazialgeol 12(2):143–156

Iken A, Echelmeyer K, Harrison WD (1989) A light-weight hot water drill for large depth: experiences with drilling on Jakobshavns glacier, Greenland. In: Rado C, Beaudoing D (eds) Ice core drilling. Proceedings of the third international workshop on ice drilling technology, Grenoble, France, 10–14 Oct 1988. Laboratoire de Glaciologie et Geophysique de l'Environnement, Grenoble, pp 123–136

Iken A, Echelmeyer K, Harrison W et al (1993) Mechanisms of fast flow in Jakobshavns Isbræ, West Greenland: Part I. Measurements of temperature and water level in deep boreholes. J Glaciol 39 (131):15–25

Kamb B, Engelhardt H (1991) Antarctic ice stream B: conditions controlling its motion and interactions with the climate system. In: Kotlyakov VM, Ushakov A, Glazovsky A (eds) Glaciers-Ocean-Atmosphere Interactions. Proceedings of the international symposium, 24–29 Sept 1990, St. Petersburg. IAHS Publ., vol 208, pp 145–154

Kamb B, Raymond CF, Harrison WD et al (1985) Glacier surge mechanism: 1982–1983 surge of Variegated Glacier, Alaska. Science 227(4686):469–479

Kaminski C, Crees T, Ferguson J et al (2010) 12 days under ice—a historic AUV deployment in the Canadian High Arctic. In: Proceedings of the 2010 IEEE/OES autonomous underwater vehicles (AUV) Conference, 1–3 Sept 2010, Monterey, California, USA, pp 1–11

Kaufman M (2012) Race is on to find life under Antarctic ice. National Geographic. Posted on 18 Dec 2012. Retrieved 10 Oct 2018 from https://news.nationalgeographic.com/news/2012/12/121218-antarctica-life-microbes-ice-science-environment/

Koci BR (1984) Hot water drilling in Antarctic firn and freezing rates in water filled boreholes. In: Holdsworth G, Kuivinen KC, Rand JH (eds) Proceedings of the second international workshop/symposium on ice drilling technology, 30–31 Aug 1982, Calgary, Alberta, Canada. USA CRREL Spec. Rep. 84–34, pp 101–103

Koci B (1989) A deep hot water drill system with potential for bottom sampling. In: Rado C, Beaudoing D (eds) Ice core drilling. Proceedings of the third international workshop on ice drilling technology, Grenoble, France, 10–14 Oct 1988. Laboratoire de Glaciologie et Geophysique de l'Environnement, Grenoble, pp 137–139

Koci B (1994) The AMANDA project: drilling precise, large-diameter holes using hot water. Mem Natl Inst Polar Res Spec Issue 49:203–211

Koci B (2002) Wotan: A drill for ice cube. Mem Natl Inst Polar Res Spec Issue 56:209–216

Koci B, Kuivinen KC (1986) PICO drilling activities at Siple Station and on the Siple Coast during 1985–1986. Antarct. J US 21(5):117

Koci B, Bindschandler R (1989) Hot-water drilling on Crary ice rise, Antarctica. Ann Glaciol 12:214

Koci B, Nagornov O, Zagorodnov V et al (1997) Hot water drilling of large diameter holes in cold ice. In: Lee Y, Hallett W (eds) Proceedings of the 5th international symposium on thermal engineering and sciences for cold regions, 19–22 May 1996, Ottawa, Canada, National Research Council Canada, pp 312–317

Kravchenko VV (1984) Bureniye ledyanikh massivov nebol'shoi moshnosti [Drilling through massive ice of small thickness]. Akademiya nauk SSSR. Institut geografii. Materialy gliatsiologicheskikh issledovanii [Academy of Sciences of the USSR. Institute of Geography. Data of Glaciological Studies] 50, pp 161–164

Kudryashov BB, Menshikov NG (1994) Ice core hot-fluid drilling. Nankyoku Shiryo (Antarctic Record) 38(3):193–198

Kuivinen KC, Koci BR (1984) Hot-water drilling on the Siple Coast and ice core drilling at Siple and South Pole Stations. Antarct. J US 19(5):58–59

Kuivinen KC, Marshall PS, Koci BR (1980) Polar Ice Coring Office (PICO) drilling activities, 1979–80. Antarct. J US 15(5):76–77

Li F (1993) An analysis of melt water freezing in the ice borehole. Polar Ice Coring Office, University of Alaska—Fairbanks, PICO TR-93-2

Lüthi M, Funk M, Iken A et al (2002) Mechanisms of fast flow in Jakobshavn Isbræ, West Greenland. Part III. Measurements of ice deformation, temperature and cross-borehole conductivity in boreholes to the bedrock. J Glaciol 48(162):369–385

Makinson K (1993) The BAS hot water drill: development and current design. Cold Reg Sci Tech 22:121–132

Makinson K (1994) BAS hot water drilling on Ronne Ice Shelf, Antarctica. Mem Natl Inst Polar Res Spec Issue 49:192–202

Makinson K (2003) Future hot water drilling on Rutford Ice Stream 2004/05. FRISP Rep 14:163–166

Makinson K, Anker PGD (2014) The BAS ice-shelf hot-water drill: design, methods and tools. Ann Glaciol 55(68):44–52

Makinson K, Pearce D, Hodgson DA et al (2016) Clean subglacial access: prospects for future deep hot-water drilling. Phil Trans R Soc A 374:20140304

Makovicka T, Strauss T, Hancock W et al (1998) The PICO hot water drill system. In: Hall J (ed) Proceedings of the seventh symposium on antarctic logistics and operations, 6–7 Aug 1996, Cambridge, United Kingdom. British Antarctic Survey, Cambridge, UK, pp 185–192

Meier M, Lundstrom S, Stone D et al (1994) Mechanical and hydrologic basis for the rapid motion of a large tidewater glacier. 1. Observations. J Geophys Res 99(B8):15219–15229

Meierbachtol T, Harper J, Humphrey N (2013) Basal drainage system response to increasing surface melt on the Greenland Ice Sheet. Science 341:777–779

Meierbachtol TW, Harper JT, Humphrey NF et al (2016) Mechanical forcing of water pressure in a hydraulically isolated reach beneath Western Greenland's ablation zone. Ann Glaciol 57(72):62–70

Mellor M (1986) Equipment for making access holes through Arctic sea ice. USA CRREL Spec Rep 86–32

Metcalfe T (2019) Photos: drilling into Antarctic Subglacial Lake Mercer. Live Science Posted on 15 Jan 2019. Retrieved on 7 Mar 2019 from https://www.livescience.com/64500-antarctica-buried-lake-photos.html

Miles K (2017) Image of the week—drilling into a Himalayan glacier. EGU Blogs. Posted on 28 July 2017. Retrieved 18 Feb 2018 from https://blogs.egu.eu/divisions/cr/2017/07/28/image-of-the-week-drilling-into-a-himalayan-glacier/

Miles KE, Hubbard B, Quincey DJ et al (2018) Polythermal structure of a Himalayan debris-covered glacier revealed by borehole thermometry. Sci Rep 8:16825

Mironov YU, Morev VA, Porubayev VS et al (2003) Study of geometry and internal structure of ice ridges and stamukhas using thermal water drilling. In: Proceedings of port and ocean engineering under arctic conditions (POAC '03), 16–19 June 2003, Trondheim, Norway, pp 623–634

Morev VA, Pukhov VA, Yakovlev VM et al (1984) Equipment and technology for drilling in temperate glaciers. In: Holdsworth G, Kuivinen KC, Rand JH (eds) Proceedings of the second international workshop/symposium on ice drilling technology, 30–31 Aug 1982, Calgary, Alberta, Canada, USA CRREL Spec Rep 84–34, pp 125–127

Morev VA, Toskin VV, Yakovlev VM (1986) Tekhnicheskie sredstva dlya teplovogo bureniya i rezaniya l'da [Devices for thermal drilling and cutting in ice]. Problemy inzhenernoi glyatsiologii [Problems of Engineering Glaciology], Nauka, Nobosibirsk, pp 37–39 (in Russian)

Müller F (1976) On the thermal regime of a high-arctic valley glacier. J Glaciol 16(74):119–133

Münchow A, Padman L, Washam P et al (2016) The ice shelf of Petermann Gletscher, North Greenland, and its connection to the Arctic and Atlantic Oceans. Oceanography 29(4):84–95

Murray T, Porter PR (2001) Basal conditions beneath a soft-bedded polythermal surge-type glacier: Bakaninbreen, Svalbard. Quatern Int 86(1):103–116

Murray T, Gooch DL, Stuart GW (1997) Structures within the surge front at Bakaninbreen, Svalbard, using ground-penetrating radar. Ann Glaciol 24:122–129

Murray T, Stuart GW, Fry M et al (2000) Englacial water distribution in a temperate glacier from surface and borehole radar velocity analysis. J Glaciol 46(154):389–398

Napoléoni JGP, Clarke GKC (1978) Hot water drilling in cold glacier. Can J Earth Sci 15:316–321

Nicholls K (2015) Arctic blog: hot water drilling on Petermann Glacier. Posted on 24 Aug 2015. Retrieved 19 Sept 2018 from https://www.bas.ac.uk/blogpost/arctic-blog-hot-water-drilling-on-petermann-glacier/

Nicholls KW, Makinson K (1998) A 'light weight' hot water drill for use on Ronne Ice Shelf. In: Hall J (ed) Proceedings of the seventh symposium on antarctic logistics and operations, 6–7 Aug 1996, Cambridge, United Kingdom. British Antarctic Survey, Cambridge, UK, pp 193–202

Nicholls KW, Østerhus S, Makinson K et al (2001) Oceanographic conditions south of Berkner Island, beneath Filchner-Ronne Ice Shelf, Antarctica. J Geophys Res 106(C6):11481–11492

Nicholls KW, Makinson K, Østerhus S (2004) Circulation and water masses beneath the northern Ronne Ice Shelf, Antarctica. J Geophys Res 109(C12):C12017

Nicholls KW, Corr HFJ, Makinson K et al (2012) Rock debris in an Antarctic ice shelf. Ann Glaciol 53(60):235–240

Nixdorf U, Mandler H, Wege C et al (1994a) Heisswasserbohrung [Hot water drilling]. In: Miller H (ed) Die Expedition ANTARKTIS-X mit FS "Polarstern" 1992: Bericht von den Fahrtabschnitten ANT-X/1a und 2 [The expedition ANTARKTIS-X of RV "Polarstern" 1992: Report of Legs ANT-X/1a and 2], Bremerhaven,

Alfred Wegener Institute for Polar and Marine Research 152, pp 191–195 (in German)

Nixdorf U, Oerter H, Miller H (1994b) First access to the ocean beneath the Ekströmisen Antarctica, by means of hot-water drilling. Ann Glaciol 20:110–114

Nixdorf U, Dunker E, Eckstaller A et al (1997) Schelfeis-Ozean-Wechselwirkung [Ice Shelf-Ocean Interaction]. In: Jokat W, Oerter H (eds) Die Expedition ANTARKTIS-XII mit FS "Polarstern" 1995: Bericht vom Fahrtabschnitt ANT-XII/3 [The expedition ANTARKTIS-XII of RV "Polarstern" in 1995: report of leg ANT-XII/3], Bremerhaven, Alfred Wegener Institute for Polar and Marine Research 219, pp 69–88 (in German)

Nøst OA (2009) The second drilling successful. Norwegian Polar Institute. Posted on 19 Dec 2009. Retrieved 16 Mar 2018 from http://www.npolar.no/en/research/ice/antarctica/fimbul-ice-shelf/expedition-0910/diary/entries/2009-12-19.html

Nøst OA, Gabrielsen PG, Smedsrud LH (2009) Hot water drilling manual. Norwegian Polar Institute, Oslo

Olesen OB (1989) A Danish contribution to the family of hot-water glacier drills. In: Rado C, Beaudoing D (eds) Ice core drilling. Proceedings of the third international workshop on ice drilling technology, Grenoble, France, 10–14 Oct 1988. Laboratoire de Glaciologie et Geophysique de l'Environnement, Grenoble, pp 140–148

Orheim O, Hagen JO, Østerhus S et al (1990) Studies on, and underneath, the Ice Shelf Fimbulisen. Meddelelser 113. In: Orheim O (ed) Report of the Norwegian Antarctic Research Expedition, Norsk Polarinstituit, Oslo, pp 59–73

Orheim O, Østerhus S, Melvold K et al (1997) Hot water drilling near Filchner Station, Ronne Ice Shelf. Meddelelser 125. In: Orheim O (ed) Report of the Norwegian Antarctic Research Expedition 1992/93. Norsk Polarinstituit, Oslo, pp 93–96

Østerhus S, Orheim O (1992) Studies through Jutulgryta. Fimbulisen in the 1991/92 season. In: Oerter H (ed) Filchner-Ronne-Ice-Shelf-Programme, Report No. 6, Bremerhaven, AWI, pp 103–109

Paren JG, Cooper S (1988) Thermal regime of George VI Ice Shelf, Antarctic Peninsula. Ann Glaciol 11:206

Penn State News (2013, 2017). Retrieved 20 Feb 2018 from http://news.psu.edu/photo/264684/2013/02/19/pig-drilling and http://news.psu.edu/photo/494005/2017/11/13/don-voigt-2

Pohjola VA (1994) TV-video observations of englacial voids Storglaciären, Sweden. J Glaciol 40(135):231–240

Poplin JP, Ralston TD, Lawrence WS (1987) A thermal ice drill for profiling thick multiyear ice. Cold Reg Sci Tech 14:1–11

Porter PR, Murray T, Dowdeswell JA (1997) Sediment deformation and basal dynamics beneath a glacier surge front: Bakaninbreen, Svalbard. Ann Glaciol 24:21–26

Rack FR (2016) Enabling clean access into Subglacial Lake Whillans: development and use of the WISSARD hot water drill system. Phil Trans R Soc A 374:20140305

Rack FR, Duling D, Blythe D et al (2014) Developing a hot-water drill system for the WISSARD project: 1. Basic drill system components and design. Ann Glaciol 55(68):285–297

Rado C, Girard C, Perrin J (1987) Electrochaude: a self-flushing hot-water drilling apparatus for glaciars with debris. J Glaciol 33(114):236–238

Rand J (1982) Developing a water well for the ice backfilling of Dye-2. USA CRREL Spec Rep 82–32

Reed L (2015) UNL drillers help make new discoveries in Antarctica. Office of University Communications, University of Nebraska–Lincoln. Posted on 21 Jan 2015. Retrieved 2 Oct 2018 from http://newsroom.unl.edu/releases/2015/01/21/UNL+drillers+help+make+new+discoveries+in+Antarctic

Return to the Glacier (2013) Arctic Research. Reports from INTERACT field sites. Posted on 26 July 2013 by chaxtell89. Available at: https://arcticresearch.wordpress.com/tag/storglaciaren/. Accessed 11 Feb 2016

Reynaud L, Courdouan P (1962) Reconnaissance du thalweg sous-glaciaire de la Mer de Glace en vue de l'établissement d'une prise d'eau. La Houille Blanche, Special Issue B-1962, pp 808–816

Roberson S, Hubbard B (2010) Application of borehole optical televiewing to investigating the 3-D structure of glaciers: Implications for the formation of longitudinal debris ridges, midre Lovénbreen, Svalbard. J Glaciol 56(195):143–156

Röösli C, Walter F, Husen S et al (2014) Sustained seismic tremors and icequakes detected in the ablation zone of the Greenland ice sheet. J Glaciol 60(221):563–575

Russell FL (1965) Water production in a polar ice cap by utilization of waste engine heat. USA CRREL Tech Report 168

Ryser C, Lüthi MP, Andrews LC et al (2014) Sustained high basal motion of the Greenland ice sheet revealed by borehole deformation. J Glaciol 60(222):647–660

SALSA Chief Scientist Reports (2019) Compiled science field reports, 15 Dec 2018–6 Jan 2019. Available on-line at: https://icedrillweb.files.wordpress.com/2019/01/salsa-science-reports-compiled-15dec-6jan-2019.pdf

Schiermeier Q (2014) Polar drilling problems revealed. Nature 505:463

Scientists drill to record depths in West Antarctica (2019) British Antarctic Survey Press release. Posted on 24 Jan 2019. Retrieved on 8 Mar 2019 from https://www.bas.ac.uk/media-post/scientists-drill-to-record-depths-in-west-antarctica/

Sediment Laden Lake Ice Drill (n.d.) U.S. Ice Drilling Program, Current Inventory. Retrieved 20 Feb 2018 from https://icedrill.org/equipment/sediment-laden-lake-ice-drill.shtml

Siegert MJ, Clarke RJ, Mowlem M et al (2012) Clean access, measurement, and sampling of Ellsworth Subglacial Lake: a method for exploring deep Antarctic subglacial lake environments. Rev Geophys 50(RG1):RG1003

Siegert MJ, Makinson K, Blake D et al (2014) An assessment of deep hot-water drilling as a means to undertake direct measurement and sampling of Antarctic subglacial lakes: experience and lessons learned from the Lake Ellsworth field season 2012/13. Ann Glaciol 55(65):59–73

Silverwood N (2018) The long haul. New Zealand Geographic. Retrieved 8 Sept 2018 from https://www.nzgeo.com/stories/the-long-haul/

Small Hot Water Drill (n.d.) U.S. Ice Drilling Program, Equipment. Retrieved 20 Feb 2018 from https://icedrill.org/equipment/small-hot-water-drill.shtml

Smeets CJPP, Boot W, Hubbard A et al (2012) A wireless subglacial probe for deep ice applications. J Glaciol 58(211):841–848

Smirnov VN, Mironov EU (2010) Issledovania prochnosti, morfometrii i dynamiki l'da v inzhenernikh zadachakh pri osvoenii shel'fa v zamerzayushchikh moryakh [Research on strength, morphometry and ice dynamics in the engineering missions during investigations of the shelf of freezing seas]. Problemy Arktiki i Antarktiki [Problems of Arctic and Anatrctica] 2(85):5–15 (in Russian)

Smith AM (2005) RABID: Basal conditions on Rutford Ice Stream, West Antarctica: Hot-water drilling and down-hole instrumentation. British Antarctic Survey Field Report, R/2004/S3, BAS Archives ref: AD6/2R/2004/S3

Smith A, Fothergill C (2016) BEAMISH initial environmental evaluation. Environment Office, British Antarctic Survey, Cambridge, United Kingdom

Spears A, West M, Meister M et al (2016) Under ice in Antarctica: the ICEFIN unmanned underwater vehicle development and deployment. IEEE Robot Autom Mag 23(4):30–41

Stanton TP, Shaw WJ, Truffer M et al (2013) Channelized ice melting in the ocean boundary layer beneath Pine Island Glacier, Antarctica. Science 341:1236–1239

Sugiyama S, Tsutaki S, Nishimura D et al (2008) Hot water drilling and glaciological observations at the terminal part of Rhonegletscher, Switzerland in 2007. Bull Glaciol Res 26:41–47

Sugiyama S, Skvarca P, Naito N et al (2010) Hot-water drilling at Glaciar Perito Moreno, Southern Patagonia Icefield. Bull Glaciol Res 28:27–32

Sugiyama S, Sawagaki T, Fukuda T et al (2014) Active water exchange and life near the grounding line of an Antarctic outlet glacier. Earth Planet Sci Lett 399:52–60

Talalay PG (2016) Mechanical ice drilling technology. Geological Publishing House, Beijing and Springer Science + Business Media Singapore

Talalay PG, Pyne AR (2017) Geological drilling in McMurdo Dry Valleys and McMurdo Sound, Antarctica: historical development. Cold Reg Sci Tech 141:131–162

Talalay P, Yang C, Cao P et al (2015) Ice-core drilling problems and solutions. Cold Reg Sci Tech 120:1–20

Talalay P, Liu G, Wang R et al (2018) Shallow hot-water ice drill: estimation of drilling parameters and testing. Cold Reg Sci Tech 155:11–19

Taylor PL (1984) A hot water drill for temperate ice. In: Holdsworth G, Kuivinen KC, Rand JH (eds) Proceedings of the second international workshop/symposium on ice drilling technology, 30–31 Aug 1982, Calgary, Alberta, Canada. USA CRREL Spec Rep 84–34, pp 105–117

Taylor S, Lever JH, Harvey RP et al (1997) Collecting micrometeorites from the South Pole water well. USA CRREL Rep 97-1

The Franklin Expedition (2017) Parks Canada. Retrieved 21 Feb 2018 from https://www.pc.gc.ca/en/culture/franklin

Thermal drilling equipment (n.d.) Kovacs ice drilling equipment. Retrieved 15 Sept 2016 from https://kovacsicedrillingequipment.com/thermal-drilling/

Thomsen HH, Olesen O, Braithwaite RJ et al (1991) Ice drilling and mass balance at Pâkitsoq, central West Greenland. Grønlands Geologiske Undersøgelse 152, Copenhagen, Denmark, pp 80–84

Thorsteinsson T, Elefsen SÓ, Gaidos E et al (2008) A hot water drill with built-in sterilization: design, testing and performance. Jökull 57:71–82

Treverrow A, Donoghue S (2010) AMISOR: understanding the ocean beneath the ice. Aust Antarct Mag 19:26–27

Treverrow A, Warner RC, Budd WF et al (2010) Meteoric and marine ice crystal orientation fabrics from the Amery Ice Shelf, East Antarctica. J Glaciol 56(199):877–890

Truffer M (2017) Drilling through the Nansen Ice Shelf. Glacier adventures: Reports from the field. Posted on 15 Feb 2017. Retrieved 8 Mar 2018 from http://glacieradventures.blogspot.ru/

Truffer M, Motyka RJ, Harrison WD et al (1999) Subglacial drilling at Black Rapids Glacier, Alaska, U.S.A.: Drilling method and sample descriptions. J Glaciol 45(151):495–505

Tsutaki S, Sugiyama S (2007) Construction of a hot water drilling system. Preprints of the annual conference, Japanese Society of Snow and Ice 2007, 4-4, 2007 (in Japanese)

Tsutaki S, Sugiyama S (2009) Development of a hot water drilling system for subglacial and englacial measurements. Bull Glaciol Res 27:7–14

Tucker WB III, Govoni JW (1987) A portable hot-water ice drill. Cold Reg Sci Tech 14:57–64

Tulaczyk S, Kamb B, Scherer R et al (1998) Sedimentary processes at the base of a West Antarctic Ice Stream: constraints from textural and compositional properties of subglacial debris. J Sediment Res 68(3):487–496

Tulaczyk S, Mikucki JA, Siegfried MR et al (2014) WISSARD at Subglacial Lake Whillans, West Antarctica: scientific operations and initial observations. Ann Glaciol 55(65):51–58

Verrall R (2001) A guide to Arctic field trips. Defence Research Establishment Atlantic, Canada

Verrall R, Baade D (1984) A simple hot-water drill for penetrating ice shelves. In: Holdsworth G, Kuivinen KC, Rand JH (eds) Proceedings of the second international workshop/symposium on ice drilling technology, 30–31 Aug 1982, Calgary, Alberta, Canada. USA CRREL Spec Rep 84–34, pp 87–94

Vogel SW (2009) On the geometry of core-catcher holders for hot-water based ice coring of sediment-laden ice. J Glaciol 55(189):188–190

Waddington BS, Clarke CKC (1995) Hydraulic properties of subglacial sediment determined from the mechanical response of water-filled boreholes. J Glaciol 41(137):112–124

Willis A, Mair D, Hubbard B et al (2003) Seasonal variations in ice deformation and basal motion across the tongue of Haut Glacier d'Arolla, Switzerland. Ann Glaciol 36:157–167

Yen Y-C, Tien C (1976) Heat transfer characteristics of melting and refreezing a drill hole through an ice shelf in Antarctica. USA CRREL Rep 76-12

Zagorodnov VS, Kelley JJ, Stanford KL et al (1992) Borehole monitoring with impulse acoustic sensors. Polar Ice Coring Office Technical Note, TN-92-4, University of Alaska Fairbanks

Steam Ice Drills

4

Abstract

Steam ice drills are used for shallow drilling to install ablation stakes, pressure transducers, and other sensors and to create large-diameter holes for sewage disposal and construction of water wells. During drilling, steam produced by boilers or steam generators flows through an insulated hose into the borehole, where it condenses and melts the ice.

Keywords

Closed boiler • Steam generator • Insulated hose • Sewage and wastewater disposal

The principle of steam ice drills is quite simple. Water is vaporized in a closed boiler, where water is heated to its boiling point. An additional energy supply turns the water into steam, which flows through an insulated hose into the borehole, where steam condenses and melts the ice. The condensed steam and meltwater travel up to the surface between the feeding hose and the ice. The high degree of latent heat contained in the steam (approximately 2.3 MJ/kg) ensures high energy flow from the boiler to the drill tip.

The drilled depths with such systems are usually very shallow (< 30–40 m). Depending on the application and the attainable depths, steam drills can be subdivided into three types: (1) portable steam ice drills for shallow drilling down to 10–20 m for installing ablation stakes, pressure transducers, and other sensors that need to be fastened to the ice and also to make access holes into lake/oceanic water; (2) large-diameter steam ice drills for sewage disposal and the construction of water walls; and (3) steam ice core-drills to recover core samples.

4.1 Portable Steam Ice Drills

4.1.1 Voyeykov's Steam Drill

The first steam ice drill was constructed by A. A. Voyeykov (1935), a member of the Lena-Khatagnskaya Expedition Glavsevmorputi (Northern Sea Route Authority), at the beginning of 1930s in order to measure the thickness of sea ice and for sub-ice sea observations (water thickness, temperature measurements, and sampling).

The drill was composed of a 5 L iron boiler (made from the body of a dry-powder fire extinguisher) heated by two kitchen kerosene stoves (Fig. 4.1a). The boiler was fixed inside a steel box and was surrounded by a tin cover with asbestos insulation. The upper surface of the steel box served as a heating plate to warm up the instruments and melt snow into a small metal container (not shown). Steam was directed onto the ice by means of an insulated 10-mm-diameter rubber hose and a drill stem (Fig. 4.1b). The 2-m-long drill stem was made of a 1.5″ steel pipe with a 6-mm-diameter copper steam pipe along the center line. Two layers of woolen cloth served as heat insulators. The hose was also covered by felt and woolen cloth.

This drill was used in Tiksi bay (Laptev sea) during the winter of 1932–1933 (Fig. 4.2). It took ~12–15 min to make a hole in 2-m-thick sea ice at an average penetration rate of 9 m/h. The diameter of the holes was 48–50 mm in the lower part and 60–65 mm in the upper part. The boiler was fed with water (melted snow/ice or seawater) every 40–50 min. The estimated steam-production capacity of the boiler was ~3 kg/h.

© Geological Publishing House and Springer Nature Singapore Pte Ltd. 2020
P. G. Talalay, *Thermal Ice Drilling Technology*, Springer Geophysics,
https://doi.org/10.1007/978-981-13-8848-4_4

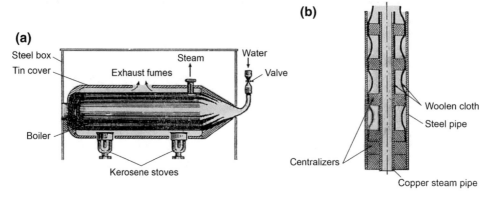

Fig. 4.1   Voyeykov's steam-operated ice drill: **a** schematic diagram of the boiler; **b** drill stem (Voyeykov 1935)

Fig. 4.2   Voyeykov's steam-operated ice drill in use on the sea ice of Tiksi bay (Laptev sea) during the winter of 1932–1933 (Voyeykov 1935)

4.1.2   Howorka and NVE Steam Drills

F. Howorka (1965), from the Institute of Meteorology and Geophysics, Innsbruck University, Austria, designed the portable steam drill to bore holes in ice down to a maximal depth of 8 m. The drill included a 3-L boiler with a burner fueled by butane cartridges and a pressure gauge mounted on top. An 8.5-m-long rubber hose (ID/OD: 13 mm/21 mm)

was connected to the vessel. To keep the borehole straight, a 2-m-long guide tube 13/11 mm ID/OD was inserted at the end of the hose. A 0.3-m-long brass tip 21/8 mm ID/OD was soldered to the guide tube. To reduce heat losses, a smaller diameter hose was inserted into the outer one, passed through the tip, and was fastened to its end, where a changeable nozzle was screwed in. The weight of the boiler, including a wind shield, the bottom plate, and the burner, was only 8.5 kg, and the weight of the hose and the drill tip was 4 kg. A handle was mounted for carrying the equipment easily from one site to another.

An excess pressure of $\sim$60 kPa was required to drive steam through the hose. During testing on Hintereisferner, Alps, 8-m-deep holes were drilled with an average penetration rate of $\sim$16 m/h, which depended on the content of the mineral inclusions in the ice; sand or sludge accumulating at the bottom of the hole slowed down penetration. The optimum nozzle diameter proved to be 2.5 mm, providing the maximum rate. One butane cartridge lasted for $\sim$110 min, which was enough to drill three 8-m-deep holes. Borehole diameter was $\sim$35 mm.

The NVE (which stands for Norges vassdrags-og energidirektorat/Norwegian Water Resources and Energy Directorate) drill was based on the Howorka ice drill and differed in that it used propane instead of butane (Hodge 1971). Propane was chosen because it is more readily available and has a higher vapor pressure at low temperatures than butane. In practice, it was found that the Norwegian drill was too fragile to be used in rough terrain and was too bulky for easy back-packing. One of the Norwegian drills was obtained in 1967 by the Glacier Project Office of the Water Resources Division, US Geological Survey, Tacoma, Washington, for the purpose of installing ablation and movement stakes on South Cascade Glacier and Nisqually Glacier, Washington.

4.1.3   Steam Drill of the University of Washington

This drill was similar to Howorka's design (Hodge 1971). The boiler was of the single-pass horizontal fire-tube type and was rectangular in shape (Fig. 4.3). It contained 24 tubes, each 19 mm in diameter, and could hold 6.9 L of water. A vertical glass tube on the front of the boiler served as a water level indicator. The top of the boiler was fitted with a quick-release coupling for the steam hose, a pressure gauge, and two safety valves, one of which was set to release at an overpressure of 350 kPa (slightly above the normal operating pressure) and the other to release at an overpressure of 520 kPa (the pressure to which the boiler was tested). The safety valves exited horizontally and did not endanger the operator when they opened.

A propane burner was fixed inside the unit and was connected to the valve on the outside. A flexible hose connected the valve to the propane tank; this allowed the tank to be easily removed and refilled. The material used was a T-6 6061 aluminum alloy, except where stainless steel was used to withstand the high temperatures of the propane flame. The basic weight of the system, with an enclosure for the propane tank and the pack frame, was 20.4 kg (considerably heavier than the Howorka and Norwegian versions). The propane tank weighed an additional 4.5 kg and could hold 1.8 kg of propane.

In much the same way as Howorka's design, the steam hose was made in 8-m lengths with an OD of 25.4 mm and consisted of an outer rubber hose and an inner Teflon hose with an air gap between them. This double-walled hose was connected to a guide tube with the same OD. The guide tube, 1.8 m long, consisted of an outer fiberglass tube and an inner Teflon tube. The bottom end of the guide tube was made of brass; the Teflon tube passed through the brass and fed the steam into a replaceable nozzle.

Experiments with blocks of clear ice showed that the spray pattern of the nozzle was very important. The optimum drilling rate was attained when ice was melted uniformly over the entire cross-section of the hole. Four nozzles with different hole configurations were tested; the one which yielded the fastest penetration rate had a 1.18-mm central hole and six holes with the same size radiating at an angle of 30° from the central hole.

This drill was used during 1969–1970 on South Cascade, Nisqually and Blue Glacier (Washington), and in the Gulkana and Wolverine glaciers (Alaska). The time necessary to bring a full tank of water from 0 °C to its boiling point and then to a normal working overpressure of 320 kPa was ∼ 12 min. An additional 12 min were necessary if snow had to be melted first. With one length of hose, an 8-m hole could be drilled at an average penetration rate of 33 m/h; in firn, the penetration rate increased to 54 m/h. Approximately four 8-m holes could be drilled in ice with one tank of water. One tank of propane lasted ∼ 180–190 min; this was sufficient to bring a full boiler to the boiling point three times and to drill approximately ten 8-m holes in ice. It was assumed that the holes had a size close to that of the drill near the

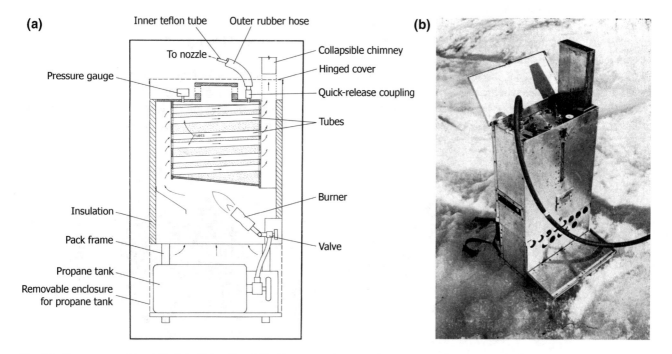

Fig. 4.3 Steam-operated ice drill of the University of Washington: **a** schematic diagram as seen from the front (safety valves and a small vent on the left side are not shown); **b** drill in use at Blue Glacier, USA, September 1970 (Hodge 1971)

bottom, which then slowly widened to ~35–40 mm at the top of the 8-m hole.

Tests were also performed on sea ice at Barrow, Alaska, in March 1970. Air temperature varied between −35 and −40 °C. The only problem encountered was that when the hose was allowed to cool down to ambient air temperature while connected to the boiler, the steam condensed and froze before reaching the drill tip. This plug of ice did not move and the drill could not be used. However, if steam was run through the hose indoors, then the hose would take 20–25 min to cool down outside. The drilling rate was ~30 m/h and fifteen 2-m holes (25.4 mm in diameter) could be drilled through sea ice with one tank of water. The holes drilled in this ice were very true to their size over their entire length and water would not start to refreeze until several minutes after the drill had been removed.

4.1.4 Steam Drill of the Laboratoire de Glaciologie

This steam drill was developed in the Laboratoire de Glaciologie in 1966 (Gillet 1975). A boiler consisting of four connected vertical tubes was heated by a propane burner that consumed 1 kg/h of gas (Fig. 4.4). The steam produced was sent to a drill tip with an OD of 32 mm that injected it directly against the ice. The total weight of the drill was 28 kg including the boiler (16 kg) and the gas tank (6 kg). The volume of gas was enough to heat water during 3.5 h.

This drill was used in the Alps especially for making holes typically 10–15 m deep and intended for installing ablation stakes; the maximal achieved depth was 30–40 m. To avoid heat losses through the feeding hose, the outer hose was replaced with soft closed-cell insulation. The penetration rate was 30–40 m/h for the first 10 m and 20 m/h for the next 10 m. The speed remained constant in ice with sandy debris. The production of water from snow was ensured using a double jacket, which allowed for drilling in

accumulation zones and when water was not available on the glacier surface. When needed, water could be injected using a small hand pump.

4.1.5 JARE Steam Drill

The JARE drill consisted of three parts: a boiler, a propane tank with a commercial burner, and steam hose with a guide tube and a nozzle (Fig. 4.5) (Naruse and Suzuki 1975). The boiler had a cylindrical shape, was 20 cm in diameter and 40 cm in length, and had 12 fire tubes. It could hold 9 L of water, of which 6 L could be safely vaporized. It was designed to withstand a pressure of 1 MPa and had a safety valve set to work at 600 kPa. The normal operating pressure was 300 kPa. The net weight of the system (without water and propane) was ~60 kg.

The propane tank could hold 20 kg of propane. The steam hose was 25 mm in diameter and 10 m long. A steel 1.5-m-long guide pipe with a brass nozzle was connected to the hose. In the same manner as the Hodge drill, the nozzle had seven holes, each 1.5 mm in diameter.

The drill was transported on a wooden sledge. After filling the boiler with warm water (20 °C), the burner was turned on. At an ambient temperature of −20 °C, it took approximately 30 min for the boiler to reach the operating pressure of 300 kPa. The mean diameter of the holes was ~30 mm. The average rate of penetration was 27 m/h in firn with temperature of −29 °C, and the amount of water consumed was ~0.2 L/m.

4.1.6 Heucke Steam Drill

The drilling device of E. Heucke, from the Commission for Glaciology, Bavarian Academy of Sciences, Munich, Germany consisted of a portable steam generator, rubber hose, and a drilling pipe with interchangeable tips (Fig. 4.6)

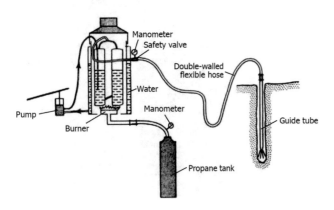

Fig. 4.4 Steam-operated ice drill of the Laboratoire de Glaciologie (Gillet 1975)

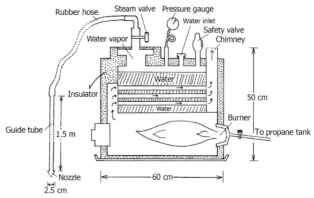

Fig. 4.5 JARE steam-operated ice drill (Naruse and Suzuki 1975)

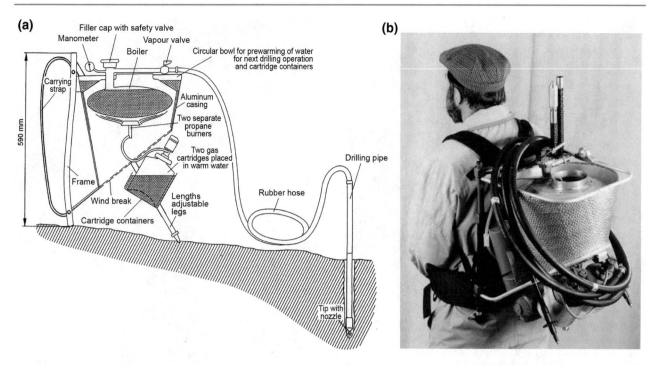

Fig. 4.6 Steam drill of the Bavarian Academy of Sciences: **a** diagram of the entire drilling device; **b** complete drilling device as a load (Heucke 1999)

(Heucke 1999). The main part of the steam generator, namely a spheroid boiler, was welded from pre-formed copper parts, and 16 heat exchange fins ensured efficient heat transmission between the gas flame and the water. The filler cap (which also served as a safety valve) was a standard part from the motor-car industry. The working pressure lied between 70 and 170 kPa, and the safety valve was opened at 180 kPa. The corresponding operating temperature was from 115 to 130 °C.

The capacity of the boiler was 4.4 L, which allowed for ~1 h of drilling operations. The boiler was heated by two gas burners to which two gas cartridges were attached, each containing 450 g. Small gas containers (cartridges in particular) cool down very quickly as the gas is consumed, and thus the burner's output soon diminishes owing to the falling vapor pressure of the gas. To avoid this, the gas cartridges were placed in warm water. It was possible to replace empty cartridges without turning off the equipment. There was alternative solution available for connecting larger gas containers.

With the help of a circular bowl above the boiler, heat was recovered from the waste gas and could be used to generate water by melting snow or to heat the water in the bowl for the next boiler filling. The entire casing was attached to an adjustable carrying frame via aluminum pipes. Two telescopic legs allowed for horizontal and safe placement even in sloped terrain. The weight of the device (without the hose, the drilling pipe, and accessories) amounted to 7.5 kg.

The rubber hose consisted of three hoses: a steam-conducting inner hose made of PTFE; a middle hose made of coiled aluminum, which reduced heat radiation from the hot inner hose; and an outer hose made of rubber reinforced with woven fabric, which was mainly responsible for the mechanical stability of the rubber hose (Fig. 4.7). Helical synthetic cords served to maintain the spacing between the individual hoses. The OD amounted to 24 mm, and the weight was ~0.4 kg/m. The combined weight of the drilling pipe and the hose couplings made it possible to drill

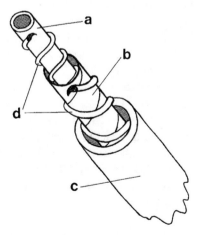

Fig. 4.7 Construction of the hose: **a** steam PTFE hose with ID/OD of 7/9 mm; **b** aluminum strap (coiled) as a heat reflector; **c** rubber hose with ID/OD of 18/24 mm; **d** helical cord to maintain the spacing (Heucke 1999)

(a)   **(b)**

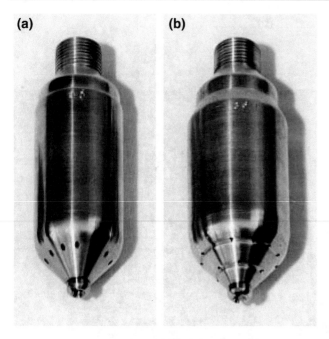

Fig. 4.8 Tips: **a** 30-mm drill tip with 13 nozzle holes set parallel to the axis, usable in ice and firn; **b** drill tip with 25 nozzle holes set parallel to the axis, designed primarily for drilling 40-mm holes in firn (Heucke 1999)

Fig. 4.9 Steam drilling to place ablation stakes into the surface of Eel Glacier, Mount Anderson, Washington, April 2015 (*Photo* B. Baccus from Irwin, n.d.)

down to depths of ∼13 m in water-filled boreholes before weight and buoyancy balanced out.

The drilling pipe was a rigid guide piece and was divided into two parts to facilitate transportation. Before drilling, they were screwed together and with the tip and were connected to the rubber hose. The total length of the guide pipe amounted to ∼1.4 m, and its weight was 1.4 kg. Various drill tips could be used. To create holes as smooth as possible with the highest penetration rate, tips with several fine nozzles, arranged in such a way so that steam could be applied evenly to the entire area to be melted, were suggested (Fig. 4.8).

Steam drills of the type described here were or are in use in Alaska, the Alps, Antarctica, Central Asia, the Himalayas, the Olympic Mountains (Fig. 4.9), Iceland, Patagonia, and Scandinavia at temperatures down to −14 °C (Antarctica) and heights up to 4000 m on ice (Abramov Glacier, Kyrgyzstan) and 4800 m on firn (Mont Blanc, France). The gas supply chosen varied according to local availability: operation with cartridges was carried out where these were available and when low overall weight was important; supply from gas bottles was used when there were enough helpers or snowmobiles.

Drilling could begin after 15 min of warm-up time. The rates of penetration decreased slightly from top to bottom. As an example, when drilling in ice down to a depth of 12.4 m at Vernagtferner, Austria, the penetration rate decreased from 35 to 20 m/h. The reason for this was that the portion of the rubber hose that was subjected to heat losses due to water cooling in the drilled hole constantly increased. The usable hole diameter obtained with a tip 21 mm in diameter ranged from 30 to 35 mm at the top and was ∼25 mm at the bottom after the completion of the drilling operation (this was measured using pipes of various diameters that were lowered on strings).

The drilling progress in firn was considerably faster, even though drilling was generally done with a larger tip. There was a gradual decrease in penetration rate as well: from 55 m/h near the surface to 15 m/h at the bottom of 27.5-m-deep holes. However, in these cases and unlike in ice, the main cause was the density of the firn, which increased with depth.

4.1.7 Danish Steam Drill

This steam ice drill was designed by C.E. Bøggild for automatic glacier ablation measurements using pressure transducers (Dyrby 2015). Sensors had to be installed into shallow holes with a minimal diameter of 25 mm. The operating pressure of the drill was 600 kPa; the working steam rate was 28 kg/h. Fuel consumption (propane) amounted to 3.2 kg/h. The guiding tube had a diameter of 38 mm. The drill weighed 68.6 kg, including the boiler (41 kg), the pump (9.2 kg), hoses (5×2.5 kg), and a guiding tube (5.9 kg).

The fastest average penetration rate for a 20-m-deep hole achieved was as high as 80 m/h, whereas the average penetration rate to make the deepest (30-m-deep) hole decreased to 30 m/h. It seems that Bøggild's equipment was the fastest portable steam ice drill because of its high operating steam pressure. However, this drill was much heavier than its predecessors.

4.2 Large-Diameter Steam Drills

Large-diameter steam ice drills are used for the construction of water-supply and sewage-disposal wells. Fresh water in Antarctica and Greenland can be found everywhere, but it in the wrong phase: it exists as either ice or snow. Ice or snow must first be changed to liquid water before it can be used. One of the most effective methods creating water supplies on the ice sheets is the construction of an in situ water well, which is simply a hole that is melted through snow until the vertical advance is intercepted by impermeable firn, where meltwater ponds.

4.2.1 Rodriguez Wells

The first experimental water well was constructed at Site 2, Greenland in 1958 (Mellor 1969). A vertical shaft was steamed down through the snow and water-impermeable firn was reached at a depth of ~ 40 m. Meltwater ponded at this level and a chamber was thawed out. Steam was supplied to the well to maintain the water level and compensate for heat loss, and water was pumped to the surface for use in the camp.

In 1959, R. Rodriguez from the USA ERDL (US Army Engineer Research and Development Laboratories) developed a water supply system which is now known as the Rodriguez Well or Rodwell. The set of equipment used

consisted mostly of modified commercial items and included (Fig. 4.10): a diesel-fired steam generator capable of producing 1.14 MPa of saturated steam at 190 °C at a rate of ~ 6 kg/min; a melting-drill bit assembly for melting a well shaft into the ice; a melting-pump bit assembly for melting glacial ice and pumping meltwater to the surface; a gasoline engine-powered cable winch for raising and lowering the bit assemblies; an A-frame and two wanigans; a 22.7-m^3 insulated and heated water storage tank; and the necessary rubber hose to convey the steam from the generator to the bit assemblies and to convey the meltwater from the pool to the storage tank.

The first water well was started at Camp Century in Greenland in July 1959, where a shaft 1.06 m in diameter was melted downwards with a 0.914-m-diameter melting bit (Schmitt and Rodriguez 1963). The thermal power delivered to the melting bit's nozzles was ~ 220 kW$_{th}$. A vertical shaft, 42.4 m deep, was melted into the glacier in ~ 30 h. Near the surface, the penetration rate was 5.5 m/h, whereas at the final depth it decreased to 0.45 m/h. At this depth, the meltwater produced no longer permeated into the surrounding firn but accumulated in the bottom of the shaft. When the pooled meltwater was ~ 1.4 m deep, the melting bit was replaced with a combined melting–pumping bit assembly. This bit assembly permitted the continuous and simultaneous injection of steam while pumping water to the surface when required.

Afterwards, steam was introduced during a total of 295 h over a 23-day period. Water was intermittently pumped to the surface via an electrically powered submersible pump at a rate of 94.6 L/min. A total of 525 m^3 of water were produced and 114 m^3 were used for the production of steam. With this arrangement, it was determined that at least 32 m^3 of water could be produced each day. Upon the completion of the melting–pumping period, an observer was lowered to physically examine the subsurface cavity that had been made. What he found was a symmetrical bell-shaped configuration in the bottom portion (Fig. 4.11a). The cavity was ~ 15 m high, with a maximum diameter of approximately 15 m that gradually tapered into the smaller cylindrical entrance shaft. The bottom of the cavity was 51.8 m beneath the glacier surface and the sidewalls of the cavity and shaft were lined with an impermeable ice coating. It was fairly conclusive that once the meltwater cavity was formed, the meltwater would be contained and the shape of the cavity could be controlled by the quantity of meltwater stored in the subsurface reservoir.

In the summer of 1960, operations similar to the original test well were resumed (Fig. 4.12); only the melting and pumping efforts was intensified. The melting equipment operated 24 h a day, six days per week. The unit was run for 102 days and a total drinking water supply of 4305 m^3 was produced. After the completion of the second summer

Fig. 4.10 Equipment of the
Rodriguez Well (Schmitt and
Rodriguez 1963)

Fig. 4.11 Rodriguez well at
Camp Century: **a** section of the
pilot after the first test season;
b water well after two seasons of
operation (Lunardini and Rand
1995 after Schmitt and Rodriguez
1963)

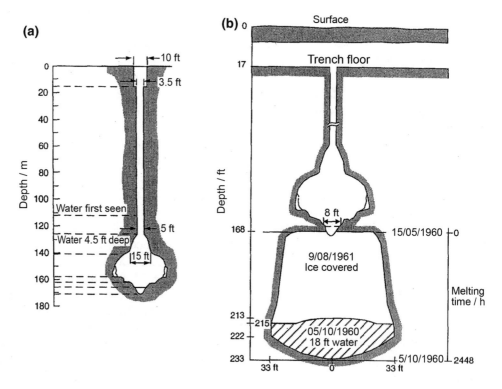

season tests, an observer again was lowered into the well, who found a cylindrical cavity immediately beneath the previous year's bell-shaped cavity (Fig. 4.11b). Its dimensions were ∼20 m in diameter, ∼20 m in height, and 70.7 m beneath the ice sheet surface.

During the first-year operations, where meltwater was pumped to the surface almost as quickly as it was produced,

the narrow part of the bell-shaped cavity was formed. During the second-year operations, in which it was determined that with the heat source available greater quantities of water could be pooled, making a cavity much larger in diameter was possible.

The drilling–melting operations were continued on the next summer. Owing to the fact that the well had reached a

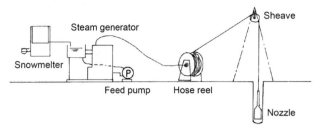

Fig. 4.13 JARE steam drilling system (Ishizawa and Takahashi 1994)

Fig. 4.12 View of the Rodriguez water well from the top at Camp Century (Clark 1965)

depth of over 150 m, which was nearing the maximum depth for the type of deep-well pump being use, the well was relocated in May 1962 after approximately 16,000 m^3 of water had been produced. A second service well reached a depth of ~92 m and a maximum pool diameter of 30–36 m. It yielded 22,730 m^3 of water over 2.25 years (Mellor 1969).

The first experimental water well at Amundsen–Scott, Antarctica station was built in 1972 for the old South Pole station (Lunardini and Rand 1995). A significant aspect of this well was that the ambient firn temperature was −51 °C, much lower than the firn temperature of −28.9 °C at Greenland sites. Despite this low temperature, the water well functioned well without significant adverse freezing problems. Its operation was halted owing to a frozen fuel line at the surface boiler; the well was then shut down and abandoned after having supplied approximately 416 m^3 of potable water. Afterwards and for many years, water at the South Pole was obtained via various energy-intensive and labor-intensive approaches for gathering and melting snow. In 1993, the Rodwell concept started to be tested again; however, hot water was used as a heating medium (see Sect. 3.2.9).

4.2.2 JARE Large-Diameter Steam Drill

This drill system was developed in relation to the construction by the JARE of Asuka station at the Japanese inland Antarctic station located on Queen Maud Land. They proposed using wells for the sewage disposal system (Ishizawa and Takahashi 1994; Ishizawa et al. 1990). This steam drill consisted of a snow melter, a steam generator, a hose reel, and a nozzle (Fig. 4.13).

First, snow was put into the snow melter by hand, and then water was supplied to the water tank of the steam generator automatically by a feed pump when the water level of the generator was lower than a set level. The steam pressure was set to 785 kPa, and the burner for steam generation stopped when the pressure reached 883 kPa. In addition, a safety valve operated when the pressure rose above a predetermined level. The steam was fed to the nozzle through a hose, which was lowered automatically by a low-speed DC motor attached to a leg of the tripod (Fig. 4.14). The lowering speed was manually set by judging whether the nozzle was touching the bottom of the borehole or not. The nozzle consisted of a skirt and 19 small nozzles (Fig. 4.15).

The drilling operation was carried out on February 9, 1987 and steam was supplied for 9 h. The lowering speed of the nozzle was 4 m/h between the snow surface and a depth of 20 m. At depths lower than approximately 20 m, speed decreased gradually. The drilled depth was 27.5 m after 9 h of operation, and the hole diameter was approximately 40 cm. The burner of the steam generator sometimes misfired during strong winds. The total amount of kerosene consumed by the snow melter and the steam generator was

Fig. 4.14 Tripod and hose winch of the JARE steam drilling system (Ishizawa and Takahashi 1994)

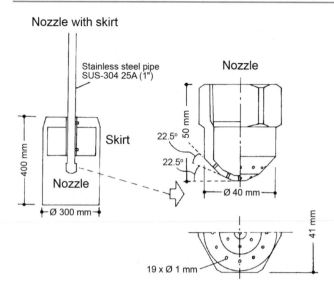

Fig. 4.15 Nozzle schematic of the JARE steam drill (Ishizawa and Takahashi 1994)

110 L. The cumulative amount of sewage was 1077 m³, which was expected to be used over 5 years.

4.3 Steam Ice-Coring Drills

To the best of our knowledge, only one type of steam ice-coring drill was developed. In an attempt at increasing efficiency when drilling dry holes in ice, the steam thermal ice-coring PTBS-112 drill was designed and tested at the LMI (Fig. 4.16) (Chistyakov et al. 1988; Solov'ev et al. 1989). At the surface, a steam generator was filled with water. At predetermined times, tubular heaters with a rated power of 4 kW were switched on and, when the pressure in the steam generator reached a preset limit, an electromagnetic valve opened, allowing the steam to pass through a steam pipe towards the thermal head.

The OD and the ID of the drill head were 112 and 98 mm, respectively. The thermal head had a spiral steam inner channel with a total length of 1.5 m. Steam was first delivered to the lower part of the head and then passed through the spiral channel, warming up the head. In the upper part of the head, steam was forwarded either to an annular space where it mixed with meltwater or directly to the water lifting pipe. The second design proved optimal.

During the melting of an annular groove at the bottom of the hole, meltwater was directed to a water tank via an air-flow created with a vacuum pump. The length of the core barrel was 2.5 m, the length of the drill was 6.5 m, and the weight of the drill was 75 kg.

In February–March 1985, the steam thermal ice-coring PTBS-112 drill was tested at Vavilov Ice Cap, Severnaya Zemlya Archipelago, and a 50-m-deep hole was drilled. The

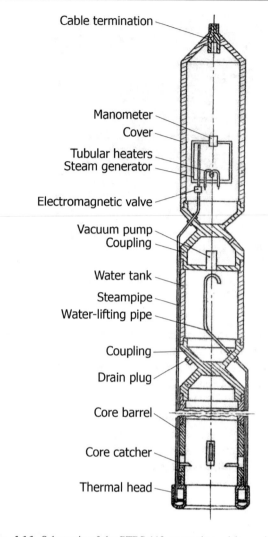

Fig. 4.16 Schematic of the PTBS-112 steam thermal ice-coring drill (Solov'ev et al. 1989)

average length of a run was 0.86 m, the core diameters were 85–92 mm, and the specific quantity of recovered water (meltwater plus condensed steam) was 5.75 L/m. The drilling achieved an average rate of penetration of 8.75 m/h, which is 2–3 times higher than that of ordinary thermal ice-coring drills. The capacity of the steam generator was enough to drill 1.5–2.0 m of ice cores.

4.4 Summary

Steam is created by heating the water in a boiler using propane or butane in portable drills and Jet A-1 fuel in large-diameter steam drills or electric heaters in steam ice-coring drill. Drills with steam jetting can be used even in ice containing small rocks, pebbles, or sand. The drilling rate may decrease by as much as a factor of 10, but holes can nevertheless be drilled. However, given the relatively high specific-heat capacity of hot water in comparison to hot air [1.01 J/(g K) for air at 100 °C and 4.22 J/(g K) for water at 100 °C (Gieck and Gieck 2006)], hot-water drills have a better capacity to melt ice than steam drills. That is to say that, near atmospheric pressure, which is essentially the pressure at the melting interface, the energy per unit volume would be over 4–5 times greater for water than for vapor.

Portable steam drills proved to be the best equipment for drilling holes with depths up to 10–15 m in ice or 30 m in firn, especially when the low weight and the low cost of the equipment are important factors (no pump is needed). These depths mentioned above can be considered as the practical limit for holes made with portable steam drills because the heat losses in the feeding hose become too large. Premature cooling causes the steam to start to condense before it reaches the nozzle rather than outside the nozzle. Any improvements in the drilling rate or the depth of holes will have to come either by increasing the operating steam pressure or by using a hose with perfect heat-insulation properties. However, this would lead to an increase in the size of the boiler and the diameter of the hoses. The heat losses through the hose completely dominate the performance of portable steam drills.

Large-diameter water and sewage-disposal wells can be successfully drilled using steam drills. After enough meltwater has accumulated in the hole, it is pumped from the well to a heat exchanger on the surface. This water is then converted into steam and returned to the well, where it is discharged through a spray nozzle. The shape and size of the ponding cavity depend on the amount of energy applied, the amount of water circulated, and the pattern of the spray discharge. Aside from the obvious benefits of not having to mechanically handle snow for surface melting operations or having to provide large insulated storage tanks for the meltwater, in situ wells provide a useful storage area for sewage after they are pumped dry.

The depth of the cavity is another dimension effecting the quantity of water that can be extracted from a well. Deeper wells require longer pipelines, larger pumps to pump against greater static heads, bigger servicing rigs to carry heavier loads, and more electrical power and a greater steam capacity to deliver the required thermal power output. With bigger equipment, complexity increases to some degree because of the increased steam and water pressures and the electrical requirements. Most importantly, the cost of the equipment and operations increase. The deepest water wells were drilled to depths slightly over 150 m.

Most Antarctic research stations are situated on the coast and the simplest solution for sewage and wastewater disposal in coastal regions is the discharge of effluents into the sea. However, sewage disposal at inland stations is a challenging task. Recently, the impacts of sewage have begun to receive more attention, as have the treatment methods required to mitigate these impacts. Although the Protocol on Environmental Protection to the Antarctic Treaty 1991 (known as the Madrid Protocol) guides all activities in Antarctica, the actual management of wastewater by the many countries operating in Antarctica varies considerably, from no treatment to advanced sewage-treatment methods. There is a growing movement towards regarding the obligations under the Madrid Protocol as the bare minimum. Storing treated sewage in ice wells could be considered as the optimum procedure for waste burial in the Antarctic and Greenland inland at the present time.

References

Chistyakov VK, Skurko AM, Zemtsov AA et al (1988) Eksperimental'niye burobie raboti na Severnoi Zemle v 1975–1985 gg. [Experimental drilling operations at Severnaya Zemlya in 1975–1985]. Geograficheskie i glyatsiologicheskie issledovaniya v polarnikh stranakh [Geographical and glaciological investigations in polar regions] Leningrad, Gidrometroizdat, pp 33–42. (in Russian)

Clark EF (1965) Camp Century evolution of concept and history of design construction and performance. USA CRREL Tech. Rep. 174

Dyrby K (2015) Recommendation of components for an Autonomous Automatic Mass-balance Station (AAMS). Bachelor's thesis. Arctic Technology Center. Technical University of Denmark, Lyngby, June 2015, 53 p

Gieck K, Gieck R (2006) Engineering formula, 8th edn. McGraw-Hill Professional, New York

Gillet F (1975) Steam, hot-water and electrical thermal drills for temperate glaciers. J Glaciol 14(70):171–179

Heucke E (1999) A light portable steam—driven ice drill suitable for drilling holes in ice and firn. Geogr Ann 81(4):603–609

Hodge SM (1971) A new version of a steam-operated ice drill. J Glaciol 10(60):387–393

Howorka F (1965) A steam-operated ice drill for the installation of ablation stakes on glaciers. J Glaciol 5(41):749–750

Irwin J (n.d.) It's been a bad year if you're a glacier. Washington's National Park Fund. Retrieved 28 May, 2018 from https://wnpf.org/its-bad-year-youre-glacier/

Ishizawa K, Takahashi A (1994) Borehole drilling for sewage disposal at Asuka Station, East Antarctica. Mem. Natl Inst Polar Res Spec Issue 49:212–217

Ishizawa K, Takeuchi S, Takahashi A (1990) Borehole drilling for sewage disposal and rise of the hole's bottom at Asuka Station East Antarctica. Nankyoku Shiryô (Antarctic Record) 34(2):145–155 (in Japanese)

Lunardini VJ, Rand J (1995) Thermal design of an Antarctic water well. USA CRREL Spec. Rep. 95–10

Mellor M (1969) Utilities on permanent snowfields. USA CRREL Monograph III-A2d

Naruse R, Suzuki Y (1975) A steam–operated drill used by the 14th Japanese Antarctic Research Expedition (1972–1974). Antarctic Record 53:53–56

Perspectives for Future Development of Thermal Ice-Drilling Technology

5

Abstract

Future development of thermal drilling systems is recommended to focus on reliable growth, safety, and environmental improvements, as well as increases in performance. Specific challenges related to improving thermal drilling technology include developing unconventional thermal ice-drilling methods; searching for new heating technologies; developing directional thermal-drilling methods; and designing automated drilling systems. The certain steps have been taken to test unconventional thermal ice-drilling methods like dissolution drilling, flame-jet drilling, and laser drilling.

Keywords

Dissolution drilling • Flame-jet drilling • Laser drilling • Automated drilling systems • Radiothermal generator

Future development of thermal drilling systems should be focused on reliable growth, safety, and environmental improvements, as well as increases in performance. The relative merits of the different ice-drilling systems can be combined on the basis of project requirements. To make systems more universal, simple thermal coring drill, hot points, and electromechanical auger drills can be reasonably included into a single lightweight set for shallow-depth drilling because the requirements of the surface equipment (e.g., mast, winch, and control system) are similar, but their missions are different. To take a different stance, in situation where difficulties while drilling the deep warm ice in the Antarctic and Greenland ice sheets may arise, thermal drills having a pumping system to remove meltwater and store it in a chamber within the drill can be considered as reasonable alternative to deep electromechanical drills (Talalay et al. 2015). In these cases, the surface drilling equipment is similar as well. As another example, it would be advantageous to use versatile hot-water drilling systems with either closed-loop (for drilling in firn) or open-circuit melting units (for drilling in solid ice).

Thermal drills can be used for some non-trivial drilling applications, such as sealing the shoe of casings in ice or to consolidate permeable near-surface snow–firn formations (Talalay 2014). Specific challenges related to improving thermal drilling technology include: (1) developing unconventional thermal ice-drilling methods; (2) searching for new heating technologies; (3) developing directional thermal-drilling methods; and (4) designing automated drilling systems. Possible ways of solving these problems are presented below.

5.1 Unconventional Thermal Ice-Drilling Methods

Common thermal ice-drilling methods employ either direct heat transfer from electric heaters or an intermediate heat-transfer agent, such as hot water or steam. In order to solve various drilling limitations and problems, such as relatively low rates of penetration, limits on depth, meltwater refreezing, the need for casing installation, and safety improvements, new drilling concepts are being continually developed and tested. Generally, these systems can be referred to as "unconventional" or "novel" drilling systems, and improvements are mainly focused on the extension of methods for melting ice.

5.1.1 Dissolution Drilling

To intensify the melting process, hydrophilic liquids (e.g., ethanol) that chemically dissolve ice can be used. In these cases, the total volume of liquid required can be reduced by up to 5–75% of the borehole volume (this depends on its temperature profile). It was found experimentally that the

© Geological Publishing House and Springer Nature Singapore Pte Ltd. 2020
P. G. Talalay, *Thermal Ice Drilling Technology*, Springer Geophysics,
https://doi.org/10.1007/978-981-13-8848-4_5

reaction between ethanol and water was endothermic and that the latent heat of ice dissolution by ethanol varied from 300 kJ/kg at a temperature of −2 °C to 100 kJ/kg at −24 °C.

Zagorodnov et al. (1994) suggested using hot points consisting of a pump, a heater, and a drilling head with a nozzle (Fig. 5.1). The pump allows for the circulation of a hydrophilic drilling liquid. This drilling liquid heats up when it passes through the heater. The schematic of this drill is similar to the small-diameter Electrochaude hot-water drill (see Sect. 3.2.1); however, drilling with a hydrophilic liquid would be faster owing to the potential energy of the solvent and also safer because it can solve the problem of water refreezing in the hole. At an ice temperature of −3 °C, the expected penetration rate of the drilling–dissolution method is 2.7 times higher than that of the drilling–melting approach.

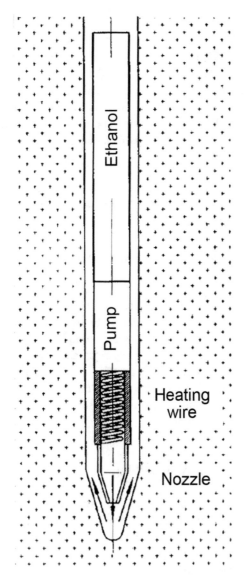

Fig. 5.1 Thermal dissolution drill (Zagorodnov et al. 1994)

5.1.2 Flame-Jet Drilling

Flame jets burn liquid hydrocarbon fuel at high rates by supplying a gaseous oxidant under pressure, typically compressed air or oxygen. Special nozzles facilitate rapid reactions, and the flame discharges at a very high velocity. Flame temperature is extremely high and provides high energy for melting ice. For a number of years in the 1960s and 1970s, flame-jet drilling machines were used in the mining industry for making vertical blast holes in hard rock or ore formations (Iagupov 1972; Maurer 1980). Even though this drilling method is not "novel" (the first attempts to use jet fire to drill in ice were made more than half a century ago), it surely is unconventional.

In 1964, a flame-jet drill designed by the Kazakh Polytechnic Institute (Alama-Ata, Kazakhstan) was used for penetrating Tuiuksu Glacier, Zailiiskii Alatau Ridge (Northern Tien Shan) (Brichkin et al. 1967). This rocket-type burner weighed 3–4 kg and was connected to a flexible pressure hose; it was hoisted in the hole with a steel cable and a hand winch (Fig. 5.2). The pressure hose was fixed to the steel cable at 2-m intervals. Kerosene and oxygen were discharged to the burner at pressures of 1.1–1.2 MPa and 1.0–1.1 MPa, respectively. Two holes were drilled to depths of 33 m and 23 m. The diameter of the first hole was 120–130 mm, and drilling it took 28 min (the average rate of penetration was 71 m/h). The diameter of the second hole was 140–150 mm, and drilling it took 26 min (the average rate of penetration was 53 m/h). During drilling, six oxygen containers (36 Ncum) and two compressed-air containers (12 Ncum) were used.

A new, modified drill was tested in August 1969 (Mikheev 1971). To decrease the weight of the drilling equipment, the oxidizing component was changed from oxygen to air, thus eliminating the need to deliver heavy oxygen cylinders to high-mountain regions. Air was provided using a modified EK7A compressor with a nominal pressure of 0.8 MPa at an output of 0.62 m³/min. The total weight of the drilling equipment was 600 kg. This new 4-m-long downhole unit included a VTBM-4 burner equipped with a special automatic check valve that ensured the optimum mixing ratio of the components and a 12-L fuel tank installed above the burner (Fig. 5.3). On the lower end of the burner, a bottom sensing gauge was attached. Four holes to depths of 28, 53.5, 72, and 53 m were drilled. The hole diameters were approximately 140 mm near the mouth and 80 mm at the bottom. The drilling times were 3.25 h for the 53.5-m-deep hole and 5.55 h for the 72-m-deep hole. The average fuel consumption was 0.15–0.17 L/m. In the early 1970's, flame-jet drilling on Tuiuksu Glacier was continued. A total of nearly 20 holes were drilled down to bedrock; the deepest hole was 102 m deep (Makarevich et al. 1984).

Fig. 5.2 Flame-jet drill used at Tuiuksu Glacier in 1964 (Brichkin et al. 1967)

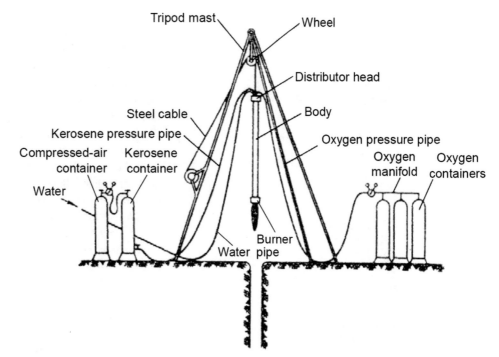

Perhaps the best-known flame-jet drilling project was accomplished at Site J-9 (82°22′S, 168°37′W), Ross Ice Shelf in the season of 1977–1978 (Browning 1978; Browning and Somerville 1978). Preliminary tests of the system took place in a 61-m-deep ice well in the CRREL, Hanover. In July 1977, the entire drilling system was air-lifted and transported to Dye 2 in Greenland, where a test hole was drilled to a depth of 183 m. Then, the flame-jet drilling system was delivered to Site J-9 and two access holes were drilled through the 420-m-thick ice of Ross Ice Shelf.

The flame-jet drill was powered by an internal burner (similar to a rocket engine) with a 16-mm-diameter exit nozzle whose reactants were compressed air and fuel oil provided at a high pressure (Fig. 5.4). The system comprised two air compressors, each with its own diesel engine. The first was of the conventional contractor type, delivering 17 m³/min at 0.96 MPa to the suction side of a booster unit capable of raising the air pressure to 8.3 MPa. However, it was found that this high pressure was not required; the initial air pressure (when starting the hole) was 2.1 MPa and the pressure during regular drilling was just over 3.4 MPa. The drill skid included a hydraulic crane and hose-handling components (Fig. 5.5). Booster diesel was also used to power a hydraulic pump, high-pressure fuel and water pumps, and various other auxiliaries. The total weight of the system was ~20 tons.

The first hole was completed on December 2, 1977 after 9 h of drilling. Hose handling was a serious problem. Three hoses and the support cable were stretched in long loops

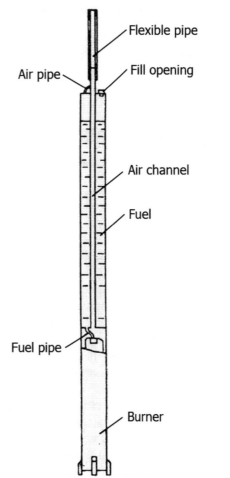

Fig. 5.3 Downhole assembly of the flame-jet drill used at Tuiuksu Glacier in 1969 (Mikheev 1971)

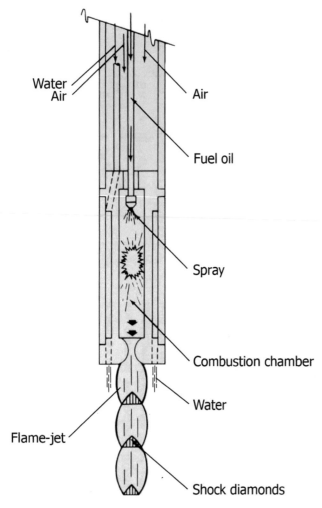

Fig. 5.4 Downhole assembly of the flame-jet drill used at Site J-9, Ross Ice Shelf in 1977 (Browning 1978)

over the ice shelf surface from the compressor to a hose wheel attached to a tractor 230 m away and then back to the drill skid positioned next to the compressor. However, the unsupported hoses sank into the snow. For example, the hydraulic hoses sank ~1 m deep into the snow, forming narrow slots. To eliminate this problem and to reduce frictional drag, the feed hoses had to be lifted and pulled by hand until the drill had reached a depth of ~90 m, the point at which gravity forces became sufficiently large to provide the required tension. This process required a crew of nearly 10 handlers.

To maximize hole size and reduce the possibility of oceanographic tools sticking in the coldest portion of the hole (−27.9 °C at a depth of 10 m), a penetration rate of only 36.5 m/h was maintained in the top 100 m of ice to produce a hole diameter of more than 450 mm. When approaching to the base of the ice shelf, the penetration rate

was increased to 110 m/h, which was its maximum value. When the hole reached the ocean cavity, the drill was forced upward as much as 15 m and the hoses and support cable reversed directions over the hose wheels above the hole. The pressure difference between the hydrostatic head in the hole and the pressure of the sea water beneath was due to the fact that the density of the water filling the hole during flame drilling was reduced appreciably by the presence of rising gases. To prevent the water in the hole from freezing and to keep the access open, a 60-kW heating cable was immediately suspended through the length of the water column contained in the hole just drilled. However, after four days, the ice had frozen the cable tightly in the hole and the hole was no longer of use.

A second hole was begun on December 14 within 1.5 m of the first one in hope that the heat already added to the ice in the adjacent hole would reduce freeze-in rates. The hole was completed in 9 h, which was the same time that it took for the drilling of the first hole. The following day, this hole was enlarged to an average diameter of 0.46 m via flame reaming, using the same drill for a second 9-h period. The diameter of the hole (measured using a large caliper) showed large irregularities over the entire hole length. There were many zones where the hole was narrow; on the other hand, hole diameters of over 0.76 m were measured at several depths. The reason for the formation of these irregularities was likely the non-uniformity of the drill feeding; if the advance rate is too low, hole diameter becomes excessive.

After successful jet-flame drilling in Antarctica, Browning Engineering Corporation secured a grant from the US National Science Foundation to develop a suspension core drill that would make it possible to use flame-jet technology for another challenging task: studying rock formations under ice caps and glaciers (Michaelides 2012). However, the results of this project are not known yet.

Flame-jet burners are compact and not extremely heavy, but the equipment that supplies the fuel and oxidant is bulky and heavy. Although the temperature of the flame may seem inappropriately high, there are significant thermal losses in the borehole, reducing the potential penetration rates. The considerable drawbacks of jet-flame drilling systems (high weight and power consumption, complexity, and safety issues) are the main reasons that this technology has not been used in glacial researches since the 1970s.

5.1.3 Laser Ice Drilling

Laser devices emit light through a process of optical amplification based on the stimulated emission of electromagnetic radiation and are widely used in the industry for

Fig. 5.5 Overview of flame-jet drill, Site J-9, Ross Ice Shelf, December 1977 (Browning 1978)

cutting different materials. Proposals for laser drilling date back to approximately five decades but, in general, lasers were limited by their relatively low power levels (Maurer 1980). However, during the last decades, intensive research has been made into the development of systems to improve the efficiency of laser-associated equipment and to transmit high-power lasers over long distances via fiber optic cables.

Zeller et al. (1989) suggested using CO_2 lasers as a field device to cut individual firn and ice cores for sample preparation. During the 1990–1991 field season at Windless Bight near Ross Island, a 25-W continuous infrared CO_2 laser was successfully used for ice-core processing. The advantage of CO_2-laser cutting systems is that the beam is emitted at an infrared wavelength, which is absorbed in a very short distance in ice.

Zeller et al. (1989) also proposed the design of a laser coring drill (Fig. 5.6), which would be constructed out of two thin-wall steel tubes arranged concentrically with the space between the tubes being used as the location of the optical waveguide fiber and the vacuum line to the scavenger pump. When the ice core breaks the beam, the laser is switched to the core-cutting mirror, which cuts the sample off in the core-barrel area. The length of the core barrel is expected to be 1.5 m. In a scaled-up version, the core barrel could be extended to 3 or 4 m. Unfortunately, a proof-of-concept prototype was not developed.

Sakurai et al. (2016) tested a CO_2 laser at 10.6 µm, a wavelength that ice strongly absorbs, to drill through ice. The rate of penetration increased nearly in proportion to the laser intensity. For an intensity of approximately 50 W/cm^2,

Fig. 5.6 Laser coring drill (Zeller et al. 1989)

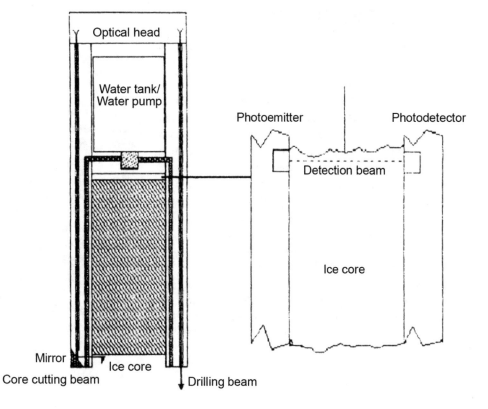

Fig. 5.7 Ice block penetrated via CO_2-laser irradiation; hole diameter was ~4 mm and hole length was ~13 cm (Sakurai et al. 2016)

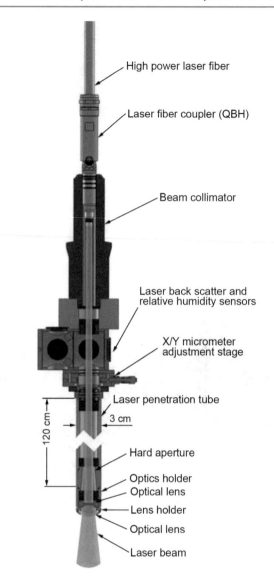

Fig. 5.8 Schematic of the DLP probe developed at Stone Aerospace, USA (Stone et al. 2018)

the melting speed was 14.4 m/h for snow with a density of 153 kg/m^3 and 2.9 m/h for solid ice. Results also showed that, in a vertical melting orientation, meltwater accumulated in the hole and lightning flashed, reducing the penetration rate. The only way to continue testing was to ensure that the test ice block was tilted down and that the beam was impinging on the side of the block so that the water generated could run out, leaving only exposed ice (Fig. 5.7). The desired behavior is obviously to have all the energy be absorbed by the ice and little, if any, by the water.

In May 2016, a series of tests were conducted in the laboratory of Stone Aerospace, USA using a 3.2-cm-diameter direct-laser-probe (DLP) operating at varying power levels from a 1070-nm ytterbium fiber laser (Stone et al. 2018). A long tube served as the body and the

collimating and focusing optics were place at its end (Fig. 5.8). A laser power coupler, a beam collimator, and an optical alignment stage (XY) were included at the top of the probe. The system was fired at successively increasing power levels from 50 W to 2.5 kW (Fig. 5.9). The rate of penetration into an ice block with a temperature of −26 °C for a laser power of 2.5 kW was greater than 12 m/h and was accelerating, but the size of the ice block limited the test to 1 m of penetration. The temperature was monitored at the nose cone and it peaked at only 30 °C at the highest power level tested (2.5 kW). Later, in August 2016, tests with a laser power of 5 kW demonstrated penetration rates of 22 m/h in ice blocks and temperatures identical to those of the earlier tests.

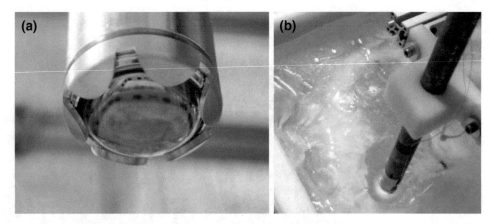

Fig. 5.9 DLP probe: **a** optics front end; **b** entering ice at a power level of 50 W; the violet glow is from the 1070-nm laser beam and is interpreted by the camera's CMOS sensor as violet owing to the lack of near-IR filtering (near-IR is not visible to the human eye) (Stone et al. 2018)

Laser drilling has high potential for ice drilling, but the effect of meltwater accumulation from the irradiation spot must be considered. The laser-beam output from the fiber should remain close to the ice (optimal distance depends on the focusing optics), thus minimizing the meltwater effect. In order to achieve the optimum performance from a laser, the rate at which the drill is lowered must be carefully controlled. If the drill is lowered too quickly, the laser comes too close to the bottom of the hole, resulting in reduced efficiency. On the other hand, if the drill is lowered too slowly, drilling speed will be reduced because of meltwater flashes. Another problem that needs to be solved is the refreezing of the meltwater and the freezing-in of the optical fiber in the borehole.

5.2   New Heating Technologies

Because issues related to energy saving and the intensification of drilling processes are always of current interest, engineers are looking everywhere for new heating technologies. New ceramic heaters constructed using aluminum nitride (AlN) can be considered as a potential alternative for designing thermal drill heads (Fig. 5.10). AlN heaters are especially suitable for applications that require a clean, non-contaminating heat source and operate at temperatures of up to 400 °C (Specialty Heaters 2009). In addition, they provide high electrical insulation and superior chemical resistance compared with traditional metal heaters. Moreover, AlN heaters can provide a high power density of up to 155 W/cm^2 in a very small package and uniform temperatures over the surface of the heater. Because AlN heaters have never been used in ice-drilling projects, it is not possible to determine their suitability for drilling in ice, especially in terms of their external pressure limit. This can only be determined after laboratory and field experiments.

Fig. 5.10 Watlow's ULTRAMIC ceramic heater constructed using aluminum nitride (AlN) (Specialty Heaters 2009)

Foil heaters are thin, flexible components consisting of an etched-foil resistive heating element laminated between layers of flexible insulation. They provide fast warmup, consistent heat distribution, and extended heater life in a broad range of applications from medical diagnostics to defense and aerospace missions. A melting probe designed by the Institut für Weltraumforschung (IWF, Space Research Institute), Graz, Austria had a melting tip with three heaters placed inside it: a ring-shaped heating foil at the top of the tip and two rectangular heating foils electrically connected in

Fig. 5.11 High-voltage water heater: a conductive fluid passes through high-voltage AC electrodes (Stone et al. 2018)

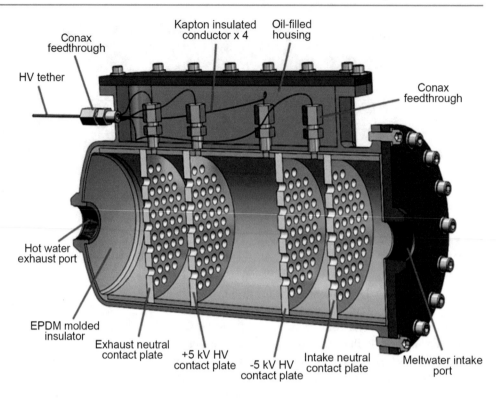

series fixed at the inner sidewall (see Fig. 1.73b). However, foil heaters do not work in water under pressure. The materials used in foil heaters are waterproof, but the edges are not sufficiently sealed for immersion. Custom designs can include increased border areas and sealed lead-wire connections that make these heaters immersible in water up to limited pressure ranges (2–3 MPa).

Carbon fiber heaters can be used to heat drill bodies. They are thin and light, have a large surface area, and can be produced in any desired shape. They are created by impregnating carbon ink (carbon in nanometer form) into a fiber glass sheet, which is sealed between two fiber glass sheets to eliminate any gases. These heaters do not use any glue or plastics. Power can be supplied to the material by attaching leads, which use either alternating or direct current, at power levels ranging from a few volts to hundreds of volts. The ability of carbon fiber heaters to work in water under high hydrostatic pressure has yet to be investigated.

Stone et al. (2018) suggested heating the water in the downhole unit by transmitting high-voltage low-AC power through a moving conducting fluid (Fig. 5.11). This generates resistive heating in the fluid with 100% efficiency without inducing electrolysis (because of the AC voltage). The resistivity of the process fluid can be tuned over a wide range by controlling the concentration of polar molecules in the fluid. This tunable resistivity allows for unprecedented power densities to be achieved. In laboratory tests, power densities of over 600 kW/L were achieved.

5.3 Directional Thermal Drilling

Directional thermal drilling is an important and challenging task to study moulins, to access cavities and sub-ice tunnels, and to bypass non-recovered downhole drilling equipment or natural obstacles in ice than cannot be re-drilled/melted. Such natural obstacles can be detected ahead of the vehicle using an acoustic sonar (Schwander et al. 2012; Zimmerman et al. 2001) or a synthetic aperture radar (Stone et al. 2018) installed at/near the melting tip. The integrity and precision of sensors that are small enough for these purposes have to be investigated; however, preliminary tests showed that a radar system can detect and model for obstacle-avoidance behaviors for 1-m objects 80 m in advance.

When a vertical well is being drilled, the axis of the borehole will not be truly vertical. There will always be a tendency for the drill to deviate from its intended course owing to joint misalignment, displacement of the drill's center of gravity from its geometrical center, manufacturing inaccuracies of the drill components, vibration of the downhole drill stem, etc. (Talalay et al. 2015). This is a well-known problem when drilling straight holes. The easiest way to deflect the trajectory of a borehole while thermal drilling is to perform asymmetrical contact heating or hot-water jetting.

Kelty (1995) suggested to individually control cartridge heaters in the hot-point thermal tip (the probe used had eight cartridge heaters spaced radially around the hot point) to steer the probe (see Figs. 1.62 and 1.63). Similarly, the IceMole maneuverable subsurface ice probe could change its melting direction via differential heating of the melting head and optional sidewall heaters (Dachwald et al. 2014).

The ability of hot-water drills to deviate was occasionally proved during tests on the Langjökull ice cap in June 2005 (Thorsteinsson et al. 2008). A nozzle with seven 0.8-mm-diameter holes (see Fig. 3.42a) was used during initial tests; however, the hole dramatically deviated from the vertical. When the drill stem was pulled to the surface, it was found that one of the holes in the nozzle was blocked, creating a deflecting force. The VALKIRIE thermal probe had five nose-mounted melting nozzles on the front of the vehicle: one nozzle directed downwards and four lateral-turning jet nozzles (Stone et al. 2018). This probe was able to run up to two lateral jets (at 90° from one another) simultaneously. Tests showed that the vehicle was able to achieve a deviation rate of $\sim 0.25°/m$ and then be brought back to vertical.

The most important matter of deviational drilling is the passableness of the thermal drill body or hot-water drill stem into an already-drilled borehole. The maximal deviation rate has to be carefully identified depending on the expected borehole diameter, the size of the drill, and the expected refreezing rate (where applicable) to ensure that the melting probe does not get stuck in the hole.

5.4 Automated Thermal Drilling Systems and Power Supply

The mean annual temperature of the Antarctic inland is $-57\ °C$, with surface elevations higher than 3000 m asl. Thus, almost all over-wintering research stations, with a few exceptions (South Pole, Vostok station), are confined to coastal or near-coastal sites. Even though the Arctic is warmer than Antarctica, long-term observations at the Greenland inland and the polar ice caps are also extremely difficult because of the harsh environment and the remoteness of these locations.

It is clear that unmanned autonomous systems that could operate in Antarctica and the Arctic unattended through the polar winter constitute an unprecedented and attractive perspective (e.g., the autonomous hot-water ice drill, Beverly and Westneat 1982; and the RECAS thermal sonde, Talalay et al. 2014). All thermal probes for getting access to the subsurface layers of extraterrestrial planetary ice sheets are intended to be autonomous (see Sect. 1.4). The capability of autonomous systems and their domains of application have expanded significantly in recent years, with high-profile successes in civilian, scientific, and military applications (Itmi and Cardon 2016). Thus, developing a well-defined,

yet modifiable, mission plan throughout drilling operations seems practicable.

The key technical challenges have been, and remain, the reliable provision of sufficient electrical power to operate the thermal probes and provide the required environmental heating and an effective data-storage retrieval system. The power system of unmanned drilling systems has to operate continuously without service or refueling for long periods between summer inspections. Over the last 50 years, there have been many attempts to develop viable unmanned power supply systems for geophysical sensors and meteorological instruments (Dudeney et al. 1998; Clauer et al. 2014). However, the power consumption of these devices (typically within 50–100 W) is by one to two orders of magnitude lower than the power required for thermal drilling. Most unmanned geophysical sensors and meteorological instruments are designed as battery-operated systems. The best power density available today using Li-ion batteries is ~ 400 Wh/kg. The present impossibility of using batteries for thermal drilling mission is readily demonstrated by simple mathematics: a 160-mm-diameter vehicle penetrating a 3-km ice sheet will require 120 kW-days of power. Using the above mentioned power density for batteries, this amounts to a battery mass alone of 22 tons. Thus, three types of power sources can be considered for autonomous thermal drilling systems: (1) solar power during the summer with short-term power storage units (batteries), (2) wind generators, and (3) automatically controlled diesel engines during the dark winter months (Fig. 5.12).

In central Antarctica, the sun shines for extended periods throughout the summer. The vertical flux of solar radiation depends on latitude, the highest possible flux being $350\ W/m^2$ during December, which is the warmest month. Solar power can be converted into electricity with an efficiency of $\sim 25\%$ using polycrystalline silicon cells (Green et al. 2012). At low temperatures, power generation increases by 5% for every 10 °C decrease in temperature compared with normal conditions (25 °C). The average summer air temperature in central Antarctica is $-30\ °C$, resulting in an expected increase in power generation of 25%. To put this into a practical perspective, to generate 10 kW of electricity, 50–60 m^2 of photovoltaic panels have to be installed. Tests in central Antarctica showed that the Conergy C167P panel (1.2 m^2, nominal power output of 167 W at 25 °C) produced a maximum of ~ 200 W and an average of ~ 2 kWh/d when the sun was above the horizon for >12 h (Lawrence et al. 2008). Solar power could be effective during the summer, but at night and in winter, an alternative power supply would have to be found.

Autonomous wind turbine generators are a practical and alternative renewable power source. The efficiency of turbines depends on wind velocity. Tests using 1-kW wind generators (of the Raum, Hummer, and Bergey types)

Fig. 5.12 Autonomous power supply options: **a** solar panels; scientific base camp erected at the coast near Crown Bay (Hubert 2012); **b** autonomous wind generator; Raum model turbine installed at the South Pole in January 2011 (Allison et al. 2012); **c** automatically controlled diesel generators; PLATO power system, Dome A (Hengst et al. 2008)

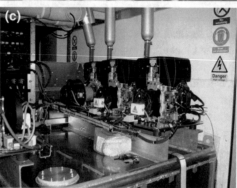

showed that the functional relationship between power output and wind velocity follows a power law with the index of about 2.8 (Allison et al. 2012). At a wind velocity of 5 m/s, the turbines generate ∼90 W. On the Antarctic plateau, katabatic winds are mild, which means low wind speed and low generator efficiency.

Therefore, solar power and wind power are not practical for high-power-load applications on the Antarctic plateau. The steady power output of diesel engine generators presents the best power-supply option. No-live-operator Hatz 1B30 diesel generators have been used during winter at the PLATO (Plateau Observatory), an Antarctic test site with an observatory (Lawrence et al. 2008). The Hatz 1B30 is an air-cooled diesel engine with a maximum power output of 1.5 kW at an atmospheric pressure corresponding to the altitude of the Antarctic plateau. Operating the engines at only 2200 rpm enhances engine durability (its nominal power at sea level is 5.4 kW/3600 rpm).

To provide higher power, the more powerful Hatz 2L41C generator can be used (Talalay et al. 2014). The standard output of this engine is 22 kW at sea level at 3000 rpm. This type of engine has an extremely long service life, which allows it to be operated reliably in remote areas or for applications without monitoring. These engines run on Jet-A1 fuel, and their estimated fuel consumption is 5–9 L/h. A bank of large fuel and oil filters ensures a stable performance, and the fuel tank to be used should have enough capacity for several months of continuous engine operation. It makes sense to use two engines alternating during continuous operation.

The use of gas turbines for generating electricity dates back to the 1930s. Recently, RIPEnergy AG, Switzerland started to produce small lightweight gas turbine generators that can be considered as a good alternative for autonomous drilling systems (10 kW Gas Turbine Generator n.d.). This 380-VAC generator with 7 kW of continuous output power weighs only 31 kg. Fuel (diesel or Jet A1) consumption is 10 L/h; the operating temperature range is from −40 to 50 °C.

Nuclear power is considered as the most reliable power supply for autonomous thermal probes for extraterrestrial investigations (Elliott and Carsey 2004; Lorenz 2012; Stone et al. 2014); the Antarctic Treaty presently forbids the presence of nuclear power sources in Antarctic areas. Nuclear materials (e.g., via the fission of uranium) yield 4.4×10^9 Wh/kg for energy storage (Stone et al. 2018). A small-diameter (<250 mm) thermal probe could easily be powered by a radiothermal generator (RTG). Stone Aerospace has done the preliminary design of two power systems for extraterrestrial missions with thermal probes: a 25-kW$_{th}$ strontium-90 RTG and an ultra-compact 100-kW$_{th}$ fission reactor that takes advantage of ice-reflected neutrons to reduce its mass. Both designs have back-power conversion to generate electricity to run the onboard guidance, navigation and control, thermal drilling, and other systems.

5.5 Summary

Drilling in ice is constantly facing the demand for higher efficiency regarding minimizing time, risk of accidents, costs, and resource consumption. Besides the technical perspective of realizing highly efficient drilling technologies, a new level of standards is being set by environmental protocols. To meet these demands, it may no longer suffice to improve already existing technologies by optimizing process parameters. Instead, advanced ice-drilling tools will be necessary and therefore have to be developed.

Besides the advanced drills described hereinbefore, such as thermal-dissolution and laser drills, new fusion methods using plasma or microwaves present some features of considerable interest. Plasma is an electrically neutral highly ionized gas composed of ions, electrons, and neutral particles. Artificial plasma can be generated via the electric arcs between electrodes. Plasma-arc torches pass gases, such as air or nitrogen, through a spark discharge system to create a plasma torch at an extremely high temperature (>3000 °C). Because the temperature of plasma is much higher than the temperature that can be survived by the mechanical parts of the torch, a flow of cooling fluid through the assembly is required (Pierce et al. 1996). Drilling systems based on this concept require a conduit with one channel for the cooling fluid (which can be used as drilling fluid) and another for the plasma gas. At least two electric lines are required for the spark discharge system, and additional lines for the control system may be needed. The microwave drilling method is based on the phenomenon of the generation of local hot spots (much smaller than the wavelength of microwaves) via near-field microwave radiation and is applicable for boring in a variety of nonconductive materials (Jerby et al. 2002). However, feasibility studies are needed to prove the concepts of plasma and microwave drilling in ice.

References

10 kW Gas Turbine Generator (n.d.) RIPEnergy AG: the power conversion company. Retrieved 26 May 2018 from http://www.ripenergy.ch/eng/gasturbine_generator/10kW_turbine_generator.html

Allison P and 48 others (2012) Design and initial performance of the Askaryan Radio Array prototype EeV neutrino detector at the South Pole. Astropart Phys 35(7):457–477

Beverly CN, Westneat AS (1982) Autonomous computer controlled ice drill. Marine Systems Engineering Laboratory, University of New Hampshire, Durhan, USA, September 1982. Report on Project NR-294–063

Brichkin AV, Mikheev SV, Boev AV (1967) Ognevoye bureniye lednikov v visokogornikh usloviakh [Flame-jet drilling of glaciers in high-mountain regions]. Izvestiya VGO [Proceedings of All-Union Geographical Society] 99(2):147–148. (in Russian)

Browning JA (1978) Flame drilling through the Ross Ice Shelf. The Northern Engineer 10(1):4–8

Browning JA, Somerville DA (1978) Access hole drilling through the Ross Ice Shelf. Antarct J U.S. 13(4):55

Clauer CR, Kim H, Deshpande K et al (2014) An autonomous adaptive low-power instrument platform (AAL-PIP) for remote high-latitude geospace data collection. Geosci Instrum Method Data Syst 3:211–227

Dachwald B, Mikucki J, Tulaczyk S et al (2014) IceMole: a maneuverable probe for clean in situ analysis and sampling of subsurface ice and subglacial aquatic ecosystems. Ann Glaciol 55 (65):14–22

Dudeney JR, Kressman RI, Rodger AS (1998) Automated observatories for geospace research in polar regions. Antarct Sci 10(2):192–203

Elliott JO, Carsey FD (2004) Deep subsurface exploration of planetary ice enabled by nuclear power. In: Proceeding of 2004 IEEE Aerospace Conference, 6–13 March 2004, vol 5. Big Sky, Montana, USA, pp. 2978–2987

Green MA, Emery K, Hishikawa Y et al (2012) Solar cell efficiency tables (version 39). Prog Photovolt: Res Appl 20(1):12–20

Hengst S, Allen GR, Ashley MCB et al (2008) PLATO power: a robust low environmental impact power generation system for the Antarctic plateau. In: Stepp LM, Gilmozzi R (eds) Proceeding of the SPIE conference on ground-based and airborne telescopes II, 27 August, 2008, vol 7012. Society of Photographic Instrumentation Engineers, Bellingham, Washington, USA, SPIE

Hubert A (2012) New solar panels and another trip to the coast. Int Polar Assoc. Posted on 2 Feb 2012. Retrieved 26 May 2018 from www.antarcticstation.org/news_press/news_detail/new_solar_panels_and_another_trip_to_the_coast/

Iagupov AV (1972) Teplovoe razrushenie gornykh porod i ognevoe burenie [Thermal destignation of rocks and jet piercing drilling]. Moscow, Nedra, 160 p. (in Russian)

Itmi M, Cardon A (2016) New autonomous systems. Wiley

Jerby J, Dikhtyar V, Aktushev O et al (2002) The microwave drill. Science 298(5593):587–589

Kelty JR (1995) An in situ sampling thermal probe for studying global ice sheets. Dissertation presented to the Faculty of the Graduate College in the University of Nebraska in partial fulfillment of requirements for the degree of Doctor of Philosophy. Major: Interdepartmental area of engineering (Electrical engineering) under the supervision of Prof. DP Billesbach, Nebraska, Lincoln, May 1995

Lawrence JS and 36 others (2008) The PLATO Antarctic site testing observatory. In: Stepp LM, Gilmozzi R (eds) Proceeding of the SPIE conference on ground-based and airborne telescopes II, 27 August, 2008, vol 7012. Society of Photographic Instrumentation Engineers, Bellingham, Washington, USA, SPIE

Lorenz RD (2012) Thermal drilling in planetary ices: an analytic solution with application to planetary protection problems of radioisotope power sources. Astrobiology 12:799–802

Makarevich RG, Vilesov EN, Golovkova RG et al (1984) Ledniki Tuiuksu [Tuiuksu Glaciers]. Gidrometeoizdat, Leningrad (in Russian)

Maurer WC (1980) Advanced drilling techniques. Petroleum Publishing Co., Tulsa, Oklahoma

Michaelides L (2012) Inventions: thermoblast flame-jet drill. Dartmouth Engineer Magazine, Winter 2012. Available on-line at: https://engineering.dartmouth.edu/magazine/inventions-thermoblast-flame-jet-drill/

Mikheev SV (1971) O burenii l'da ognevym sposobom [About ice drilling by flame-jet drilling]. Akademiya nauk SSSR. Institut geografii. Materialy gliatsiologicheskikh issledovanii [Academy of Sciences of the USSR. Institute of Geography. Data of Glaciological Studies] 18, pp 160–163. (in Russian)

Pierce KG, Livesay BJ, Finger JT (1996) Advanced drilling systems study. Report SAND95-0331, Sandia National Laboratories, Albuquerque, USA, 163 p

Sakurai T, Chosrowjan H, Somekawa T et al (2016) Studies of melting ice using CO_2 laser for drilling. Cold Reg Sci Tech 121(2016):11–15

Schwander J, Walther R, Moret H (2012) Downhole bedrock sonar. In: First open science conference international partnerships in ice core sciences, 1–5 October, 2012. Presqu'île de Giens, Côte d'Azur, France. Booklet of Abstracts, p 185

Specialty Heaters (2009) MarComm sensor and controller catalog, pp 457–474

Stone WC, Hogan B, Siegel V et al (2014) Progress towards an optically powered cryobot. Ann Glaciol 55(65):1–13

Stone W, Hogan B, Siegel V et al (2018) Project VALKYRIE: laser-powered cryobots and other methods for penetrating deep ice on ocean worlds. In: Badescu V, Zacny K (eds) Outer solar system. Springer, Cham, pp 47–165

Talalay PG (2014) Perspectives for development of ice drilling technology: a discussion. Ann Glaciol 55(68):339–350

Talalay PG, Zagorodnov VS, Markov AN et al (2014) Recoverable autonomous sonde (RECAS) for environmental exploration of Antarctic subglacial lakes: general concept. Ann Glaciol 55(65):23–30

Talalay P, Yang C, Cao P et al (2015) Ice-core drilling problems and solutions. Cold Reg Sci Tech 120:1–20

Thorsteinsson T, Elefsen SÓ, Gaidos E et al (2008) A hot water drill with built-in sterilization: design, testing and performance. Jökull 57:71–82

Zagorodnov VS, Kelley JJ, Nagornov OV (1994) Drilling of glacier boreholes with a hydrophilic liquid. Mem Natl Inst Polar Res Spec Issue 49:153–164

Zeller E, Dreschhoff G, Laird CM (1989) Development of laser ice-cutting apparatus. Antarct J U.S. 26(5):89–91

Zimmerman W, Bonitz R, Feldman J (2001) Cryobot: an ice penetrating robotic vehicle for Mars and Europa. In: IEEE aerospace conference, vol 1, 10–17 March 2001. Big Sky, Montana, USA, pp 311–323

Appendix A: Records of Thermal Drilling in Ice

0.8 m	Largest-diameter ice hole drilled with a "plain" hot point to study Antarctic lakes
18 mm	Smallest-diameter hot points: Department of Atmospheric Sciences, University of Washington (USA) and the LGGE (France)
756 m	Deepest hole drilled using an electrically heated hot point, Granduc Mines Ltd., Salmon Glacier, Canada (1956)
952.5 m	Deepest open hole drilled with a TELGA-14 M thermal drill at Vostok Station, East Antarctica (1972)
200 mm	Largest-diameter ice core recovered with the ANARE large-diameter electrothermal drill (known colloquially as the "Errol drill")
2201.7 m	Ice-core drilling record achieved with the TBS-112VCh thermal drill at Vostok station, Antarctica in August 1985 (it was beaten by the GRIP hole in Greenland in August 1991)
871.5 m	Deepest borehole drilled with an ethanol–water solution at Komsomolskaya Station, Antarctica using the ETB-5 thermal drill (1981–1983)
2755.3 m	Record depth for ice-core thermal drilling, achieved using the TBZS-132 thermal drill at Vostok station, Antarctica in September 1993
2500 m	Record depth for hot-water drilling within the IceCube project at the South Pole (2004–2011)
125–200 m/h	Fastest (in clear ice) hot-water drilling system of the Geological Survey of Greenland

© Geological Publishing House and Springer Nature Singapore Pte Ltd. 2020
P. G. Talalay, *Thermal Ice Drilling Technology*, Springer Geophysics,
https://doi.org/10.1007/978-981-13-8848-4

Appendix B: Abbreviations of Institutes, Organizations and Projects

AARI	Arctic and Antarctic Research Institute, St. Petersburg, Russia
ANARE	Australian National Antarctic Research Expedition
ANDRILL	ANtarctic geological DRILLing project
AINA	Arctic Institute of North America, University of Calgary, Canada
AMISOR	Amery Ice Shelf Ocean Research project
APL-UW	Applied Physics Laboratory, University of Washington, Seattle, USA
ARA	Askaryan Radio Array, large neutrino detector at South Pole
AWI	Alfred Wegener Institute, Helmholtz Centre for Polar and Marine Research, Bremerhaven, Germany
BAS	British Antarctic Survey, Cambridge, UK
BGR	Bundesanstalt für Geowissenschaften und Rohstoffe (Federal Institute for Geosciences and Natural Resources), Hannover, Germany
BPRC, BPCRC	Byrd Polar Research Center at The Ohio State University, USA in 2014 renamed to the Byrd Polar and Climate Research Center
Caltech	California Institute of Technology, Pasadena, California, USA
CRREL	US Army Cold Regions Research and Engineering Laboratory, Hanover, USA
DLR	Deutsches zentrum für Luft-und Raumfahrt (German Aerospace Center), Cologne, Germany
DREP	Defence Research Establishment Pacific, Esquimalt, Canada, research agency of the Department of National Defence (agency was closed in 1994)
EGIG	Expédition Glaciologique Internationale au Groenland (1967–1968)
ETH	Swiss Federal Institute of Technology Zurich (German: Eidgenössische Technische Hochschule Zürich)
FISP	Filchner Ice Shelf Project
GISP	Greenland Ice Sheet Program (1971–1981)
GRIP	Greenland Ice Core Project (1989–1995)
IGSFE	Institute of Geography of Siberia and Far East, Irkutsk, USSR/Russia (in 2005 renamed to V.B. Sochava Institute of Geography)
IceCube	IceCube Neutrino Observatory, Amundsen–Scott Station, Antarctica
IDDO	Ice Drilling Design and Operations group, University of Wisconsin-Madison
IGAS	Institute of Geography, USSR Academy of Sciences (since 1991 IGRAS-Institute of Geography Russian Academy of Science)
IGSFE	Institute of Geography of Siberia and Far East, USSR Academy of Sciences (now Institute of Geography named after V.B. Sochava, RAS Siberian branch), Irkutsk
IWF	Institut für Weltraumforschung (Space Research Institute), Graz, Austria
JLU	Jilin University, China
JPL	Jet Propulsion Laboratory, California Institute of Technology, Pasadena, USA
ILTS	Institute of Low Temperature Science, Hokkaido University, Japan
JARE	Japanese Antarctic Research Expedition
HSD	Hydrological Service Division, National Energy Authority, Iceland
KOPRI	Korea Polar Research Institute

(continued)

© Geological Publishing House and Springer Nature Singapore Pte Ltd. 2020
P. G. Talalay, *Thermal Ice Drilling Technology*, Springer Geophysics,
https://doi.org/10.1007/978-981-13-8848-4

LGGE	Laboratoire de Glaciologie et Géophysique de l'Environnement, CNRS, Grenoble, France
LMI, SPSMI	Leningrad Mining Institute (since 1991 SPSMI—St. Petersburg State Mining Institute; now—St. Petersburg Mining University), Russia
NRC	National Research Council of Canada
NSF	US National Science Foundation
PICO	Polar Ice Coring Office was established in March 1974 at the University of Nebraska-Lincoln University. In August 1989 PICO contract is awarded for University of Alaska-Fairbanks; PICO ended activities at University of Alaska-Fairbanks in March 31, 1995
RAE	Russian Antarctic Expedition
RISP	Ross Ice Shelf Project (1973–1979)
SAE	Soviet Antarctic Expedition (1955–1992); in the beginning, three first expeditions (1955–1959) were referred as Complex Antarctic Expedition (CAE); since 7 August 1992 SAE became Russian Antarctic Expedition (RAE)
SAFIRE	Subglacial Access and Fast Ice Research Experiment
SakhalinNIPImorneft	Sakhalin Research, Design, and Survey and Institute of Oil and Gas, Russia
SALSA	Subglacial Antarctic Lakes Scientific Access
SCINI	Submersible Capable of under Ice Navigation and Imaging project
SIRG	Snow and Ice Research Group, University of Nebraska-Lincoln, USA
TRS	Tarfala Research Station, Sweden
UAF	University of Alaska-Fairbanks, USA
UI	University of Iceland
USGS	US Geological Survey
UW	University of Washington, USA
WISSARD	Whillans Ice Stream Subglacial Access Research Drilling

Printed in the United States
By Bookmasters